The Climate of the Arctic

ATMOSPHERIC AND OCEANOGRAPHIC SCIENCES LIBRARY

VOLUME 26

Editors

Robert Sadourny, *Laboratoire de Météorologie Dynamique du CNRS, École Normale Supérieure, Paris, France*
Lawrence A. Mysak, *Department of Atmospheric and Oceanographic Sciences, McGill University, Montreal, Canada*

Editorial Advisory Board

L. Bengtsson	Max-Planck-Institut für Meteorologie, Hamburg, Germany
A. Berger	Université Catholique, Louvain, Belgium
P.J. Crutzen	Max-Planck-Institut für Chemie, Mainz, Germany
J.R. Garratt	CSIRO, Aspendale, Victoria, Australia
G. Geernaert	DMU-FOLU, Roskilde, Denmark
K. Hamilton	University of Hawaii, Honolulu, HI, U.S.A.
M. Hantel	Universität Wien, Austria
A. Hollingsworth	European Centre for Medium Range Weather Forecasts, Reading, UK
H. Kelder	KNMI (Royal Netherlands Meteorological Institute), De Bilt, The Netherlands
T.N. Krishnamurti	The Florida State University, Tallahassee, FL, U.S.A.
P. Lemke	Alfred-Wegener-Institute for Polar and Marine Research, Bremerhaven, Germany
P. Malanotte-Rizzoli	MIT, Cambridge, MA, U.S.A.
S.G.H. Philander	Princeton University, NJ, U.S.A.
D. Randall	Colorado State University, Fort Collins, CO, U.S.A.
J.-L. Redelsperger	METEO-FRANCE, Centre National de Recherches Météorologiques, Toulouse, France
R.D. Rosen	AER, Inc., Lexington, MA, U.S.A.
S.H. Schneider	Stanford University, CA, U.S.A.
F. Schott	Universität Kiel, Kiel, Germany
G.E. Swaters	University of Alberta, Edmonton, Canada
J.C. Wyngaard	Pennsylvania State University, University Park, PA, U.S.A.

The titles published in this series are listed at the end of this volume.

The Climate of the Arctic

by

Rajmund Przybylak

*Department of Climatology,
Nicholas Copernicus University,
Toruń, Poland*

SPRINGER-SCIENCE+BUSINESS MEDIA, B.V.

A C.I.P. Catalogue record for this book is available from the Library of Congress.

DOI 10.1007/978-94-017-0379-6

Layout and Composition:
Studio KROPKA dtp – *Piotr Kabaciński*
Tel. 048 (56) 66 01 737; 048 (602) 303 814
Kozacka Street 13/1; 87-100 Toruń; Poland

Printed on acid-free paper

All Rights Reserved
© 2003 Springer Science+Business Media Dordrecht
Originally published by Kluwer Academic Publishers in 2003
MyCopy version of the original edition 2003

No part of this work may be reproduced, stored in a retrieval system, or transmitted
in any form or by any means, electronic, mechanical, photocopying, microfilming, recording
or otherwise, without written permission from the Publisher, with the exception
of any material supplied specifically for the purpose of being entered
and executed on a computer system, for exclusive use by the purchaser of the work.
www.springer.com/mycopy

*To wife Dorota
and our daughters Anna and Julia*

CONTENTS

Preface ix
Acknowledgements xi

1. Introduction 1
   1.1 Boundaries of the Arctic 1
   1.2 Main Geographical Factors Shaping the Climate 5

2. Atmospheric Circulation 13
   2.1 Development of Views on Atmospheric Circulation in the Arctic 13
   2.2 Large-scale Atmospheric Circulation 14
   2.3 Synoptic-scale Circulation 21
   2.4 Winds 25
   2.5 Local Circulation and Mesoscale Disturbances 28

3. Radiation Conditions 33
   3.1 Sunshine Duration 35
   3.2 Global Solar Radiation 39
   3.3 Short-wave Net Radiation 43
   3.4 Long-wave Net Radiation 47
   3.5 Net Radiation and Other Elements of the Heat Balance 49

4. Air Temperature 63
   4.1 Mean Monthly, Seasonal, and Annual Air Temperature 63
   4.2 Mean and Absolute Extreme Air Temperatures 81
   4.3 Temperature Inversions 91

5. Cloudiness 97
   5.1 The Annual Cycle 98
   5.2 Spatial Patterns 101
   5.3 Fog 104

6. Air Humidity 109
   6.1 Water Vapour Pressure 109
   6.2 Relative Humidity 112

7. Atmospheric Precipitation and Snow Cover 117
   7.1 Atmospheric Precipitation 120
   7.2 Number of Days with Precipitation 134
   7.3 Snow Cover 136

8. Air Pollution 141

9. Climatic Regions 149
   9.1 The Atlantic Region 150
   9.2 The Siberian Region 152
   9.3 The Pacific Region 153
   9.4 The Canadian Region 154
   9.5 The Baffin Bay Region 156
   9.6 The Greenland Region 157
   9.7 The Interior Arctic Region 158

10. Climatic Change and Variability in the Holocene 161
    10.1 Period 10-11 ka–1 ka BP 161
    10.2 Period 1 ka–0.1 ka BP 173
    10.3 Period 0.1 ka–Present 181

11. Scenarios of the Arctic Climate in the 21st Century 203
    11.1 Model Simulations of the Present-day Arctic Climate 203
    11.2 Scenarios of the Arctic Climate in the 21st Century 211

References 225
Copyright Acknowledgements 259
Index 263

PREFACE

Towards the end of the 19th century some researchers put forward the hypothesis that the Polar regions may play the key role in the shaping of the global climate. This supposition found its full confirmation in empirical and model research conducted in the 20th century, particularly in recent decades. The intensification of the global warming after about 1975 brought into focus the physical causes of this phenomenon. The first climatic models created at that time, and the analyses of long observation series consistently showed that the Polar regions are the most sensitive to climatic changes. This aroused the interest of numerous researchers, who thought that the examination of the processes taking place in these regions might help to determine the mechanisms responsible for the „working" of the global climatic system. To date, a great number of publications on this issue have been published. However, as a review of the literature shows, there is not a single monograph which comprises the basic information concerning the current state of the Arctic climate. The last study to discuss the climate of the Arctic in any depth was published in 1970 (*Climates of the Polar Regions*, vol. 14, ed. S. Orvig) by the World Survey of Climatology, edited by H. E. Landsberg. This publication, however, does not provide the full climatic picture of many meteorological elements. The issue of climatic changes is raised only cursorily and the information provided is now long outdated. As far as the Antarctic is concerned, the situation is far better for there are numerous synthetic works on Antarctic climatology in many languages.

It was the rather astonishing paucity of academic studies concerning the climate of the Arctic that prompted my decision to embark on a work which aimed at filling this gap. This sort of compilation work, which sums up the present state of knowledge on the subject, can only be successfully accomplished if it is done in a well-equipped library. This is why most of the present book was written in the Scott Polar Research Institute library, Cambridge, UK, which, in all likelihood, contains the most comprehensive collection of Polar literature.

The primary aim of the publication is to present the current state of knowledge concerning the Arctic climate using, whenever possible, the latest meteorological data. In view of the importance of climatic changes, this issue has been given more attention than is customary in similar studies.

It is now commonly accepted that the mean physical state of the atmosphere is one of the key elements of the Arctic climatic system. Consequently, a variety of climatic data is indispensable not only for climatologists, but also for other researchers of the Arctic environment (glaciologists, oceanographers, botanists, etc.). Up-to-date and reliable climatic data are also requisite to validate climatic models. The author hopes that the book will be of particular interest to all researchers who represent the above scientific disciplines in their research.

The present work should also be helpful to students of geography and related disciplines, both in the didactic process and in research. It may also be of use to all those who are interested in this part of the world.

Finally, I would like to express my hope that the reader will find the book gratifying in terms of readability and the usefulness of the information it contains. I would also like to apologize for any mistakes in the text that went unnoticed in the publication process.

Rajmund Przybylak
Toruń, June 2002

ACKNOWLEDGEMENTS

It would not have been possible to carry out the research for the present volume without the financial support provided by the Nicholas Copernicus University in Toruń. For their assistance in securing this support I would like to thank the Vice-Rector for Research and International Relations Prof. Marek Zaidlewicz, the Dean of the Faculty of Biology and Earth Sciences Prof. Andrzej Tretyn, the Director of the NCU Institute of Geography Prof. Jan Falkowski, and the Administrative Director Dr. Krzysztof R. Lankauf. I would also like to register my appreciation for the opportunity to make two research visits to the Scott Polar Research Institute in Cambridge – the months I spent there allowed me to gain access to the materials on which the greater part of the present work is based. In particular I would like to thank the library staff at the Institute under the direction of William Mills, whose assistance contributed greatly to the project.

I would sincerely like to thank the anonymous reviewers of the study, whose comments and criticisms were extremely helpful in drawing my attention to improving certain sections of the text. Obviously any faults which remain in the text are my own.

I am very grateful to Krystyna Czetwertyńska, Anita Krawiec, and Zsuzsanna Vizi for contributing their knowledge and computer expertise in reproducing the graphics for the book.

Special warm thanks to John Kearns for improving my English, and to my editor at Kluwer, Marie Johnson, for her assistance in the preparation of this book.

Last but not least, I would like to thank my wife Dorota for assisting me in many ways and for all her personal support, particularly during the periods which I had to spend away from home.

Chapter 1

INTRODUCTION

The word "Arctic" is derived from the Greek word *Arktos* ('bear'). In its Latin equivalent, this occurs in the names of two constellations – *Ursa Major* and *Ursa Minor* – which circle endlessly around the one fixed point in the heavens: *Polaris*, the North Polar Star.

1.1 Boundaries of the Arctic

The Arctic is not an easily definable geographic entity similar to, for example, Iceland, Lake Baykal, or even the Antarctic. Therefore, until recently, it has not been possible to arrive at any single definition of the area. Since the 1870s a large number of researchers representing different disciplines such as geography, climatology, and botany have tried to establish a widely accepted criterion to delimit the Arctic boundary (Figure 1.1). In almost all the geographical monographs and other books dealing with Arctic or Polar regions one can find a variety of attempted definitions (e.g. Bruce 1911; Brown 1927; Nordenskjöld and Mecking 1928; Baird 1964; Sater 1969; Sater *et al.* 1971; Baskakov 1971; Petrov 1971; Barry and Ives 1974; Weiss 1975; Sugden 1982; Young 1989; Boggs 1990; Stonehouse 1990; Barry 1995; Bernes 1996; Przybylak 1996a; Niedźwiedź 1997; Mills and Speak 1998). However, the most comprehensive reviews have been given by Petrov (1971) and Baskakov (1971). The oldest conception of the Arctic is one which considers it to be a region of the Northern Hemisphere lying north of the Arctic Circle ($\varphi = 66°33'N$). The majority of the above authors agree that this astronomically distinguished line of latitude cannot be considered to be the real Arctic boundary. This fact was noted as early as 1892 by Bruce (Bruce 1911) and later in 1927 by Brown, who wrote, "The Arctic and Antarctic circles merely mark the equatorial limits of the zones in which the sun is never more than 23°30' above the horizon. [...] The circles are astronomical lines without climatic significance." The careful reader will note that here Brown gives the wrong value of the height of the sun. The correct value is 47° and can be ascertained using the formula $h = 90° - \varphi + \delta$ where φ is the geographical latitude and δ is the declination of the sun.

A more meaningful and more frequently used definition of the Arctic is a climatological one. Among the many known climatic criteria, the most popular is still the older proposition given by Supan (1879, 1884), i.e. the 10°C mean isotherm of the warmest month. This criterion was later modified, first by Vahl (1911) and then by Nordenskjöld (1928). Vahl did not determine the precise borders of the Polar regions, but, as he seems to have let it coincide with the tree line, he regarded the equation $V < 9.5° - 1/30$ K to be the most favourable for the determination of the position of this boundary. In this formula V and K denote the mean temperature of the warmest and coldest months, respectively. Nordenskjöld (1928) found that the role of the coldest month in determining the Arctic boundary should be greater than was assumed Vahl (1911). Therefore, he proposed a new formula: $V < 9° - 0.1$ K. In addition, he also extended it to the seawater areas (see Figure 1.1). According to this criterion the Arctic includes regions in which the temperature of the warmest month ranges from 9°C (when the temperature of the coldest month is 0°C) to 13°C (when the temperature of the coldest month is –40°C).

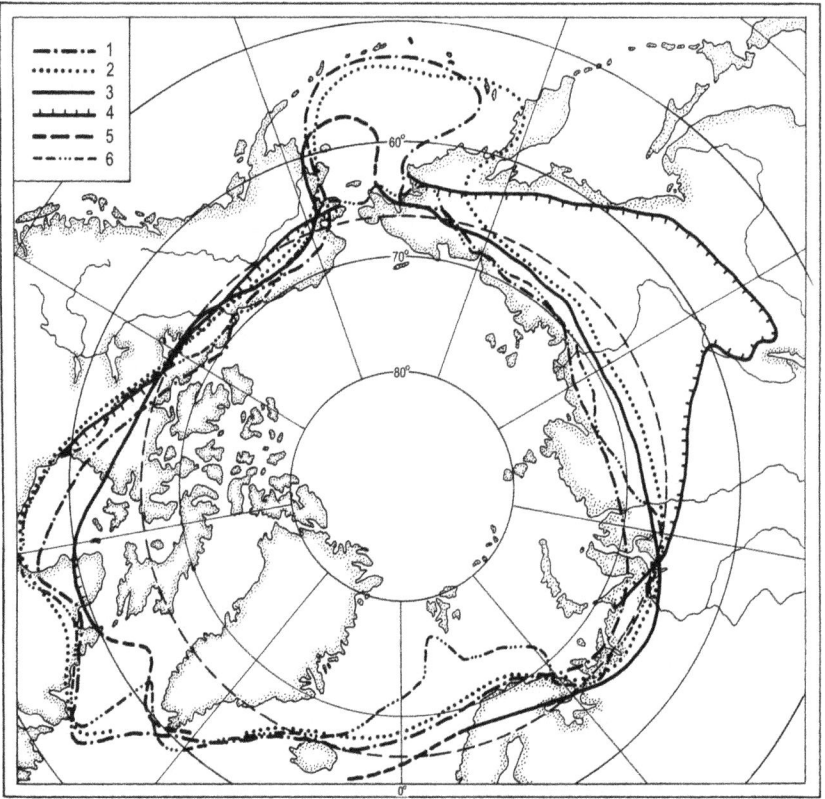

Figure 1.1. Boundaries of the Arctic. 1 – isotherm of the warmest month 10°C, 2 – boundary of the Arctic after Nordenskjöld, 3 – line denoting net radiation of 62.7 kJ/cm² /year (15 kcal/cm²/year), 4 – boundary of the permafrost, 5 – Arctic Circle, 6 – boundary of the Arctic after *Atlas Arktiki* (1985).

The boundary of the Arctic can also be drawn using the criterion proposed by Gavrilova (1963) and Vowinckel and Orvig (1970). According to them, all areas where the net radiation balance is lower than 62.7 kJ/cm² /year (15 kcal/ cm² /year) may be considered to belong to the Arctic (Figure 1.1).

The authors of the *Atlas Arktiki* (1985) have recently presented a new, very good, proposition. The southern Arctic boundary has been delimited using mean long-term values of almost all meteorological elements. Thus, the concept of climatic regionalisation is employed. The Arctic perimeter on the continents lies mostly between the boundaries of the 10°C mean isotherm of the warmest month and the so-called Nordenskjöld line (Figure 1.1). In addition, the authors of the *Atlas* have also distinguished seven climatic regions within the Arctic (Figure 1.2). These facts have persuaded me to adopt their definition of the Arctic for the purposes of this monograph.

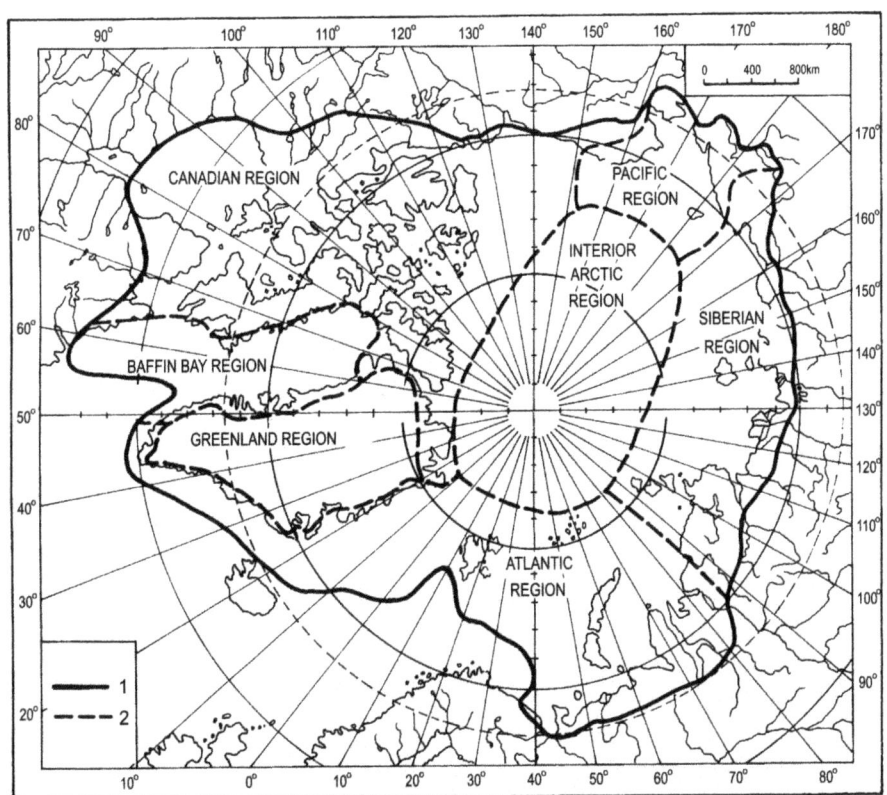

Figure 1.2. Boundaries of the Arctic (1) and climatic regions (2) (after *Atlas Arktiki* 1985).

The third criterion quite often used (aside from astronomical and climatological criteria) is a (geo)botanical one. The southern boundary of the tundra or the northern boundary of the tree line is considered to be the natural boundary of the Arctic. Supan (1879, 1884), in his classification of climates, was probably the first to distinguish the Arctic area using both climatological and geobotanical criteria. The areas distinguished show a good correlation and most other later analyses confirm Supan's finding. Sugden (1982) presents several advantages of using the tree line to delimit the southern land boundary of the Arctic. He writes, "Not only does it represent a fundamentally important vegetation boundary, but it is also important in terms of animal distributions. It coincides approximately with a mean July temperature isotherm of 10°C and thus is also of climatic significance." However, one must be aware that there are also some disadvantages of this criterion. For example, as many as three possibilities to define the boundary of the Arctic can be used: a northern limit of continuous forest, a northern limit of erect trees, or a northern limit of species. For more details see Hare (1951) and references therein.

Almost all the above-presented criteria should be used exclusively with reference to land Arctic regions. However, in the present analysis we also need to establish boundaries over sea areas. A number of researchers, mainly oceanographers, have suggested replacing the boundary at sea delimited using the above criteria with a boundary delimited using more appropriate criteria for a water environment. For example, Baskakov (1971) suggests that the boundary of the Arctic in the sea areas should be drawn according to oceanological characteristics, i.e. hydrological, ice, and geomorphologic. He has provided us with the most comprehensive definition of the oceanic Arctic region and I cite here only the most important fragment: "Those water areas may be considered part of the oceanic Arctic region which, during the cold period of the year, are generally (average outflow over a several year period) covered by sea ice of various ages including perennial ice, and in which the upper layer of water under the ice (of a depth of not less than 30 m) has negative temperature and low salinity (less than 34.5‰)." Simply speaking, as a sea boundary of the Arctic one can accept the southernmost extent normally reached by the Arctic Waters.

At the end of this section one should also mention the opinion of some researchers (Armstrong *et al.* 1978; Sugden 1982; Stonehouse 1990) who are convinced that it is practically impossible to achieve a delimiting of the precise boundary of the Arctic which will gain the acceptance of scientists from different disciplines. Sugden (1982) has written, "...the boundaries should remain flexible. Some boundaries seem appropriate for some purposes and other boundaries for others." To a certain extent it is possible to agree with this view. However, I think that an Arctic boundary should be at least agreed on among scientists of the same discipline, e.g. among climatologists. In an era of global warming, this is becoming more and more urgently needed. Otherwise, our estimations of mean Arctic climatic trends may be equivocal (see Przybylak 1996a, Przybylak 2000a, 2002a).

1.2 Main Geographical Factors Shaping the Climate

Undoubtedly, geographical latitude is the main factor determining the weather and climate both in the Arctic and elsewhere. For the purpose of this work, the Arctic has been defined after *Atlas Arktiki* (1985) (Figure 1.2). From Figure 1.2 it can be seen that the southern boundary of the Arctic thus defined ranges between about 54°N (the Labrador Peninsula) to about 75°N near Spitsbergen. No matter how we define the Arctic, its location in high latitudes limits significantly the magnitude of receiving energy from the sun. In regions lying beyond the Arctic Circle, the most unusual feature is occurrence of seasonal day and night. As we know, the length of both polar night and polar day varies from one day at the Arctic Circle to about six months at the North Pole (Figure 1.3). In addition, because of the atmospheric refraction, the total time when the sun is visible over the horizon during the year is greater at high latitudes than in more temperate latitudes. Also, since the sun crosses the horizon at a shallow angle in the Arctic, dawn and dusk persist for long periods before and after the sun is visible. As a result, winter days are much longer here than summer nights. The elevation of the sun in noon anywhere cannot be higher than about 47°. This fact is mainly responsible for the lower income of the solar energy (on an annual basis) here than in lower latitudes. However, the total solar radiation in June, which the Arctic receives at the top of the atmosphere, is even higher than in equatorial areas. For example, the solar irradiance flux reaching the upper boundary of the earth-atmosphere system is equal to 129 kJ/cm^2 (31 kcal/cm^2) and 98.2 kJ/cm^2 (23.5 kcal/cm^2) at the 80°N and the equator, respectively (Budyko 1971). From a climatological point of view, however, it is only the solar radiation absorbed by the surface which is important. Due to the high albedo of the earth-atmosphere system in the Arctic, this component of the radiation balance is markedly lower than in the rest of the globe.

Looking at the map of the Arctic, one can easily see that the Arctic, in contrast to the Antarctic, consists of an ocean encircled by land. The central main part of the ocean is called the Arctic Ocean and is ice-covered year-round, while snow and ice are present on the land for almost all the year. The land encompasses the northern parts of two major land masses – Eurasia and North America – as well as quite a large number of islands, especially on the American side. Of these the largest are Greenland (2,175,600 km^2), Baffin Island (476,070 km^2), Ellesmere Island (212,690 km^2) and Victoria Island (212,200 km^2) (*The Times Atlas of the World* 1992). The area of Greenland, including the islands, is 2,186,000 km^2 (Putnins 1970). A substantial break in the ring of land exists only between Greenland and Norway (Figure 1.2). Other breaks between Asia and America (the Bering Strait) and between the islands of the Canadian Arctic Archipelago, are of marginal significance. The highest mountains are to be found in south-eastern Greenland, where two sum-

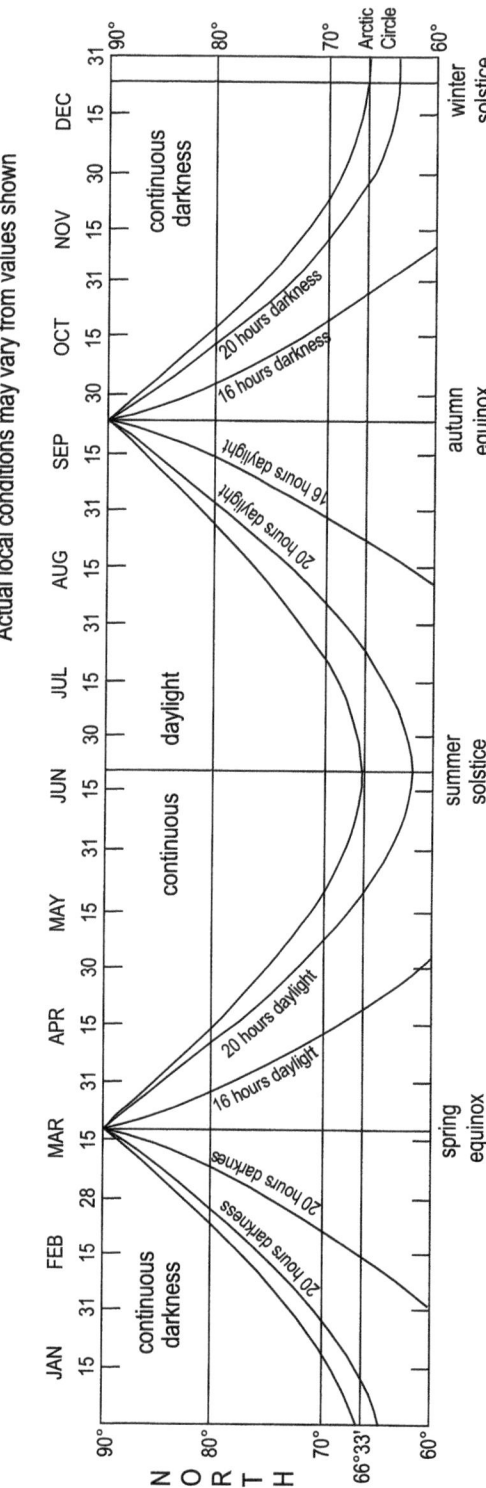

Figure 1.3. Duration of daylight and darkness in the latitude band 60°–90°N (after CIA 1978).

mits rise over 3000 m a.s.l.: Gunnbjörn Fjeld (3700 m, $\varphi = 68°54'N$, $\lambda = 29°48'W$) and Mt. Forel (3360 m, $\varphi = 67°00'N$, $\lambda = 37°00'W$) (*The Times Atlas of the World* 1992). Much of the Arctic is low lying, except for the Greenland ice sheet, the ice-covered mountains of Ellesmere and Axel Heiberg islands, and the mountains in the northern part of the Beringia region. The differentiated influence of land and sea areas on the climate of the Arctic is significantly lower than in the moderate latitudes. This is true, particularly in winter, when the land and most of the sea areas are covered by snow. The long-term mean depths of snow cover for May, calculated from measurements taken mainly in Russian drifting stations NP3 – NP31 over the period 1954–1991, vary from 30–40 cm in the central part of the Arctic to more than 80 cm in the mountainous regions. The maximum snow-cover depth is most often observed in April or May except in the Canadian Arctic, where it is observed in March. The decay of the snow cover begins in the south of the Arctic in the first ten days of June, and in the vicinity of the Pole in mid-July. The number of days with snow cover is greatest in the central Arctic (more than 350 days). This number decreases towards the south and is equal to about 280–300 days across those Arctic islands which have a continental climate (for more details see sub-section 7.3.).

In general, three physical characteristics of snow – high reflectivity, high infrared emissivity and high insulating property – mean that it plays a very important climatic role. The high albedo of the snow surface significantly reduces the net radiation balance of the surface and low troposphere. The high infrared emissivity of snow is one of the most important factors, which causes near-surface atmospheric temperature inversions, especially in the cold half-year. In addition, it helps in the development and stabilising of the anticyclones. Snow cover, as one of the best insulators of all known natural surfaces, is a very important element in the atmosphere-cryosphere-ocean system, and thus significantly influences heat transport. A snow cover of more than 15 cm in depth may completely stop the heat transport between the atmosphere and land or sea ice.

The Arctic Ocean and its bordering seas occupy an area of 14 million km^2 (Barry 1989). In late winter (February–March) almost all this area is covered by sea ice. During the summer (August–September), the sea ice is at its minimum extent (approximately 8 million km^2). The role of the sea ice in shaping the climate of the Arctic, and indeed that of the whole globe, is crucial. Generally, four main properties of the sea ice contribute to this. The first property is the significantly higher albedo of sea ice (0.5 to 0.7) in comparison to an open ocean (0.1). As a result, water covered by sea ice absorbs much less radiation than do open waters. A second property is the insulating role of the sea ice, restricting the exchange of heat and moisture between ocean and atmosphere. Maykut (1978) reported that measurements of wintertime sensible heat flux showed that between 10 and 100 times more heat is transferred from a calm open-water ocean to the atmosphere than from an ocean covered by a 2-metre layer of sea ice. A third property is the large latent heat of freezing and melting, which makes sea ice act as a thermal reservoir delaying the seasonal temperature cycle. These processes also alter the salinity content of the upper layers of the ocean. During freezing, a sea salt is forced out of the sea ice (resulting in an increase of salinity in the water); on the other hand, during melting, the fresh water transferred to the upper layers reduces its salinity. Recently, Wadhams (1995) has drawn our attention to the fact that sea-ice motion (a fourth property), driven mainly by wind stress, is also very important for climate and climate-change studies. Processes connected with this motion, such as divergence and convergence of sea ice, create leads and pressure ridges, respectively. The latter forms contain about half of the total Arctic ice volume (Wadhams 1981). As a result of these processes, atmosphere-ocean heat and moisture fluxes are highly time- and space-dependent.

Sea ice in the Arctic never exists as an unbroken cover or as a floating ice cap. Three categories of sea ice can be distinguished here: Polar Cap Ice, Pack Ice, and Fast Ice (Pickard and Emery 1982). Polar Cap Ice covers about 70% of the Arctic Ocean. It occurs in the vicinity of the Pole near the 1000-m isobath, and consists of ice which is several years old. In winter, the average thickness of undisturbed ice is about 3–4 m, but hummocks can increase the height locally up to 10 m a.s.l. In summer the average thickness decreases to about 2.5 m. The Pack Ice lies outside the polar cap and covers about 25% of the Arctic area. Its areal extent is greatest in May and lowest in September. The Fast Ice grows seawards from the coast to the pack. It is most often anchored to the shore and extends out to about the 20–30-m isobath. The Fast Ice occurs only in wintertime and its thickness reaches 1 to 2 m. Sea ice in the Arctic is continually in motion as a result of the effects of wind, tide, and ocean currents. The same factors create open-water areas known as leads and polynyas. Leads are cracks in the ice which are a few kilometres in width and tens of kilometres long, though which are often short lived. On the other hand, polynyas are large open-water areas in the frozen sea and range in size from a few hundred square meters to thousands of square kilometres. Polynyas appear in winter when the air temperature is well below the freezing point of seawater. The role of open-water areas in the Arctic climate system is sufficiently important to be studied more seriously by climatologists. Through these areas the Arctic surface loses huge amounts of heat because sea-surface temperature in winter can be up to 20°C higher (as in the case of the so-called North Water polynya in the northern part of Baffin Bay) than that of the surrounding areas and because there is no sea-ice cover, which significantly reduces the heat exchange between the ocean and the atmosphere, as was mentioned above.

Another type of ice which occurs in the Arctic takes the form of icebergs and originates as a result of the "calving" of tidewater glaciers. Each year a highly variable number of these navigational hazards (about 1000 across the 55°N latitude) move southward into the Atlantic together with the cold water of East Greenland and Labrador Currents.

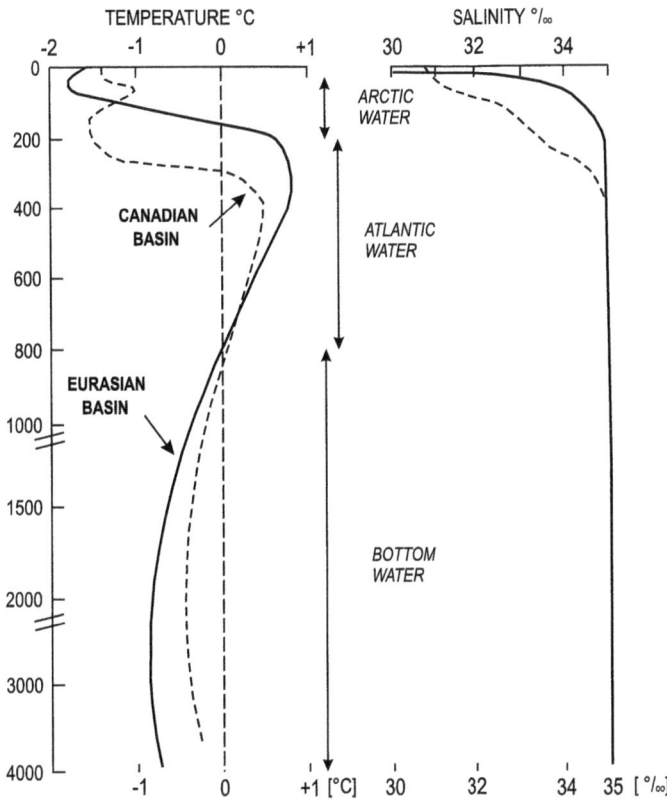

Figure 1.4. Typical temperature and salinity profiles for the Arctic Sea (the Eurasian and Canadian basins) (after Pickard and Emery 1982).

Coachman and Aagaard (1974) distinguished three main water masses in the Arctic Ocean: the surface or Arctic Water from the sea surface to a depth of 200 m, the Atlantic Water from 200 m to 900 m, and the Bottom Water below 900 m. For the study of the Arctic climate knowledge of the Arctic Water is most important and this can be divided into three layers: the Surface Arctic, the Sub-surface Arctic and the Lower Arctic Waters. The physical characteristics of these types of water masses are provided in Table 1.1 and Figure 1.4. Surface waters extend from the surface to depths of about 25 and 50 m. Both the salinity and temperature of the water is strongly controlled by melting and freezing. As a result, the temperature oscillates near the freezing point of seawater, which varies only from −1.5°C at a salinity of 28‰ to −1.8°C at a salinity of 33.5‰. Throughout the year both salinity and temperature show rather small changes, which range up to 2‰ and 0.1–0.2°C, respectively.

Table 1.1. Arctic Sea water masses (after Pickard and Emery 1982)

Water mass Name (circulation direction)	Boundary depth	Properties Temperature (T) and Salinity (S)	Seasonal variation
ARCTIC SURFACE	Surface	T: Close to F.P., i.e. −1.5 to −1.9°C S: 28 to 33.5‰	DT: 0.1°C DS: 2 ‰
	25 to 50 m		
ARCTIC SUB-SURFACE		T: Canadian Basin −1 to −1.5°C Eurasian Basin −1.6°C to 100 m, then increase S: Both basins 31.5 to 34‰	Small
	100 to 150 m		
ARCTIC LOWER		Intermediate between Sub-surface and Atlantic	
	200 m		
(all above masses circulate clockwise)			
ATLANTIC (anticlockwise)		T: Above 0°C (to 3°C) S: 34.85 to 35‰	Negligible
	900 m		
BOTTOM (uncertain, small)		2000 m Bottom T: Canadian Basin −0.4°C −0.2°C Eurasian Basin −0.8°C −0.6°C S: Both Basins 34.90 to 34.99‰	(rise adiabatic)
	Bottom		

The surface water and sea-ice circulation in the Arctic has been largely known from the observed drift of camps on the ice, floe stations, and ships. The earliest information comes from the famous "Fram" drift (1893–1896) and from the icebreaker "Sedov" drift (1937–1940). Observational evidence together with theoretical calculations of upper-layer circulation based on water density distribution, give a consistent picture of circulation in the Arctic (Figure 1.5). In the Beaufort Sea the surface waters have a clockwise movement in agreement with the anticyclonic pattern of blowing winds and lead out to the East Greenland Current. From the Eurasian side of the Arctic Ocean the surface waters move towards North Pole and exit the Eurasian Basin as the East Greenland Current. This current is known as the Transpolar Drift Stream. The speeds of these waters are of the order of 1cm/s to 4 cm/s (300 km/year to 1200 km/year). It is worth to add here that a sea-ice circulation in the Arctic Ocean is similar to the described above circulation of the surface water currents.

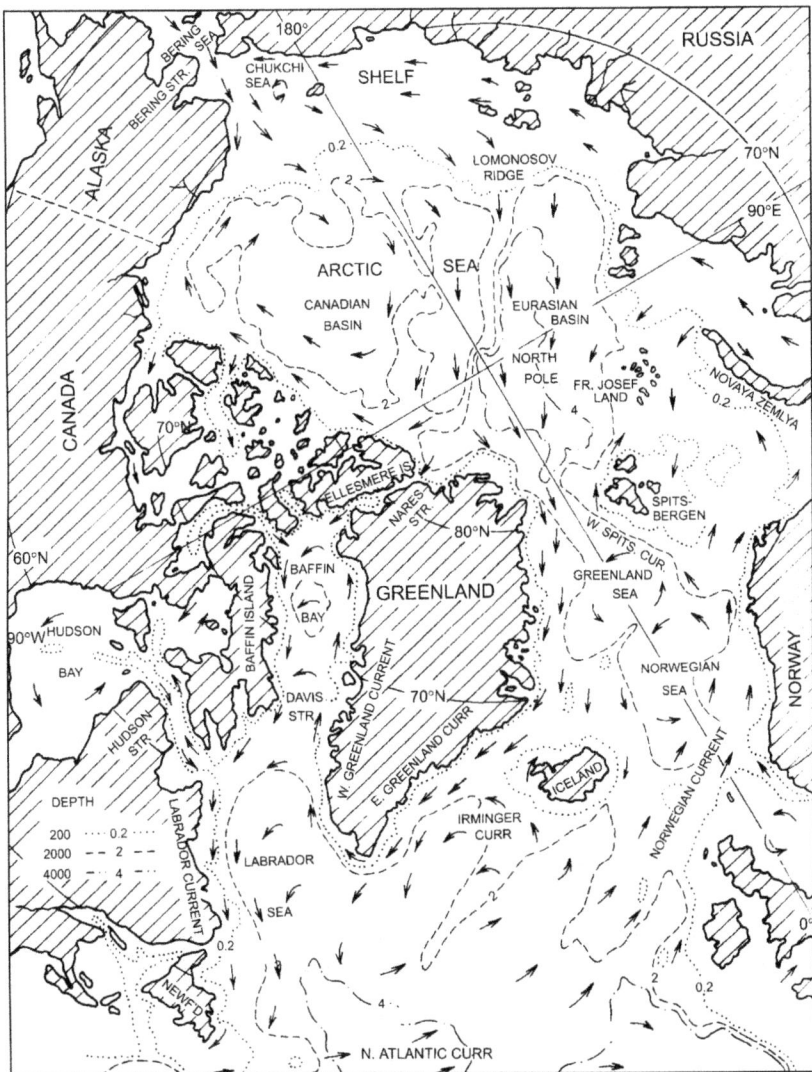

Figure 1.5. Arctic Sea and North Atlantic adjacent seas: bathymetry and surface currents (after Pickard and Emery 1982).

The circulation of the Atlantic Water is basically counter-clockwise around the Arctic Ocean, i.e. in a direction opposite to that of the Arctic Water above it (Pickard and Emery 1982). The Atlantic Water (West Spitsbergen Current) enters the Eurasian Basin from the Greenland Sea and flows further east along the edge of the Eurasian continental slope. Some waters branch off to the north and leave the Arctic as part of the East Greenland Current. The remainder flow across the Lomonosov Ridge into the Canadian Basin. The mixed Arctic (East Greenland Current) and Atlantic (Irminger Current, southwest of Iceland) Waters mass round the southern tip of Greenland and reach the Labrador Sea. Further, they flow as the West Greenland Current to Baffin Bay. This inflow of water is balanced by the southward flow of the Baffin Island Current and Labrador Current. There is also evidence that significant quantities of water of Atlantic origin enter the Arctic Ocean via the Barents and Kara shelves, where they may be considerably modified. Some warm water comes to the Arctic from the Pacific through the Bering Strait (see Figure 1.5). The principal outflows from the Arctic Ocean are through the Fram Strait and the Canadian Arctic Archipelago.

According to research conducted by Alekseev *et al.* (1991), the advection of warmth from the lower latitudes supplies more than 50% of the annual heat supply to the Arctic climate system. Most of this warmth (95%) is, however, transported by atmospheric circulation, with the remainder (5%) being transported by oceanic circulation. In winter, during the polar night, only these two fluxes of warmth reach the Arctic and protect it from significant radiation cooling.

Chapter 2

ATMOSPHERIC CIRCULATION

2.1 Development of Views on Atmospheric Circulation in the Arctic

In the late 19th century scientists undertook attempts to construct various schemes of atmospheric circulation in the Arctic on the basis of theoretical considerations. Ferrel (1882, 1889) argued that the mid-latitude westerlies circulate around a large low pressure system occurring in the Arctic with its centre above the Pole. His idea was preserved until the publication of the meteorological observations from the "Fram" drift (Mohn 1905), and according to Hobbs (1926) even until 1920. Mohn, analysing data from "Fram", confirms an opposite conception to that which had been presented by Ferrel, and earlier by Helmholtz (1888), of the predominance of high pressure and anticyclonic circulation in the Arctic. This concept, further developed by Hobbs (1910, 1926) in his "glacial anticyclone theory", was generally accepted by the majority of meteorologists and climatologists, and was very popular in the 1920s and 1930s (e.g. Brown 1927; Shaw 1927, 1928; Bergeron 1928; Clayton 1928; Baur 1929; Schwerdtfeger 1931; Sverdrup 1935; Vangengeim 1937). It was known as the "permanent Arctic anticyclone" hypothesis. Even Sverdrup (1935), having strong evidence from the "Maud" expedition that deep cyclones also enter the Arctic Ocean in winter, did not challenge this opinion. The supporters of the Arctic anticyclone hypothesis argued that, on average, the atmospheric pressure at sea level in the central Arctic was higher than at temperate latitudes, and the pressure maximum coincided with the temperature minimum. Nothing changed until the publication of the meteorological and aerological observations carried out on the Soviet drifting station NP-1 (Dzerdzeevskii 1941–1945), which confirmed and significantly supplemented the observations of the "Maud" expedition. The observations provided extensive evidence of the absence of a permanent anticyclone in the Arctic. Dzerdzeevskii (1941–1945) showed that both in winter and summer various types of isobaric systems occur in the Arctic, including intense cyclones. In addition, he calculated that the number of days with cyclones in summer was equal to, or exceeded, the number of days with anticyclones. This and other strong evidence presented in his work eventually led to the rejection of the erroneous hypothesis that a stable polar anticyclone occurs in the Arctic. The delay in reaching this conclusion was caused, among others, by the ideas and assertions of Hobbs (1926), which were highlighted again in two papers published in 1945 (Dorsey 1945 and Hobbs 1945). Even up to 1950, Hobbs (1948) maintained that a semi-permanent Arctic high in the Arctic should be present (Jones 1987). The most comprehensive review is presented in Dzerdzeevskii's work and in its English translation (Dzerdzeevskii 1954).

All synoptic charts prior to about 1931 (the U.S. Historical Weather Map series), as well as both the NCAR and UKMO grid-point pressure data sets constructed using these maps, show excessively high values of the mean sea level pressure in the Arctic (up to 8 hPa over the central Arctic away from the North Atlantic sector) (Jones 1987). Jones further writes that the reason for this is "...a lack of basic station data, and the belief amongst many North American meteorologists of the 1920s and 1930s of the existence of a polar or glacial anticyclone."

2.2 Large-scale Atmospheric Circulation

The Arctic must import heat from southerly latitudes due to the net radiation loss to space from the top of the atmosphere. Investigations have also shown that almost the entire deficit of energy is supplemented by atmospheric circulation (see previous chapter). This fact ascribes considerable importance to atmospheric circulation as a climatic factor.

Schemes of general atmospheric circulation which are frequently published in handbooks show the occurrence of the so-called 'polar cell' in the Arctic. In this cell cold dense air flows out from a polar high pressure centre towards a belt of low pressure located about 60–65°N. As a result, easterly and north-easterly winds should dominate in the Arctic. However, in reality, as we know from the previous section, the Arctic high is by no means a quasi-permanent feature of the arctic circulation. Thus, easterly winds are a characteristic phenomenon only for the Atlantic and Pacific sectors of the Arctic. Calculations of the average wind directions for the whole Arctic have not shown any coherence system of polar latitudes either (Barry and Chorley 1992).

Clearly simpler is the circulation in the middle and upper troposphere between about 3 km and 10 km, where a circumpolar cyclonic vortex occurs (Barry and Hare 1974). In these layers the general westerly air circulation is present as a result of the large-scale equator-to-pole temperature gradient and the Earth's rotation (Figure 2.1).

Atmospheric Circulation 15

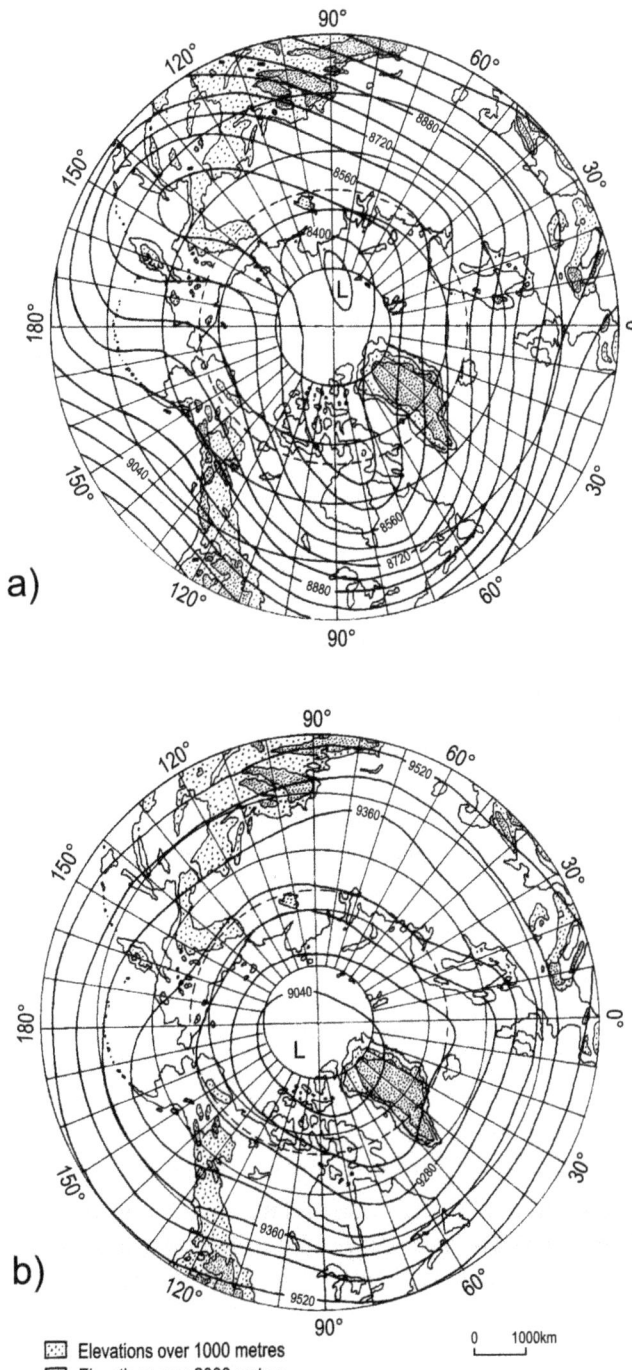

Figure 2.1. Mean height (gpm) of the 300 hPa surface for (a) January and (b) July (after Crutcher and Meserve 1970).

Investigations of the atmospheric pressure field in the Arctic were, and still are, limited by an observation network which is generally too sparse. As has been mentioned earlier, the earliest views presented were based on theoretical considerations. Mohn (1905) published probably the first maps of atmospheric pressure distribution in the Arctic. The next, improved analysis, was provided by Baur (1929). He used all the available data from different expeditions (e.g. the Fram expedition) and data collected during the First International Polar Year of 1882/1883. According to Rodewald (1950), Baur's mean annual chart can be accepted as representative for the years 1874–1933. Later maps became increasingly detailed (e.g. Sverdrup 1935; Dzerdzeevskii 1941–1945; Dorsey 1949 (unpublished) presented by Petterssen *et al.* 1956; Prik 1959; Baird 1964; Crutcher (unpublished) presented by Barry and Hare 1974; Colony and Thorndike 1984; Gorshkov 1980; *Atlas Arktiki* 1985; Serreze *et al.* 1993; Rigor and Heiberg 1997). The authors of the first two maps assumed the existence of a permanent Arctic anticyclone in the central Arctic. As Baur (1929) wrote "...In all months of the year the pressure at the North Pole is higher than at 70° north latitude..." Similar conclusions were presented by Sverdrup (1935), whose research resulted in corrections and supplementations to Baur's maps. Since the second half of the 1930s Soviet scientists have intensified the investigation of the Arctic climate, mainly using data obtained from drifting stations NP-1 (1937) – NP-31 (1991) floating on the Arctic Ocean. As has been mentioned earlier, the results of meteorological observations carried out during the drift of the first station (Dzerdzeevskii 1941–1945) helped to change the view of the distribution of atmospheric pressure in the Arctic. Later on, these investigations were conducted mainly by Prik, who in 1959 published her very well known work (where air temperature was also presented) which has been cited by many authors (Stepanova 1965; Vowinckel and Orvig 1970; Barry and Hare 1974; Sugden 1982). Generally, Prik (1959) confirmed the results presented by Dzerdzeevskii, but, of course, also introduced some changes. She showed that the Arctic anticyclone could only be found as a bridge of high pressure connecting the Siberian high with the Canadian high in the winter months, while in some cases it appeared in the form of a small anticyclone over the Canadian Arctic Archipelago. In the 1960s (Baird 1964) and 1970s (Crutcher, unpublished, after Barry and Hare 1974), maps presenting the mean sea-level pressure were published. However, they did not give any information about the data used in the process of map construction. In comparison with Prik's maps, they are less detailed. Generally, the January pressure distribution in the Arctic is similar in all maps, except those of Greenland. Over Greenland, we may notice the occurrence of anticyclones (Prik 1959) or at least wedges of high pressure (Baird 1964; Barry and Hare 1974). On the other hand, large differences occur in the summer distribution of air pressure. Baird's map (1964) shows the presence of high pressure in the vicinity of the North Pole. Crutcher presents quite a similar pattern. On his map, instead of a high-pressure centre, a wedge of high-pressure covers the Pole, while on Prik's map a low-pressure centre surrounds the North Pole. In the 1980s two atlases were published in Russia, in which syntheses of the Arctic climate (among others) are presented (Gorshkov 1980; *Atlas Arktiki* 1985). The charts of the distribution of atmospheric air pressure in both of these atlases were prepared by Prik. In the first atlas, the mean air pressure distribution for the period 1881–1970 for each month is presented. In the second one, only the mean air pressure for January and July 1881–1965 is shown. It seems to me that, at present, these atlases are the best sources of information about the mean sea-level air-pressure distribution in the Arctic. Therefore, they have been used to describe in detail the patterns of this element in the area studied. However, I must mention here a new possibility which supplements and improves our knowledge concerning pressure distribution in the central Arctic. Since 1979, a network of Arctic drifting buoys has been operated through the University of Washington Polar Science Center (Thorndike and Colony 1980; and subsequent reports through Rigor and Heiberg 1997). The pressure analysis published by Thorndike and Colony for the period 1979–1985 uses data from a dozen buoys and approximately 70 coastal and island stations around the Arctic Ocean. According to McLaren *et al.* (1988) they are more accurate than earlier Arctic pressure fields. This conclusion, it seems to me, is rather untrue, because Prik (for example) used data from 290 stations and 20 drifting stations in constructing the maps published in *Atlas Arktiki* (1985). In any case, data from buoys should significantly improve our knowledge concerning the distribution of air pressure in the Arctic. Maps presenting the mean sea-level pressure for January and July (1979–1996) published by Rigor and Heiberg (1997) show a generally similar pattern to those published in the *Atlas Okeanov* and *Atlas Arktiki* (Figures 2.2a–d). Finally, the results from a recently published work by Serreze *et al.* (1993), examining climatological patterns of Arctic synoptic activity for the period January 1952 – June 1989, should be briefly summarised. The authors have used National Meteorological Center sea-level pressure data set, which, since 1979, have incorporated data from arrays of drifting buoys from the Arctic Ocean Buoy Program. For winter Serreze *et al.* (1993) received results similar to those from most of the other works cited here, including *Atlas Arktiki* (1985). In summer, a much greater difference between these sources exists, although the general pattern is also quite similar. This concerns, in particular, the location of anticyclone pressure centres. In contrast to the map published in *Atlas Arktiki* (Figure 2.2c), Serreze *et al.* (1993) found the lack of a mean summer low both in the central Arctic and in the eastern part of the Canadian Arctic, although a decrease in sea-level pressure is evident in their map.

18   *The Climate of the Arctic*

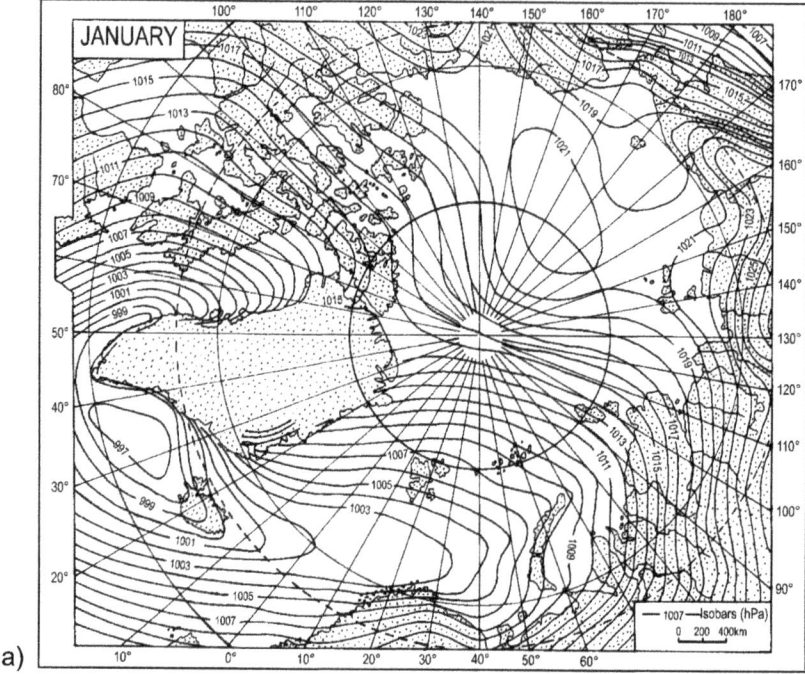

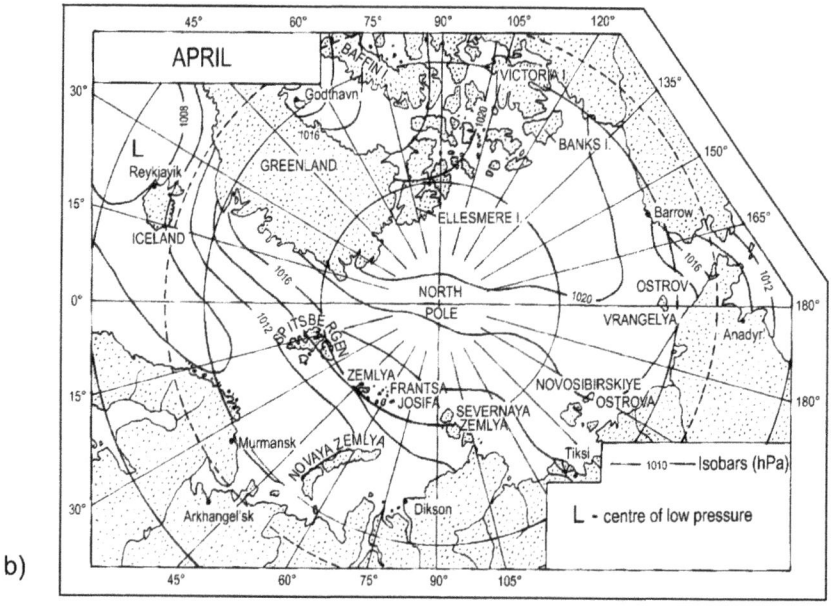

Figure 2.2. Spatial distribution of average air pressure for January (a), April (b), July (c), and October (d) in the Arctic (after *Atlas Arktiki* 1985 (January and July) and Gorshkov 1980 (April and October)).

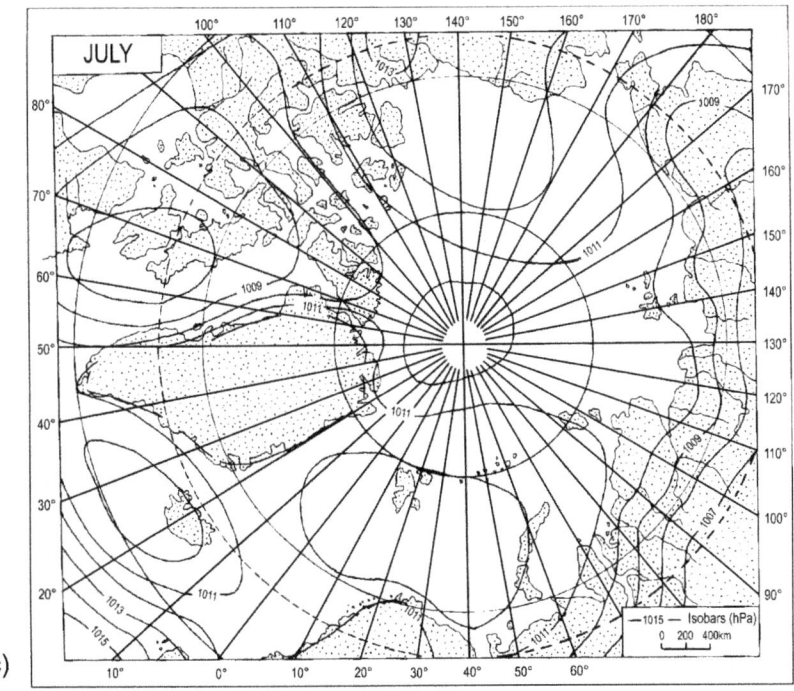

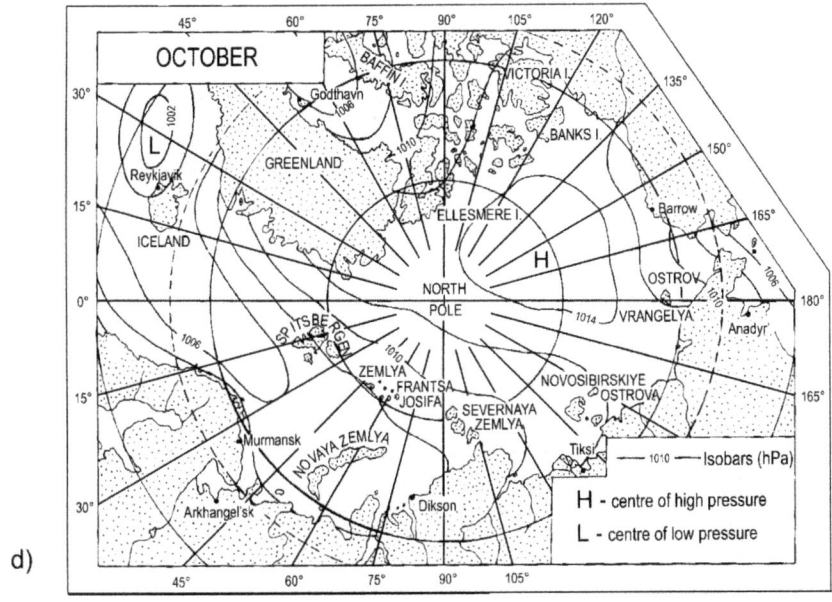

Figure 2.2. cont.

The winter pressure field consists of two main belts. The first one, with low air pressure, encompasses the entire Atlantic sector of the Arctic up to the North Pole (Figure 2.2a), and is mainly controlled by the dynamic of the Icelandic low. Near Iceland on the Arctic front cyclones are generated, which then move into the Arctic, reaching as far as the Kara Sea. As a consequence of this process, in the mean pressure field, the so-called Iceland-Kara Sea trough can be seen. Another extensive trough covers the Baffin Bay region. The second belt with high air pressure encompasses almost all other parts of the Arctic, excluding the Bering Sea and Bering Strait regions. In this part of the Arctic, half way between the Siberian high and the Canadian high, a small centre of high pressure (> 1021 hPa) is present (Figure 2.2a).

In spring (Figure 2.2b), represented by the mean pressure field for April, high pressure may be found to dominate in the whole Arctic, with the maximum (> 1020 hPa) in the northern parts of the Canadian Arctic and Greenland, in the Beaufort Sea, and in the part of the Arctic Ocean neighbouring these areas. The lowest air pressure (< 1012 hPa) is only over the Norwegian and Barents seas. One can agree with the assertion made by Vowinckel and Orvig (1970) that in spring the anticyclonic activity occurs most often over the central Arctic and that in this season the old concept of "Arctic anticyclone" is closest to being fulfilled.

The summer pressure field (Figure 2.2c) shows two centres of high pressure: the first one covers the central part of the Atlantic region (from Novaya Zemlya to Jan Mayen Island), and the second is located over the Beaufort Sea, Alaska, and the MacKenzie Basin. The data for Greenland are not presented on this chart. However, there is some evidence that in the northern part of this island a third centre of high pressure exists (see Serreze *et al.* 1993, their Figure 9) A small low-pressure centre (< 1010 hPa) in the vicinity of the North Pole separates the first two high-pressure centres. A trough of low pressure spreads from here to the eastern part of the Canadian Arctic (where a clear low-pressure centre, < 1008 hPa, exists) on the one hand, and to the central part of the continental Russian Arctic on the other. It is worth noting that the differentiation of air pressure in summer in the Arctic is significantly lower than in other seasons, especially winter. According to Serreze *et al.* (1993), this may be due to "a more even distribution of cyclonic activity than observed in winter, the general lack of spatial variations in mean cyclone and anticyclone pressures, as well as the tendency in other regions for alternation between cyclonic and anticyclonic regimes."

In autumn (Figure 2.2d), the pattern of air-pressure distribution in the Arctic is quite similar to that of winter. However, neither high- nor low-pressure centres are as strong as in winter. On the other hand, the area covered by the Iceland-Kara Sea trough in autumn is greater than in winter and reaches the vicinity of Severnaya Zemlya. The same is true of the low pressure occurring in the region of the Baffin Bay and Pacific sectors of the Arctic.

One can see that the charts do not present the distribution of atmospheric pressure over Greenland. However, from other sources (e.g. Prik 1959; Rigor and Heiberg 1997) we can say that over the year as a whole there is the occurrence of a semi-permanent high-pressure centre (or at least a wedge of high pressure).

2.3 Synoptic-scale Circulation

Similarly to the mid-latitudes, synoptic-scale disturbances control the daily weather events in the Arctic. Vangengeim (1952, 1961) showed that changes of synoptic processes in the Arctic are about 1.5 times faster than in moderate latitudes. Working from this and other facts given earlier, it can be concluded that the climate of the Arctic is significantly more sensitive to atmospheric circulation variability than the climates in both moderate and low latitudes.

The first analysis of the frequency of cyclones and anticyclones, as well as their tracks, based on measurements made on the drifting station North Pole-1 (11 May 1937 – 19 Feb. 1938) was given by Dzerdzeevskii (1941–1945, 1945). These publications are very important for our understanding of the pressure pattern in the central Arctic, and were translated into English by the University of California, Department of Meteorology, appearing in Scientific Report No. 3, 1954. Dzerdzeevskii was the first to draw a map presenting the frequency of the number of days with cyclonic activity for each month from the period May–October. He also distinguished six types of cyclonic activity in the Arctic.

However, the first real cyclone and anticyclone climatologies could not be made until the turn of the 1950s and 1960s, when synoptic charts were more reliable due to a denser network of stations (Keegan 1958; Ragozin and Chukanin 1959; Reed and Kunkel 1960; Gaigerov 1962, 1964). From the more recent works, one can mention McKay *et al.* (1970), Gorshkov (1980), LeDrew (1983, 1984, and 1985), Whittaker and Horn (1984), *Atlas Arktiki* (1985), Serreze and Barry (1988) and Serreze *et al.* (1993). The last two works present an updated climatology of the synoptic systems in the Arctic, taking into account a new data set of air pressure available since 1979 within the Arctic Ocean Buoy Program.

The various sources presented here generally show similar patterns of sea-level cyclonic activity in winter in the Arctic (see Figure 2.3a). The cyclones are most frequent in the Atlantic region of the Arctic and in the Baffin Bay region. On the maps by Ragozin and Chukanin (1959) and Gaigerov (1964), the local maximum of high frequency of cyclones also occurs over the East Siberian Sea. The mean number of passing cyclones oscillates between 4–6 over the Norwegian-Barents-Kara seas and about 4 over the East Siberian Sea (Ragozin and Chukanin 1959). The lowest occurrence of cyclone frequency is noted in the northern part of the Canadian Arctic and in the neighbouring Arctic Ocean. A more detailed analysis made by Stepanova (1965) has shown that the highest frequency of cyclones in the central Arctic is observed with a meridional or eastern type of circulation. A low number of cyclones is noted when the western type of circulation is well developed. The majority of winter cyclones enter the Arctic from the North Atlantic and the Barents Sea, and then track north-eastwards, rarely reaching the Western Arctic. Cyclones from other regions are not often observed.

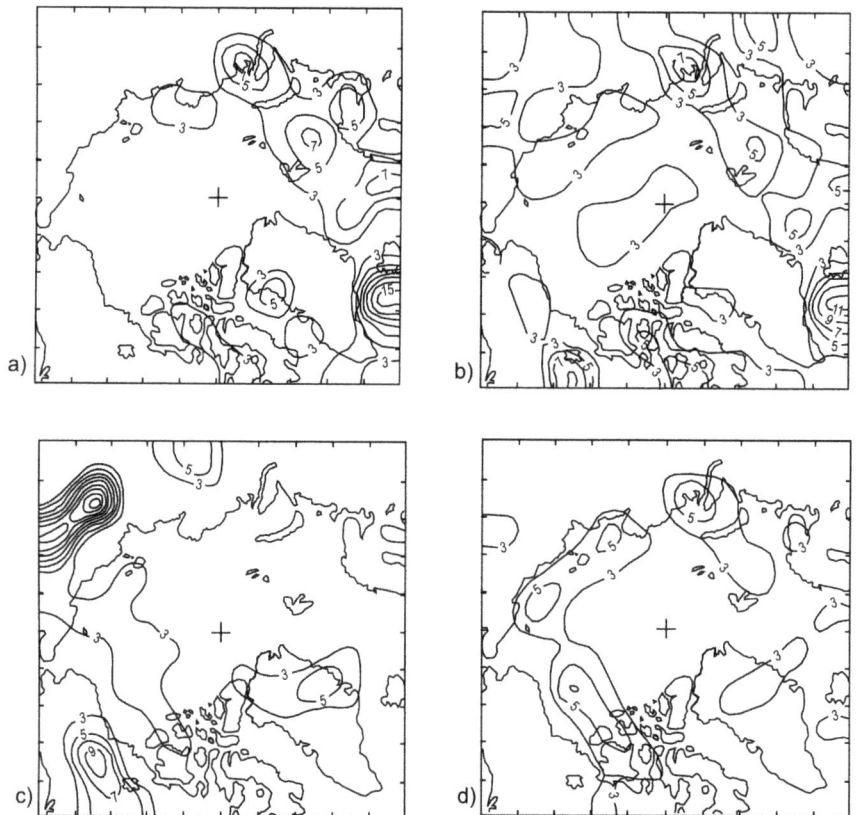

Figure 2.3. Cyclone (a, b) and anticyclone (c, d) % frequencies for winter (a, c), 1952/53 – 1988/89 and summer (b, d), 1952–1989 in squares of 306,000 km², for systems with central pressure < 1012 hPa (cyclones) and >1012 hPa (anticyclones). A cutoff contour of 3% frequency is chosen to accentuate areas of frequent synoptic activity (after Serreze et al. 1993).

Winter (January) anticyclones are almost entirely restricted to the Canadian Basin on the map presented by Serreze and Barry (1988). On Gaigerov's (1964) map this maximum is shifted to the western part of the Canadian Arctic Archipelago. In addition, another maximum occurs over the Barents Sea. In turn, according to Ragozin and Chukanin (1959), the frequency of winter anticyclones is clearly highest in the northern part of Greenland and in the southern part of the Arctic Ocean neighbouring the Beaufort and Chukchi seas. Recent calculations of winter anticyclone frequencies (Serreze *et al.*, 1993) for the period 1952/53 – 1988/89 showed a similar pattern to that presented by Ragozin and Chukanin (1959) (Figure 2.3c). In any case, the highest frequency is clearly seen in the Western Arctic.

In summer, a general decrease in atmospheric pressure occurs. As a result, the mean occurrence frequency of cyclones in this season is similar to that of winter. However, the pattern of cyclone frequency is quite different (compare Figure 2.3b with Figure 2.3a). Moreover, there is a significant divergence between different sources (compare maps in Ragozin and Chukanin 1959; Reed and Kunkel 1960; Gaigerov 1964; Stepanova 1965; Gorshkov 1980; Serreze and Barry 1988; Serreze *et al.* 1993). According to the most recent results (Serreze *et al.* 1993), summertime cyclone frequencies show maxima centred in the Barents and Kara seas as well as over the southern part of the Canadian Arctic (Figure 2.3b). Secondary maxima occur over the central Arctic in the vicinity of the North Pole and over the Laptev and East Siberian seas. Serreze and Barry (1988) note the lack of lows in the Baffin Bay region. It is a surprising result, because the other cited sources (also Serreze *et al.* 1993) show a high frequency of lows here. This means the period 1979–1985 analysed by Serreze and Barry (1988) does not provide a good representation of mean synoptic conditions in the Arctic, at least in the Baffin Bay region. Cyclones move into the Arctic Ocean from various directions. However, they usually arrive from the Siberian coast (Kara, Laptev, and Chukchi seas), the North Atlantic, and the Baffin Bay regions (Figure 2.4). Few cyclones enter the Arctic from the Bering Strait or the Canadian Arctic Archipelago. From the presented map it can be seen that the cyclone tracks meet in the central Arctic, especially over the Canadian Basin. This is contrary to what we observe in winter (see Ragozin and Chukanin 1959).

The summer frequency of anticyclones (Figure 2.3d) shows that they are most common over the following three regions: 1) the western part of the Canadian Arctic and the Beaufort Sea; 2) the East Siberian and Laptev seas; and 3) the Kara and Barents seas. A smaller centre also occurs over northern Greenland. The results of research by Reed and Kunkel (1960) and Ragozin and Chukanin (1959) show a similar pattern. The mean speed of both cyclones and anticyclones is significantly higher in the cold half-year, reaching a maximum in March (Ragozin and Chukanin 1959). The lowest speed of cyclones and anticyclones occurs in August and September, respectively. The mean speed of anticyclones is greater than that of cyclones and oscillates from 43 km/h in March to 35 km/h in September. Analogous values for cyclones are equal to 40 km/h (March) and 34 km/h (August).

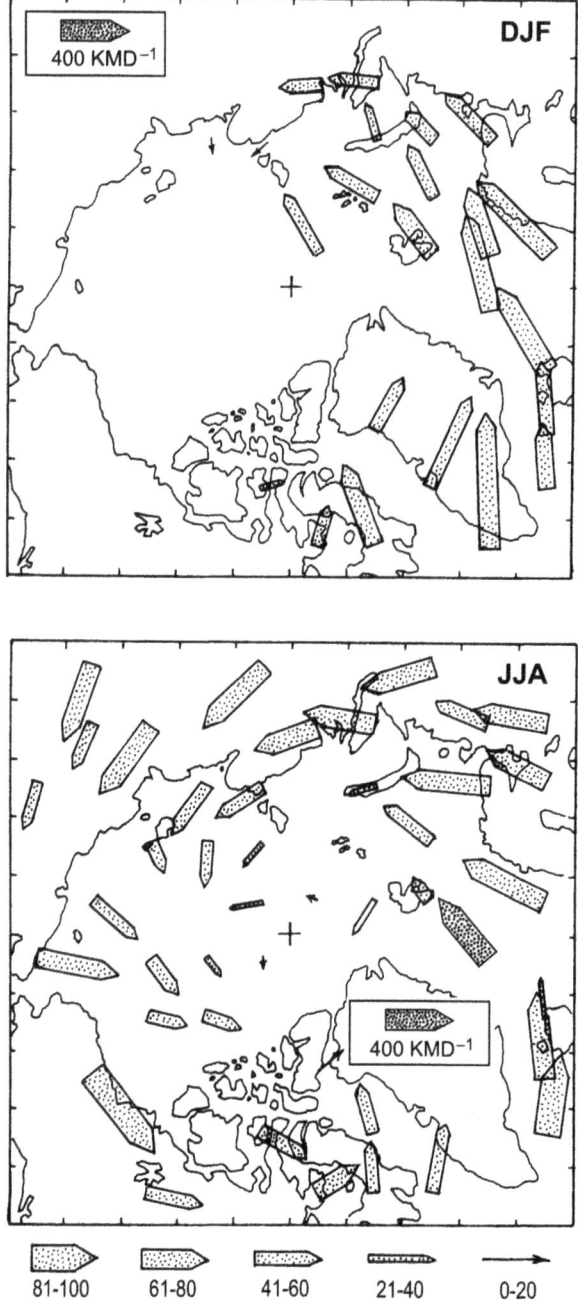

Figure 2.4. Mean cyclone motion vectors for winter (DJF) and summer (JJA), 1975–1989. The length of each arrow is the mean vector magnitude, with the width proportional to the index of motion constancy. Vectors are only plotted for grid cells with at least a 3% frequency of cyclones (after Serreze *et al.* 1993).

2.4 Winds

Winds, as we know, are the result of both large-scale and synoptic-scale atmospheric circulation. In addition, local factors such as geography, orography, and topography (altitude and relief) can sometimes significantly influence the direction and speed of winds (Rae 1951; Wagner 1965; Markin 1975; Maxwell 1980, 1982; Ohmura 1981; Pereyma 1983; Wójcik and Przybylak 1991). There is a paucity of scientific literature describing winds in the Arctic in general. Some information may be found in the following sources: Mohn 1905; Sverdrup 1933; Dzerdzeevskii 1941–1945; Petterssen *et al.* 1956; Prik 1960; Gaigerov 1962; Stepanova 1965; Vowinckel and Orvig 1970; Sater *et al.* 1971; Maxwell 1980, 1982). However, the best sources of information about winds in the Arctic are two atlases (Gorshkov 1980 and *Atlas Arktiki* 1985).

In winter, the "polar easterlies" exist most markedly over the Norwegian and Barents seas and in the Pacific region (Figure 2.5). The main air stream over the western and central Russian Arctic flows from the southern sector (Siberian high) and is then directed towards the North Pole. After passing the North Pole vicinity, the air masses leave the Arctic Ocean through the gate between Spitsbergen and Greenland. Greenland, being a very significant orographic barrier, causes quite a sharp turn of air masses flowing from the Pacific and the eastern Siberian region area of the Pole. After crossing the Arctic Ocean, the air masses are directed south-eastwards and flow over the north-eastern part of the Canadian Arctic, reaching the Labrador Sea. From Figure 2.5 some great differences may be seen between the main air streams and local winds. This is especially true for Greenland stations, where katabatic winds prevail.

Mean wind speed in the Arctic is strongly negatively correlated with the magnitude of atmospheric pressure and simultaneously it is also highly positively correlated with the intensity of cyclonic activity. Mean wind speed in January (Figure 2.6) in the regions characterised by low atmospheric pressure and high cyclonic activity (Atlantic, Baffin Bay and Pacific regions) oscillates between 6 m/s and 10 m/s. On the other hand, the regions with high atmospheric pressure and high anticyclonic activity (almost the whole Arctic Ocean and the northern and western parts of the Canadian region) have the lowest wind speeds (4–6 m/s). The highest wind speeds observed in the central Arctic rarely exceed 25 m/s. The maximum wind speeds in the most windy part of the Arctic (i.e. in the Atlantic region) are twice as strong (up to 50 m/s) (Gorshkov 1980). These storm winds are probably connected with vigorous moving cyclones or with such mesoscale phenomena as "polar lows" (for details see the next section).

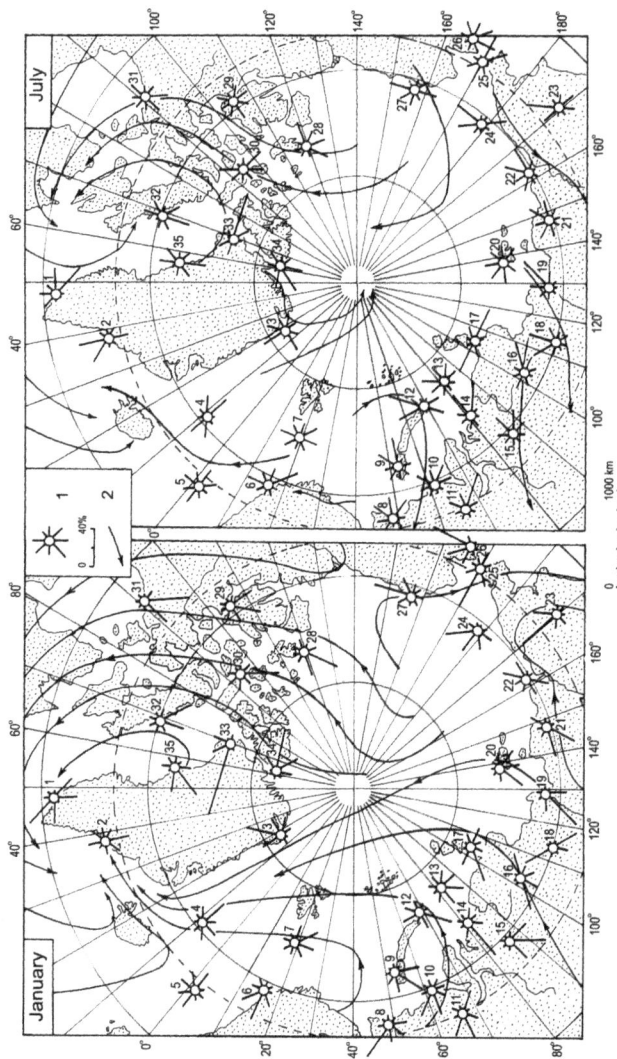

Figure 2.5. Frequency (in %) of the occurrence of winds from eight main directions (1) and prevailing main air streams (2) in the Arctic (after *Atlas Arktiki* 1985). Meteorological stations: 1 – Ivigtut, 2 – Angmagssalik, 3 – Nord, 4 – Jan Mayen, 5 – Ship M, 6 – Annenes, 7 – Björnöya, 8 – Indiga, 9 – Malye Karmakuly, 10 – Amderma, 11 – Salekhard, 12 – Mys Zhelaniya, 13 – Ostrov Yedineniya, 14 – Ostrov Dikson, 15 – Dudinka, 16 – Khatanga, 17 – Mys Chelyuskin, 18 – Zilinda, 19 – Tiksi, 20 – Ostrov Kotelny, 21 – Chokurdakh, 22 – Ostrov Chetyrekhstolbovoy, 23 – Markovo, 24 – Ostrov Vrangel, 25 – Uelen, 26 – Nome, 27 – Barrow, 28 – Mould Bay, 29 – Cambridge Bay, 30 – Resolute, 31 – Chesterfield Inlet, 32 – Clyde, 33 – Thule, 34 – Alert, 35 – Upernavik.

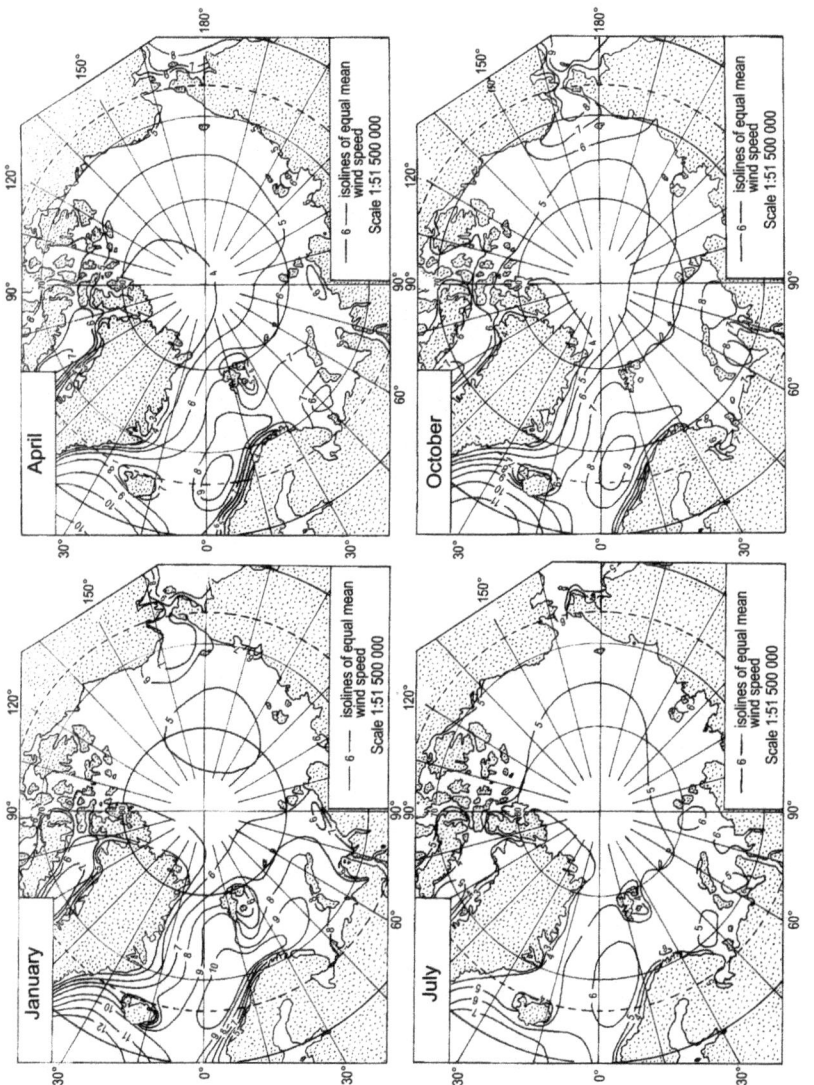

Figure 2.6. Average monthly wind speed (in m/s) in January, April, July, and October in the Arctic (after Gorshkov 1980).

In the course of the year, the maximum anticyclonic activity and the minimum cyclonic activity occur in spring, especially in the Western Arctic. In the area of the greatest occurrence frequency of anticyclones (> 15%) between the North Pole and the northern parts of Greenland and the Canadian Arctic there are no clear dominant wind directions. Simultaneously, mean (Figure 2.6) and maximum wind speeds are the lowest in these areas (< 4 m/s and < 20 m/s, respectively). In stations Alert and Isachsen the frequency of calms was noted in 59% and 32% of all observations, respectively. Southern and south-western winds prevail on the Russian Arctic coast, with the exception of the Chukchi Peninsula. In the rest of the Arctic the main air streams are similar to winter. In the Western Arctic, mean wind speeds rarely exceed 6 m/s. Higher speeds are only recorded in the southern part of the Canadian region and in the Baffin Bay region. As in winter, the strongest winds occur in the southern part of the Atlantic region.

In summer, the main air streams in the Canadian Arctic and over the Norwegian Sea show generally the same pattern as in winter (Figure 2.5). In the northern part of the Barents Sea, south-western and western winds prevail. Further east, in most parts of the Russian Arctic, northern and eastern winds dominate. From the Pacific region the main air stream flows northwards towards the pressure depression around the Pole. Almost from the opposite side of the Arctic (the Greenland Sea) large air streams also reach that depression. This means that there is a marked convergence zone of air streams in the central Arctic Ocean in summer. Wind speeds in this season in the Canadian Arctic and in the Arctic Ocean are mostly slightly higher than in spring (Figure 2.6). On the other hand, in the Atlantic region and in the Baffin Bay region the wind speeds are significantly lower in summer, rarely exceeding 6 m/s. Generally, the distribution of the maximum wind speeds also follows the same pattern.

In autumn, the patterns of distribution of atmospheric pressure and synoptic activity are roughly similar to those of winter (see Figures 2.2a and 2.2d and maps in *Atlas Okeanov*). As a consequence, maps showing the distribution of mean wind speed (Figure 2.6) in the Arctic in the seasons analysed are also similar. In autumn, however, wind speeds are generally slightly lower, with the exception of the Pacific region, where the strongest wind speeds are probably connected with greater number of cyclones in this season entering the Chukchi Sea through the Bering Strait. Stronger cyclonic activity also causes greater wind speeds over the Kara, Laptev and East Siberian seas.

2.5 Local Circulation and Mesoscale Disturbances

As mentioned at the beginning of previous section, local factors can sometimes significantly change the surface wind speed and direction. This change in many cases is so great that little or any connection with the large-scale circulation exists. In addition, local circulation and other mesoscale phenomena such as polar (also called Arctic) lows can also markedly change the parameters of wind. Thus to describe large-scale atmospheric circulation, we cannot use the observations of wind speeds and directions from the stations (especially land stations) where local influences are great. In such cases it is better to use geostrophic winds as an indicator of surface winds.

2.5.1 Local Winds

According to the nomenclature proposed by Barry (1981), three kinds of so-called "fall" winds can be distinguished: the bora, the foehn, and (mesoscale) katabatic winds. Katabatic winds are treated here in the strict sense as local downslope gravity flows caused by nocturnal radiative cooling near the surface under calm clear-sky conditions. It seems to me that in the Arctic, where polar night and polar day occurs, distinguishing between the bora and the katabatic winds can be difficult. Thus, I propose to treat them as one type of wind, more or less cold, dry and gusty.

The bora winds are very well known from Greenland, where they are generally directed from the interior towards the coast. These downslope winds result from the presence of cold air over the ice cap, which subsequently flows down the slopes of ice cap under the influence of gravity. The speeds of these winds mainly depend on the steepness of the slope and on the pressure gradient between the summits of the ice cap and the coasts. On the ice cap it is mostly katabatic winds which occur (Putnins 1970). According to measurements taken by Loewe (1935), katabatic winds were recorded at Weststation and at Eismitte for three-quarters of the observation period. They are more common in the winter half-year (see the directions of winds in Greenland stations, Figure 2.5). The katabatic winds are better developed on clear days when the temperature at the centre of ice cap is below normal (greater radiation cooling). Bora winds are also present in other Arctic regions. They can occur everywhere else where a sufficient area of high elevation exists and allows the accumulation of air masses. Boras were observed, for example, in Novaya Zemlya (Vize 1925; Kanevskiy 1962; Shapaev 1959; Barry and Chorley 1992), Svalbard, Zemlya Frantsa Josifa, Ostrov Vrangelya (Shapaev 1959), on the coast of the Kara, Laptev, East Siberian, and Chukchi seas (Shapaev 1959), and on some islands of the Canadian Arctic Archipelago (Maxwell 1980).

The foehns also occur quite often in the Arctic. They are warm, dry, and gusty winds occurring on the lee side of mountain ranges. The warmth and dryness of the air is due to adiabatic compression on descending the mountain slopes. In Greenland their advection is generally towards the sea on both coasts (Putnins 1970). The vertical extents of the foehns are not great. On the other hand, temperature increases can be very large. Exceptionally strong foehns can raise the temperature by more than 30°C (Schatz 1951). Their duration is not long and usually does not exceed two to five days. Generally speaking, foehns can also occur everywhere else in the Arctic, where flowing air masses have to cross sufficiently high mountain ridges. They were identified in, among other locations, Spitsbergen (e.g. Rempp and Wagner 1917; Pereyma 1983; Wójcik et al. 1983; Kalicki 1985; Marciniak et al. 1985; Gluza and Piasecki 1989), in Novaya Zemlya (Vize 1925; Shapaev 1959; Kanevskiy 1962; Kanevskiy and Davidovich 1968), in Zemlya Frantsa Josifa (Krenke and Markin 1973b), on the coasts of the Kara, Laptev, East Siberian, and Chukchi seas (Shapaev 1959); in Alaska (Gledonova 1971), and in the Canadian Arctic (Defant 1951; Andrews 1964; Barry 1964; Müller and Roskin-Sharlin 1967; Jackson 1969; Gledonova 1971).

Other types of local winds, such as land and sea breezes, occur more rarely in the Arctic. Shapaev (1959) notes that breeze circulation occurs in July and August on the coasts of Arctic seas and on the coasts of largest islands. Jackson (1969) has noted the importance of the sea breeze influence for the high frequency of south-westerly winds during the summer at Tanquary Fiord in the Canadian Arctic. Sea breezes occur mainly in summer in the Arctic when there is a strong thermal gradient between the warm land and the relatively cold seawater. A detailed investigation of wind directions in Hornsund (Spitsbergen) shows that their frequency is greatest near noon and in the afternoon hours (Wójcik and Kejna 1991). On the other hand, in winter the land breezes occur in places where warm open waters come into contact with the cold land coasts or pack ice (Maxwell 1980).

In the hilly and mountainous regions of the Arctic in the warm half-year, mountain (katabatic) and valley (anabatic) winds may also occur. On warm sunny days, the heated airs in a valley go up the axis of the valley. At night the process is reversed: the cold denser air from higher elevations flows down. When the glacier covers the upper part of the valley, then the glacier wind merges with the mountain wind. Glacier winds, which are special cases of mountain wind, occur both at night and during the day. During the day, however, the wind is weak and present only in a shallow layer over the glacier (usually < 2 m).

2.5.2 Polar Lows

As Turner et al. (1991) have stated, the investigation of polar lows does not have a long history – only about 30 years. British meteorologists were the first to use the term "polar lows" to describe cold air depressions that affect the British Isles (Meteorological Office 1962). Businger and Reed (1989) formulated a rough definition of polar lows. This phenomenon may "denote any type of small synoptic or subsynoptic scale cyclone that forms in a cold air mass polewards of major jet streams or frontal zones and whose main cloud mass is largely of convective origin." Typical horizontal resolutions of polar lows vary from a few hundred kilometres to more than 1000 km in diameter. Similarly, their intensities may range from moderate breezes to hurricane force winds (e.g. Reed 1979; Locatelli *et al.* 1982; Rasmussen 1981, 1983; Shapiro *et al.* 1987). They form over the oceans and decline rapidly on reaching land. In the Northern Hemisphere polar lows occur only in winter. More information about polar lows may be found in a book entitled *Polar and Arctic Lows* (Twitchell *et al.* 1989).

From this description it may be ascertained that polar lows can significantly change the weather. In the Arctic, polar lows occur mainly in the Barents and Norwegian seas (Rasmussen 1985a, b; Shapiro *et al.* 1987), but have also been noted (although more rarely) in the Greenland Sea (Fett 1989), the Beaufort and Chukchi seas (Parker 1989), and in the Bering Sea (Businger 1987).

Chapter 3

RADIATION CONDITIONS

In the history of actinometric measurements in the Arctic, five phases can be distinguished:

Phase 1. The 19th century. During this period, measurements of solar radiation were made using ordinary thermometers, i.e. the difference between the readings of thermometers with shaded and exposed bulbs, placed in the sun and in the shade, was used to estimate the intensity of radiation. According to Gavrilova (1963) the first such measurements were made during the expeditions of John Franklin to the polar sea in the years 1825, 1826, and 1827 (Franklin 1828). Later on, using the same method, measurements were conducted during different expeditions to the Arctic (*Solar Radiation...*, 1876; *Report of the International...*, 1885; *Observations of the International...*, 1886).

Phase 2. The end of 19th century – the Second International Polar Year (1932/1933). For actinometric measurements, the date of the construction of the first pyranometer by Ångström at the end of the 19th century was very important. Westman conducted the first measurements using this instrument at Treurenberg Bay in Spitsbergen in 1899–1900 (Westman 1903). A greater number of measurements during this period were carried out in the 1920s and at the beginning of 1930s (e.g. Kalitin 1921, 1924, 1929; Götz 1931; Mosby 1932; Georgi 1935; Kopp 1939; Wegener 1939). All these actinometric measurements were, however, of a temporary and episodic character. At the end of this period most scientists knew that the establishment of a network of stations was necessary in order to determine the radiation regime of the Arctic.

Phase 3. The Second International Polar Year – 1950. In this period in the years 1932–1933 the first continuous actinometric observations were made simultaneously at a number of stations. The second important event was the organisation of an actinometric network in the Soviet Arctic (six stations).

Phase 4. 1950 – the start of the satellite era (ca. 1972). This period (mainly in the 1950s) saw the establishing of most of the actinometric stations which now exist in the Arctic. The majority of them (about 20) are located in the Russian Arctic (see Figure 1 in Gavrilova 1963). In addition, since 1950, regular observations have been carried out in the central Arctic by the drifting stations "Severnyy Polyus" (Volkov 1958; Sychev 1959).

Phase 5. Satellite era – the present. In this period, besides standard in situ actinometric observations, different remote sensing techniques have become increasingly popular, especially those using satellites. These new methods are especially important for the Arctic, where the network of meteorological stations, as we know, is scarce. They permit the calculation of incoming solar radiation, albedo, and outgoing long-wave terrestrial radiation loss – all the necessary components of the net radiation of the earth-atmosphere system (see Cracknell 1981; Lo 1986; Harris 1987; Schweiger *et al.* 1993). On a global scale, several satellite-based data sets exist (International Satellite Cloud Climatology Project, Earth Radiation Budget Experiment), which have been used recently for projects such as the calculation of Surface Radiation Budget (Whitlock *et al.* 1993). In recent years, satellite-based methods have also been used to retrieve surface radiative fluxes for small areas (Parlow 1992; Scherer 1992; Duguay 1993). For this purpose Landsat satellites are best, due to satellite orbit and sensor resolution (Haefliger 1998). For greater areas the data received from NOAA AVHRR (Advanced Very High Resolution Radiometer) and SSM/I (Special Sensor Microwave/Imager) are most useful for calculations of short-wave and long-wave surface radiative fluxes. Such studies have been conducted for the Fram Strait area (Kergomard *et al.* 1993) and the Greenland Ice Cap (Haefliger 1998).

Although, it is possible to find a great quantity of literature and points of view on the different aspects of the radiation regime in the Arctic, our knowledge about the climatology of the radiation balance and its components is still meagre. This is especially true of the central Arctic and Greenland Ice Sheet regions. The main reason for this is the sparse network of actinometric stations in the Arctic. From the different sources presenting radiation conditions in the study area, especially those in the form of charts, the reader should know that the isolines drawn give only a rough approximation of the real situation. In many cases, the charts were based on theoretical calculations. It is hoped, however, that in the future, satellite remote sensing techniques will help to collect data sets of surface (and other) radiative fluxes for the Arctic with much greater temporal and spatial resolution than we have today.

Gavrilova (1963), Marshunova and Chernigovskii (1971), and Ohmura (1981, 1982) gave excellent reviews of the historical development in radiation studies up to about 1980. Therefore, there is no need to repeat this here again. However, some more important elaborations, from a climatological point of view, should be mentioned here, together with some new works which appeared after Ohmura's reviews. It is rather unquestionable that the role played by Soviet (Russian) climatologists in investigations into the radiation regime in the Arctic was, and probably still is, the greatest. The most important works published by Russians are the following: Kalitin (1940, 1945), Marshunova (1961), Budyko (1963), Gavrilova (1963), Chernigovskii and Marshunova (1965), Stepanova (1965), Marshunova and Chernigovskii (1971), Gorskhov (1980), Makshtas (1984), *Atlas Arktiki* (1985) and Khrol (1992).

From the non-Russian authors, without doubt the greatest contribution in investigations of radiation and heat balance of the Arctic and their components has been made by the staff of the Arctic Meteorology Research Group, Department of Climatology, McGill University in Montreal, and in particular Vowinckel and Orvig. Their results were first published in the McGill University Scientific Reports, Publications in Meteorology (Larson and Orvig 1962; Vowinckel and Orvig 1962, 1963, 1964a, 1965; Vowinckel 1964a, b; Vowinckel and Taylor 1964) and in shorter versions in a series of papers published in Archiv für Meteorologie, Geophysik und Bioklimatologie (Vowinckel and Orvig 1964b, c, d, 1966; Vowinckel and Taylor 1965). All these papers were later used by Vowinckel and Orvig in the preparation of some fragments of their best known work published in the series of the World Survey of Climatology (Vowinckel and Orvig 1970). A good summary of the heat budget of the earth-atmosphere system in the Arctic, based mainly on above-mentioned sources, has been provided by Fletcher (1965). Of the other important works, in particular those which give the spatial distribution of radiation balance and its components only for the Canadian Arctic, one should mention the publications of Maxwell (1980), McKay and Morris (1985), and Woo and Young (1996), among others. In recent years intensive investigations of radiation conditions on the Greenland Ice Sheet have been mainly carried out by the Institute for Atmosphere and Climate at the Swiss Federal Institute of Technology in Zürich using satellite techniques (Ohmura *et al.* 1991, 1992; Konzelmann 1994; Konzelmann and Ohmura 1995; Haefliger 1998).

3.1 Sunshine Duration

Knowledge about sunshine duration, aside from being important theoretically, is also of practical significance. The study of sunshine duration enables improved calculations of global solar radiation (e.g. Spinnangr 1968; Dahlgren 1974; Markin 1975). Such a possibility is very important for the Arctic, where only infrequent numerous and short series of actinometric observations are available. Sunshine duration, having a strong relationship with cloudiness, can also supplement our information, especially concerning the changes of cloudiness in the daily cycle. Registration of the duration of sunshine is a routine observation carried out in most meteorological stations. Bearing in mind all the above-mentioned facts, it is very strange that the distribution of sunshine duration in the Arctic is not better known. Bryazgin (1968) relates this to the surprisingly little attention which climatologists have paid to this element. During the last thirty years situation has not changed.

Some information about sunshine duration may be found in papers describing the climates of different (mainly small) parts of the Arctic (e.g. *Meteorology...*, 1944; Petterssen *et al.* 1956; Gavrilova 1963; Spinnangr 1968; Krenke and Markin 1973a, b; Markin 1975; Maxwell 1980; Pereyma 1983). Only Bryazgin (1968) has made a synthesis for the entire Arctic. In this paper he described the mean monthly and annual sums of sunshine duration, but only the maps for April and for the year as a whole are presented. April was chosen because, according to Bryazgin, the amounts of sunshine duration are greatest in this month. Later on, Bryazgin was a leading author of maps presenting sunshine duration in the Arctic, which are published in the *Atlas Okeanov* (Gorshkov 1980) and the *Atlas Arktiki* (1985). It is important to add that Bryazgin, in constructing these maps, used observational data from the period 1936–1970 (in *Atlas Arktiki*) from only 42 Russian meteorological stations. For the rest of the Arctic, the monthly and annual amounts of sunshine duration were computed based on a significant correlation between sunshine duration and cloudiness. So, for this part of the Arctic, the maps present only rough approximations of sunshine duration in reality.

Marshunova and Chernigovskii (1971), having data about sunshine duration for the whole Arctic, challenged Bryazgin's (1968) assertion that in the entire Arctic the maximum sunshine duration occurs in April. Bryazgin (1968) based his claim only on data from Russian stations and generalised the results from them for the whole of the Arctic. It turned out, however, that in the Canadian and Pacific regions the maximum duration of sunshine is observed in June or July. Even in Spitsbergen, as reported Spinnangr (1968), the highest values do not occur in April, but in May.

In January, sunshine duration does not occur above 70°N (polar night). In the lower latitudes of the Arctic, the monthly mean amounts of sunshine duration rarely exceed 10 hours.

In April, the amounts of sunshine duration, as mentioned above, are considerable (Figure 3.1). The maximum values of sunshine duration may be found in the vicinity of the North Pole and in the central part of Greenland Ice Sheet (> 400 h). A high duration of sunshine is generally typical of the whole Arctic Ocean, Greenland (excluding the southern coastal parts), and the northeastern parts of the Canadian Arctic (> 300 h), where anticyclonic activity prevails (low cloudiness) – see Figure 5 in Gavrilova (1959). On the other hand, the lowest amounts of sunshine duration (< 200 h) occur in the areas characterised by intensive cyclonic activity (the Atlantic region, southern part of the Baffin Bay region, and the southern part of Chukchi Sea and Bering Strait regions).

Radiation Conditions 37

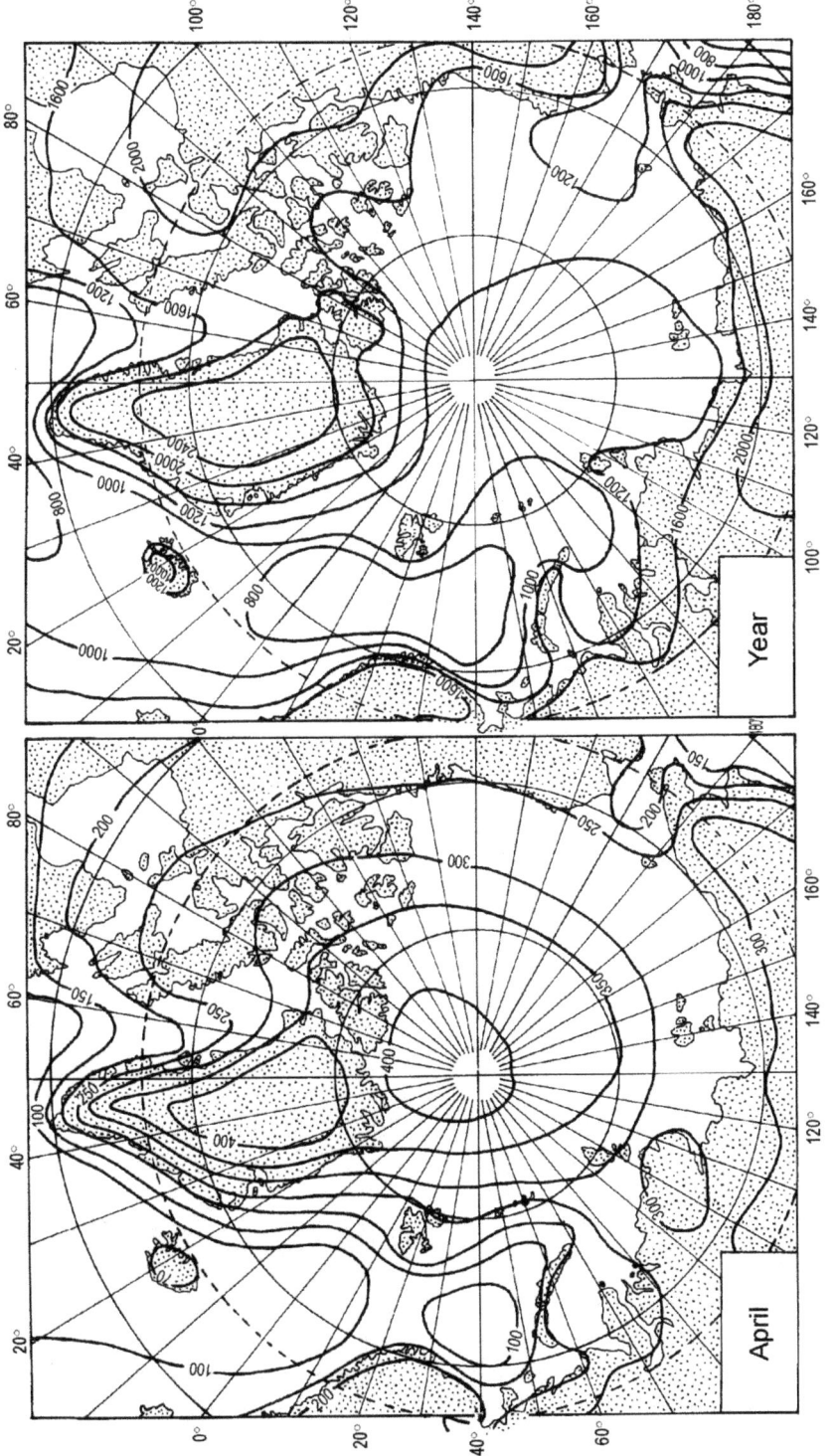

Figure 3.1. Sunshine duration (in hours) in the Arctic (after *Atlas Arktiki* 1985).

In July, the situation in comparison with April is dramatically worse, mainly in the Arctic Ocean, where, at this time, very high low-level cloudiness is observable (see Figure 5.4 or Vowinckel and Orvig 1970). Mean amounts of sunshine duration only vary about 100 h (four times less than in April). The same values, or even lower, are noted in the Norwegian and Barents seas. The Arctic shelves seas and coastal parts of the Russian Arctic receive about 150 h. Further to the south, the duration of sunshine rises to about 200–250 h. Bryazgin (1968) did not give the sunshine duration for the Canadian Arctic and Greenland. Dahlgren (1974) reported that the mean value of sunshine duration (from two years) at Devon Island in July was equal to 316 h. In Greenland, similar or (more probably) greater amounts occur.

In mid-October the polar night is present above 82°N. The Arctic islands above 80°N have only 1 h of sun. Further to the south the situation is better and coastal parts of the Arctic seas have about 20 h to 40 h and along the Arctic Circle about 60 h to 80 h (Bryazgin 1968). On Devon Island the sunshine duration is equal to 25 h (Dahlgren 1974) and is almost twice as great as in Spitsbergen (13 h, Isfjord Radio, 1951–1960 (Spinnangr 1968)). According to data published by Gavrilova (1963), it seems that the coastal parts of the Kara and Laptev seas have lower sunshine duration amounts in October (< 10 h) than were given by Bryazgin.

On an annual basis, the inner parts of Greenland receive the greatest amounts of sunshine duration (> 2400 h) (Figure 3.1). Aside from Greenland, the south-western part of the Canadian Arctic and the western part of the Chukotka region receive more than 2000 h of sun. Almost the whole Canadian Arctic also has high values (> 1600 h), along with some southern parts of the Russian Arctic and Alaska. The lowest duration of sunshine occurs between Jan Mayen Island and Novaya Zemlya (< 800 h). Bryazgin (1968) distinguished a small area between Jan Mayen and Björnöya islands where annual amounts of sunshine duration do not exceed 300 h. Of all the Arctic stations, the lowest values are found in Björnöya (annual mean only 249 h), then at Hopen and Jan Mayen (both 444 h). This is the region with the highest cyclonic activity and, as a consequence, also the highest cloudiness in the Arctic.

The average mean relative sunshine duration for the whole Arctic, which indicates the percentage of the astronomically possible sunshine, oscillates between 20% and 25%. Throughout the year, the highest fraction of actual to possible duration of sunshine is recorded in March and April (30%–40%) and the lowest in October and November (3%–5% in the marginal Arctic seas and 15%–20% in the continental Arctic) (Gavrilova 1963).

There is a considerable fluctuation in the duration of sunshine in some years. From the comparison of the maximum and minimum annual values of this element given in *Atlas Arktiki* (1985), it is evident that the maximum values may be two to three or more times greater than the minimum values.

3.2 Global Solar Radiation

Global solar radiation is one of the important factors in the formation of the radiation regime, weather, and climate. Its role in both net radiation and energy balance is crucial. Luckily, this component of the radiation balance is very easy to measure. From the reasons mentioned above, global solar radiation measurements are most often conducted in actinometric stations, not only in the Arctic. In spite of this, the network of stations is still insufficient for analysing the field of global solar radiation. Therefore, authors presenting the spatial distribution of this (or any other) component of radiation balance have to obtain additional information from calculations. There are two methods of determining the values of the global radiation (Marshunova and Chernigovskii 1971). The first method uses existing radiation data from some stations and established relationships between radiation and other meteorological elements such as cloudiness and sunshine duration. The second method, the so-called analytical method, uses knowledge concerning processes influencing the radiation both outside the atmosphere and on its way through atmosphere to the Earth's surface.

The most comprehensive and detailed information about global solar radiation for the Arctic in general or for specific parts of it is contained in the following works: Gavrilova 1963; Vowinckel and Orvig 1964b, 1970; Chernigovskii and Marshunova 1965; Marshunova and Chernigovskii 1971; Gorshkov 1980; Maxwell 1980; *Atlas Arktiki* 1985; McKay and Morris 1985 and Khrol 1992. These authors present various maps showing both monthly and annual mean sums of this element. Below, the main features of the distribution of global solar radiation in the Arctic for four months representing all seasons, and for the year as a whole, are described based on the recently published *Atlas of the Energy Balance of the Northern Polar Area* (Khrol 1992). The authors of these maps are Girdiuk and Marshunova.

In January, the polar night occurs in the greater part of the Arctic and the radiation flux is naturally zero. According to Gavrilova (1963), the zero isoline approximately follows the 71°N latitude, while on Girdiuk and Marshunova's map this isoline is shifted to about 68°N. It should be noted that all captions under figures presenting global radiation in the English translation of Gavrilova's book (Gavrilova 1966) are wrongly positioned. Most southern parts of the Arctic receive no more than 2 kJ/cm^2.

In April, the mean sums of global solar radiation oscillate from 50 –53 kJ/cm^2 in the southern part of central Canadian Arctic and in the southern inner fragment of Greenland to 25–29 kJ/cm^2 over the Norwegian and Barents seas, where the cloudiness is highest. In the central part of the Arctic Ocean the incoming radiation changes from 35 kJ/cm^2 (in the vicinity of the North Pole) to about 38 kJ/cm^2 at the 80–85°N latitude (except in the part of the Arctic neighbouring the Norwegian and Barents seas) (Figure 3.2). Please note the similarities in pattern distribution of sunshine duration and global radiation in the Arctic (compare Figure 3.1 and Figure 3.2). The contribution of April to the annual influx of radiation is substantial at around 13–15% (Marshunova and Chernigovskii 1971).

In July, with the decreasing altitude of the sun, the global radiation fluxes are about 1.2–1.4 times lower than in June. Moreover they are reduced by the considerable increase of cloudiness in July. Clearly the highest sums of global radiation (84–85 kJ/cm^2) are received in the northern half of Greenland. The secondary maximum occurs in the Canadian Arctic (except its western part) and in the Arctic Ocean neighbouring the Beaufort and Chukchi seas (59 –63 kJ/cm^2). Almost the entire Atlantic region receives < 50 kJ/cm^2. The absolute minimum (40–42 kJ/cm^2) of incoming radiation occurs in the areas to the south and south-west of Spitsbergen (Figure 3.2), where mean cloudiness is the highest and clouds are most dense in the Arctic. The contribution of July to the annual flux of radiation is 17–19%.

In October, the pattern of global radiation distribution is very simple and depends mainly on the length of the days. Therefore, the run of the isolines is more or less zonal. For example, in the region surrounding the North Pole, where the polar night has already begun, the zero isoline passes close to 83°N. The latitudinal band (73–75°N) receives about 4 kJ/cm^2. The greatest sums of global radiation (> 10 kJ/cm^2) are received in the southern parts of the Canadian Arctic and Greenland.

On an annual basis, the global radiation distribution pattern closely resembles the atmospheric circulation and cloudiness distribution patterns. The parts of the Arctic which have the greatest cyclonic activity and cloudiness (mainly Atlantic region) receive the lowest totals of global radiation (< 250 kJ/cm^2). On the other hand, the southern part of the Canadian Arctic and Greenland (central part), where anticyclonic circulation prevails and the lowest cloudiness occurs, receive more than 350 kJ/cm^2 and even 400 kJ/cm^2 (Figure 3.3a).

The values of global radiation in particular years may be different from the average conditions presented here. However, as reported Marshunova and Chernigovskii (1971), the mean deviations of the monthly sums of global radiation oscillate mainly between 8% and 12%. Only extreme deviations sometimes reach up to 30%. To reliably describe the radiation regime in the Arctic, at least five years of observations is needed (Marshunova and Chernigovskii 1971).

Radiation Conditions 41

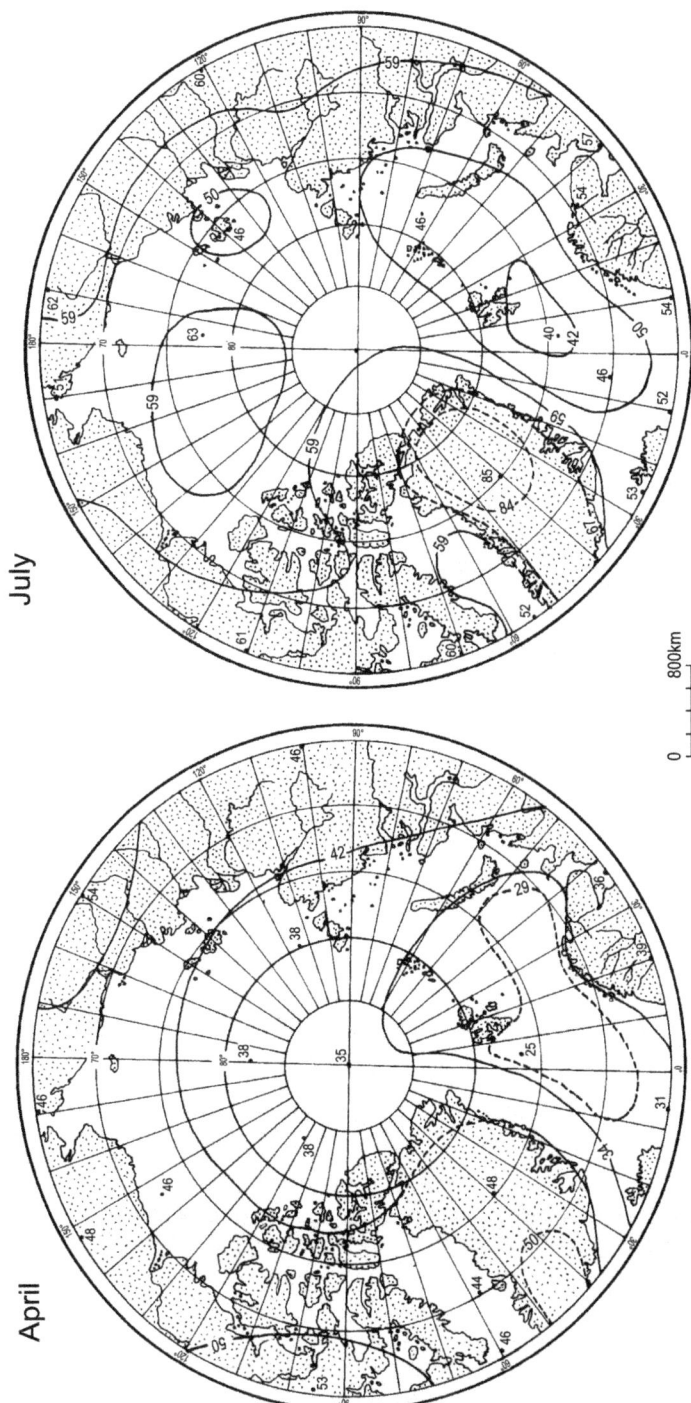

Figure 3.2. Global radiation (in kJ/cm^2/month) in the Arctic in April and July (after Khrol 1992).

42   The Climate of the Arctic

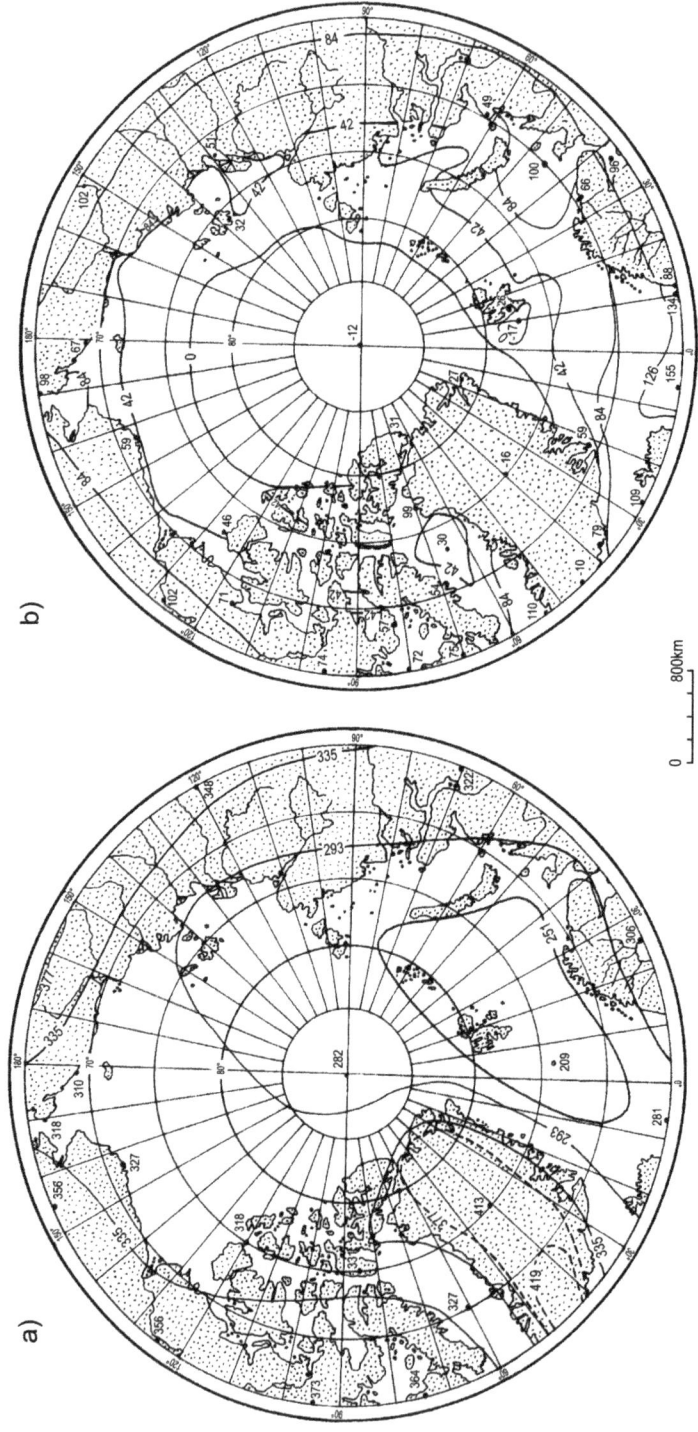

Figure 3.3. Annual totals (in kJ/cm^2) of global radiation (a) and of net radiation (b) in the Arctic (after Khrol 1992).

3.3 Short-wave Net Radiation

From a climatological point of view, knowledge about short-wave net radiation (or the absorbed solar radiation) is more important than about the potential global solar radiation reaching the earth's surface. Short-wave net radiation depends mainly on the declination of the sun and surface albedo. Along the same latitude only the albedo determines the amount of absorbed energy by the earth's surface. On a local scale (in mountainous areas), however, differences in the elevation of the land, its aspect and inclination also significantly control the amount of solar radiation which is received.

3.3.1 Albedo

As mentioned earlier, albedo is a very important factor in the short-wave balance of the surface. The computation of mean monthly values of albedo for the entire Arctic is a very difficult task. Problems arise for different reasons, e.g. the lack of insufficient in situ measurements or measurements taken from aircraft, dynamical changes of the area and physical characteristics of vegetation, snow, and ice covers, which mainly influence the albedo. In recent years, however, the chance of receiving the real distribution of albedo changes (and other components of the radiation balance) in the Arctic has markedly grown thanks to the possibilities provided by satellite techniques.

At present there are only a few publications which give the mean monthly distribution of albedo in the Arctic. Larsson and Orvig (1961, 1962) and Larsson (1963) published their results first in the form of maps and then in the form of stereograms. These maps were compiled from different kinds of information about natural vegetation, large scale physiographic features, snow cover, sea-ice cover, and glacierised areas etc. Marshunova and Chernigovskii (1971), using the same method as Larsson and Orvig, also constructed maps showing mean albedo values for March, May, July, and September. It seems to me that the best record of the albedo in the Arctic can be obtained from satellites such as those recently presented by Robinson *et al.* (1992), Schweiger *et al.* (1993), and for Greenland by Haefliger (1998).

Albedo in March in the Arctic Ocean and seas covered by sea ice and snow cover is, according to Marshunova and Chernigovskii (1971), about 82% (Figure 3.4). The northern parts of Russia and North America, including the Canadian Arctic Archipelago and probably Greenland, have similar albedo values. A significant drop in albedo (from 82% to only 20%) is observed between the regions covered by sea ice and open waters. For May it is

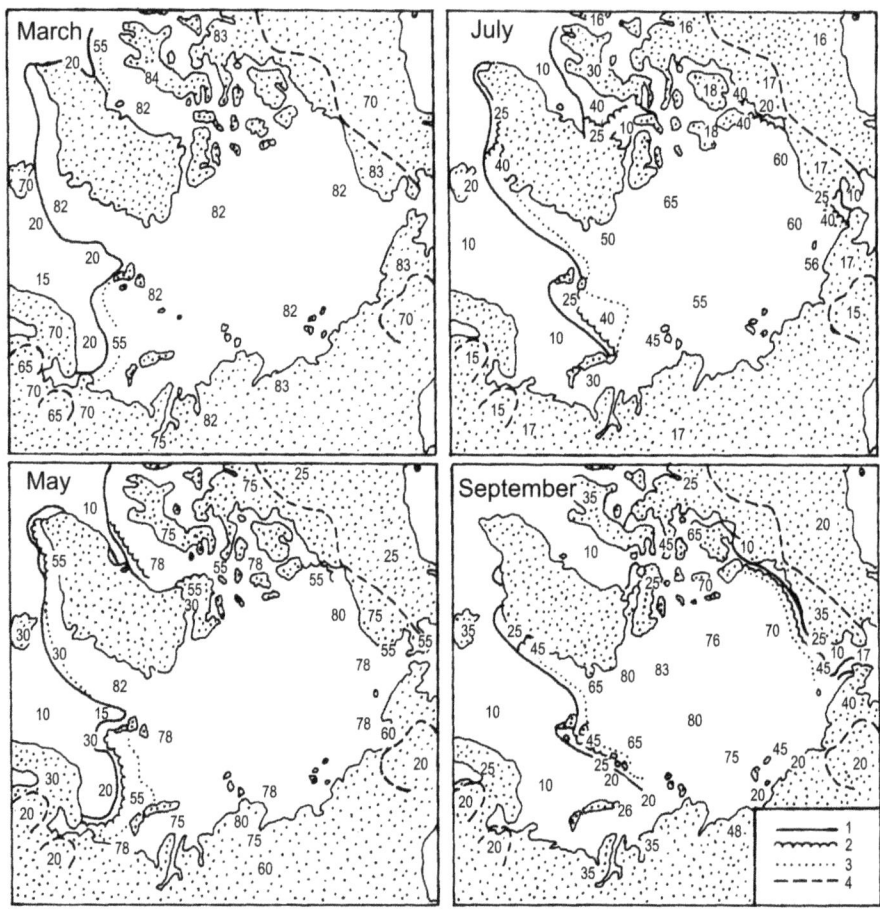

Figure 3.4. Mean monthly albedo (March, May, July, and September) (in percentages) in the Arctic (after Marshunova and Chernigovskii 1971). 1 – boundary between ice and water, 2 – boundary between ice of 1–5 and 5–8 points, 3 – boundary between ice of 5–8 and 8–10 points, 4 – forest limit.

possible to compare albedo values received by traditional methods and satellite methods. Marshunova and Chernigovskii's (1971) maps show the greatest correspondence with maps published by Robinson *et al.* (1992). This correspondence is surprisingly high because nowhere does the difference exceed 3%. Albedo, according to Marshunova and Chernigovskii (1971), in the areas of the Arctic Ocean and the seas surrounding the Arctic covered by sea ice, varies between 78% and 82% (Figure 3.4), while according to Robinson *et al.* (1992) it varies from 75% to 80%. In July, the correspondence is only a little worse than in May, but the differences rarely exceed 5%. In the central part of the Arctic the albedo, according to Marshunova and

Chernigovskii, oscillates between 60–65%, while on Robinson *et al.*'s map these values vary from 55–60%. The albedo near the sea-ice edge is equal to 50–55% (Marshunova and Chernigovskii 1971) and about 45–50% (Robinson *et al.* 1992). Albedo of the drifting ice (Barents, Norwegian, Greenland seas and in Baffin Bay) oscillates between 25–40%. In July the albedo is at its lowest in the whole year and on the tundra it reaches a minimum value of 16% to 18%.

In September, the surface reflectivity of the central Arctic is 70–83% (Figure 3.4). The albedo of the tundra increases to 25–35% (Marshunova and Chernigovskii 1971). The highest mean monthly values of the surface albedo of the Arctic seas occur in February and March (82%), and the lowest in July (55%).

3.3.2 Absorbed Global Solar Radiation

The magnitude of absorbed global solar radiation on every point of the Earth is determined by the incoming global radiation and by the reflective characteristics (albedo) of the underlying surface. Both these quantities change significantly in the annual cycle. Moreover, as may be seen from previous sections, the existing network of actinometric stations in the Arctic is very scarce. Therefore, the drawing of maps presenting the distribution of absorbed radiation in the whole Arctic is rather difficult. Reviewing the literature we only find a few teams of authors who have attempted to present such a distribution. Gavrilova (1959, 1963) was the first to publish maps presenting the monthly amounts of absorbed radiation in the Arctic. A little later, Vowinckel and Orvig (1962) also presented their results. Some of the maps from this paper were also later included in their better known articles (Vowinckel and Orvig 1964b, 1970). The third known attempt was made by Marshunova and Chernigovskii, first only for the Soviet Arctic (Chernigovskii and Marshunova 1965) and then also for the whole Arctic (Marshunova and Chernigovskii 1971). In the last work, all material (fortunately aside from maps) is limited to the non-Soviet Arctic, in accordance with the title. It is worth noting, however, that all these published geographical distributions of mean monthly and annual amounts of absorbed radiation are more rough approximations of the reality than in the case of incoming radiation.

In January, only areas to the south of 68°N receive solar radiation. However, within the Arctic these fluxes of solar radiation are small. Moreover, they are almost entirely (80–85%) reflected back by the snow cover. As a result, the zero isoline of the absorbed radiation more or less passes near the Arctic Circle (see Gavrilova 1963).

In April (Figure 3.5), the whole area covered by sea ice and snow (the Arctic Ocean, the Laptev Sea, and the central and northern parts of other Arctic seas, as well as the northern part of the Taymyr Peninsula and Greenland, and the Canadian Arctic Archipelago) absorbs radiation at a rate of 1.5–2.0 kcal/cm² (6.3–8.4 kJ/cm²). On the other hand, the highest values are absorbed by the open water areas in the Norwegian and Barents seas, as well as in Baffin Bay. In the southern parts of the continental Arctic, the absorbed radiation oscillates between 2 kcal/cm² and 3 kcal/cm² (8.4–12.5 kJ/cm²).

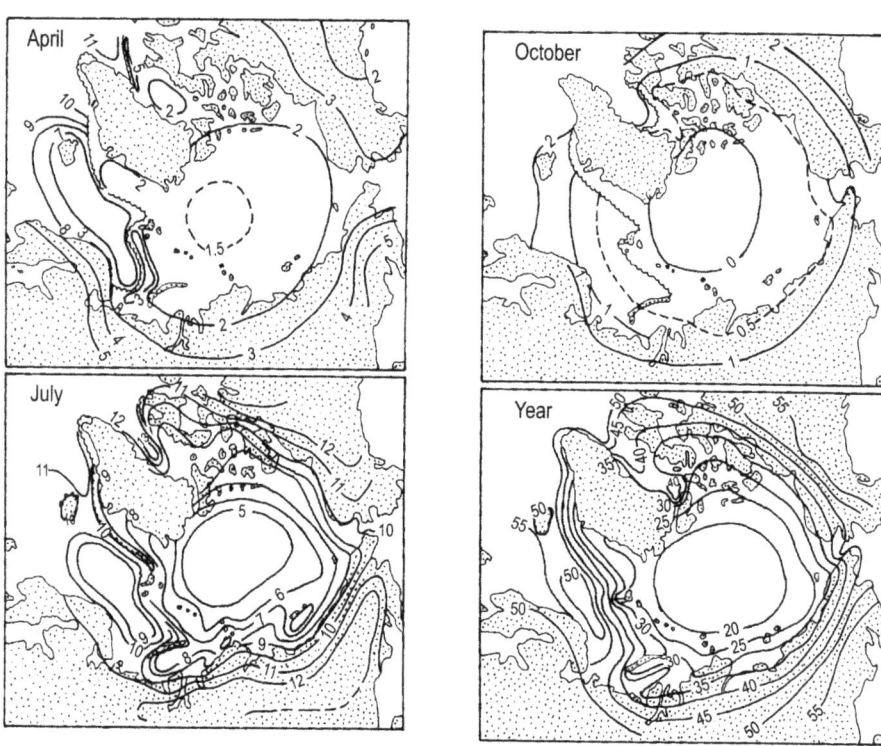

Figure 3.5. Mean totals of absorbed radiation in the Arctic for April, July, and October (in kcal/cm²/month) and for the year (in kcal/cm²/year) (after Marshunova and Chernigovskii 1971).

In July, most Arctic regions absorb their highest values of solar radiation (see Marshunova and Chernigovskii 1971). The Arctic Ocean receives from 5 kcal/cm² (20.9 kJ/cm²) in the vicinity of the North Pole to about 6 kcal/cm² (25.1 kJ/cm²) along the latitude 80°N. Further to the south, the absorbed radiation systematically rises to about 10 kcal/cm² (41.8 kJ/cm²) in the northern continental part of the Russian Arctic and Alaska, and in the southern part of the Canadian Arctic Archipelago. Similar values and even greater, up to 12 kcal/cm² (50.2 kJ/cm²), occur in the Norwegian Sea, in the pure water of the Barents and Greenland seas, and in Baffin Bay. In the central part of

Baffin Bay, values exceeding even 13 kcal/cm² (54.3 kJ/cm²) are observed (Figure 3.5).

In October (Figure 3.5), the central parts of the Arctic, up to about the latitude 80°N, do not absorb any solar radiation (polar night). The 0.5 kcal/cm² (2.1 kJ/cm²) isoline runs between mainly 70°N and 75°N. The highest values of absorbed radiation (> 2 kcal/cm² [8.4 kJ/cm²]) are in the southern parts of the Canadian Arctic and probably in the coastal parts of southern Greenland (the southernmost parts of the Arctic).

On an annual basis (Figure 3.5), the maximum values of absorbed radiation (50–55 kcal/cm² [209–230 kJ/cm²]) occur in the southernmost parts of the Arctic (the southern Canadian Arctic) and in the Norwegian and Barents seas, where, for the greater part of the year, an open water or thin drifting ice is observed. The sums of the absorbed radiation systematically decrease in a northerly direction and oscillate between 17 kcal/cm² and 20 kcal/cm² (71.1–83.6 kJ/cm²) in the vicinity of the North Pole. Vowinckel and Orvig (1962, 1964b) give significantly higher values for this area of the Arctic: c. 28–30 kcal/cm² (117.0–125.4 kJ/cm²). However, Badgley (1961) received similar results to those of the Russian authors. The mean July and August absorption of solar radiation in years with slight ice formation, in comparison with years with heavy ice conditions in the Arctic, is 1.4–1.5 times greater (see Marshunova and Chernigovskii 1971, their Table 21).

3.4 Long-wave Net Radiation

Long-wave net radiation (so-called effective radiation) is a residual of two fluxes: terrestrial radiation (upward infrared radiation) and the "counter-radiation" of the atmosphere (downward infrared radiation). The main factors determining effective radiation are air temperature and humidity, temperature of the surface, stratification of the atmosphere, and cloudiness (cloud amount and type, height and physical properties of clouds). Counter-radiation plays a very important role in the Arctic, especially in the winter when insolation becomes negligible. Vowinckel and Orvig (1970), however, have also shown that this component is dominant in summer, too (Table 3.1).

Table 3.1. Per cent contribution by insolation and counter-radiation to total surface radiation income in June (after Vowinckel and Orvig 1970)

Type of radiation	Latitude (°N)					
	65	70	75	80	85	90
Long-wave (%)	60	64	68	69	69	69
Short-wave (%)	40	36	32	31	31	31

Long-wave net radiation and its elements are very rarely measured in Arctic actinometric stations, and, as Marshunova and Chernigovskii (1971) write, such observations were not made in the non-Soviet Arctic at all. Therefore, our knowledge about effective radiation comes mainly from computations.

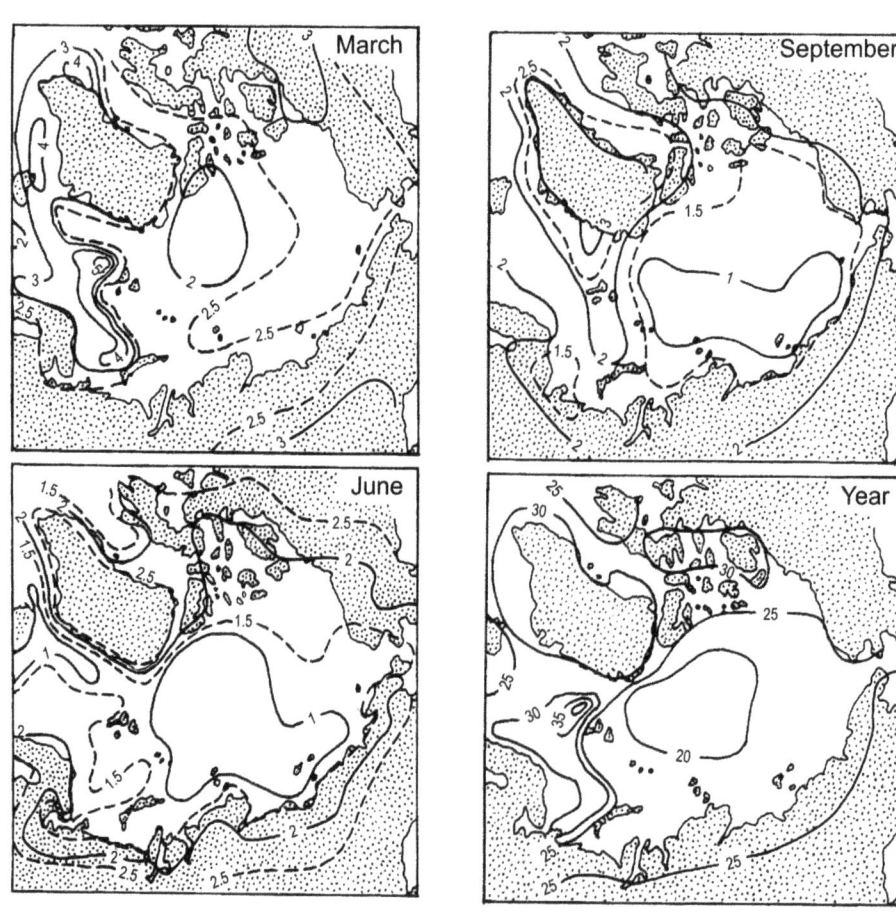

Figure 3.6. Mean totals of effective radiation in the Arctic for March, June, and September (in kcal/cm^2/month) and for the year (in kcal/cm^2/year) (after Marshunova and Chernigovskii 1971).

From November to March, the effective radiation over the open water surface in the Barents and Norwegian seas and on the eastern coast of Greenland is twice that in the coastal regions (4–5 kcal/cm^2/month [16.7–20.9 kJ/cm^2/month] and 2–2.25 kcal/cm^2/month [8.4–9.4 kJ/cm^2/month]). In March, the effective radiation near the North Pole is only a little lower than in the southern part of the Arctic (Figure 3.6). In the warm half-year, the effective radiation in the central part of the Arctic is twice as low as in winter, while in the rest of the Arctic the decrease is not so big (compare especially the continental parts of the Arctic). On an annual basis (Figure 3.6), the effective radiation is lowest in the vicinity of the North Pole (20 kcal/cm^2 [83.6 kJ/cm^2]). According to Marshunova and Chernigovskii (1971) this is a consequence of the low surface temperature in winter and high cloudiness in summer. In the rest of the Arctic, the effective radiation rarely exceeds 30 kcal/cm^2 (125.4 kJ/cm^2) reaching a maximum (> 40 kcal/cm^2 [167.2 kJ/cm^2]) in a small area located to the west of Spitsbergen (all-year open water connected with the warm West Spitsbergen Current). Positive values of effective radiation mean here that the terrestrial radiation is greater than the counter-radiation of the atmosphere.

3.5 Net Radiation and Other Elements of the Heat Balance

3.5.1 Net Radiation

The net radiation balance of the surface is a result of the subtraction of its long-wave component from its short-wave component. The net short-wave radiation in the Arctic is always positive or equal to zero (polar night). The effective radiation exists throughout the whole year and it is mainly positive in the sense given in the previous section. For the mean monthly and annual averages which we have analysed, it is always positive, as was shown in the previous section (see Figure 3.6). Vowinckel and Orvig (1970) distinguished two types of radiation regime in the Arctic: the Norwegian Sea and the pack-ice types (Figure 3.7). The first type occurs over open ocean areas north of the Arctic Circle. The characteristic feature of this type is the occurrence of a large negative balance (lower than one can expect) during winter, which almost completely reduces the positive balance during summer. In comparison with the pack-ice type, the radiation balance of the first type shows significantly greater changes in the annual march from –2 cal/cm^2/day to 3 cal/cm^2/day (–8.4 J/cm^2/day to 12.5 J/cm^2/day) versus –1 cal/cm^2/day to 2 cal/cm^2/day (–4.2 J/cm^2/day to 8.4 J/cm^2/day). From Figure 3.7 it may be seen that the net radiation balance is a very small residual of large components of incoming and outgoing radiation fluxes. Thus, as Vowinckel and Orvig (1970) notice, the balance will be highly sensitive to slight inaccuracies in the estimated incoming and outgoing radiation. For more details see Vowinckel and Orvig (1970).

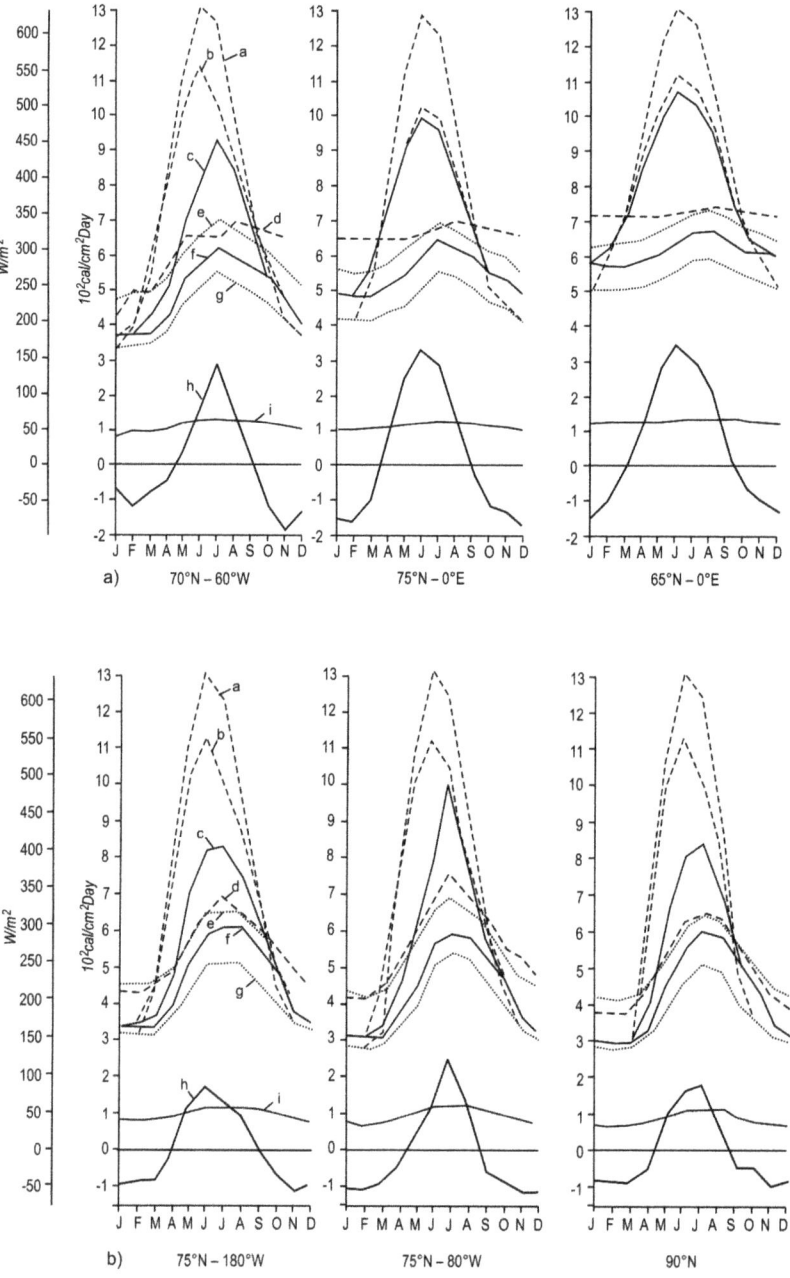

Figure 3.7. Radiation regimes in the Arctic: (a) Norwegian Sea type and (b) pack ice type after Vowinckel and Orvig (1970). a – total incoming radiation, cloudless sky; b – actual total incoming radiation; c – actual total radiation absorbed on the ground; d – long-wave radiation from the ground; e – long-wave incoming radiation, overcast sky; f – actual long-wave incoming radiation; g – long-wave incoming radiation, cloudless sky; h – actual radiation balance; i – long-wave radiation by CO_2.

In January (Figure 3.8), solar radiation is not present in the greater part of the Arctic, so in this month the radiation balance is caused by effective radiation. The net balance is equal to -8 kJ/cm^2 almost over the whole Arctic, except the open water in the Greenland, Norwegian, and Barents seas and Baffin Bay. The highest negative values, up to -25 kJ/cm^2, occur in the eastern part of the Greenland Sea, especially near the western coast of Spitsbergen. Some polynyas in the Kara and Laptev seas also have values (-13 kJ/cm^2) which are lower than normal.

In spring (April), the radiation balance is still slightly negative in most of the Arctic. Small positive values occur only in the southern parts of the Canadian Arctic and Pacific regions (up to about 3 kJ/cm^2). Significantly greater values of the net radiation balance (up to 13–14 kJ/cm^2) are noted in the southern part of the Atlantic region. However, the highest values (up to 21 kJ/cm^2) are recorded in Baffin Bay (Figure 3.8).

In July (Figure 3.9), the radiation balance reaches its highest positive values. In the central part of the Arctic it varies between 15 kJ/cm^2 and 17 kJ/cm^2, near the sea-ice edge it is equal to 34 kJ/cm^2, and in the open water of the Arctic seas it is at its highest, reaching as much as 42 kJ/cm^2. Continental parts of the Arctic receive 30–35 kJ/cm^2.

In October (Figure 3.9), the radiation balance of the surface becomes negative again over the entire Arctic. In the Arctic Ocean and the Arctic seas covered by sea ice, the values of the balance oscillate mainly from -5 kJ/cm^2 to -6 kJ/cm^2. Similar values are also observable in the continental Arctic. Open water near the sea-ice edge (Greenland, Norwegian, Barents, and Chukchi seas, and Baffin Bay) has the highest negative radiation balance (from -8 kJ/cm^2 to -13 kJ/cm^2).

The annual values of the net radiation balance (Figure 3.3b) are negative in the central part of the Arctic, mainly above 77–82°N reaching -12 kJ/cm^2 at the North Pole. The lowest observable values have been noted, however, in the centre of the northern part of the Greenland Ice Sheet (-16 kJ/cm^2) and in the Greenland Sea near the coast of Spitsbergen (-17 kJ/cm^2). In the continental parts of the Arctic (< 70°N), especially in the Canadian Arctic, the net radiation values exceed 70 kJ/cm^2 and probably reach more than 100 kJ/cm^2 in the southernmost fragments. High values of radiation balance (100–110 kJ/cm^2) also occur in the southern parts of the Barents Sea, the Denmark Strait, and Baffin Bay.

Comparison of the distribution of the net radiation balance in the Arctic (presented here after Khrol (1992)) with other sources reveals significant differences in many cases. These differences are greatest in the warm half-year and are probably caused by the different methods used for net radiation balance calculations.

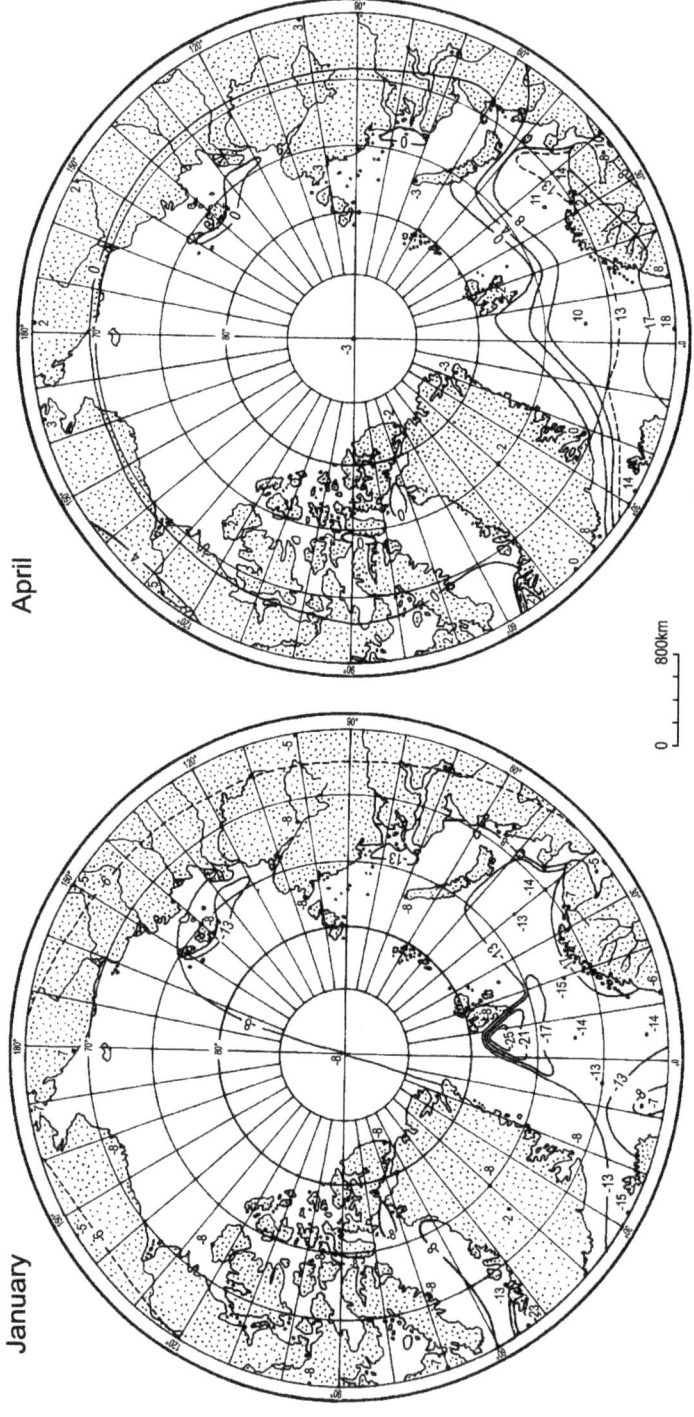

Figure 3.8. Average monthly (January and April) totals (in kJ/cm^2) of the net radiation in the Arctic (after Khrol 1992).

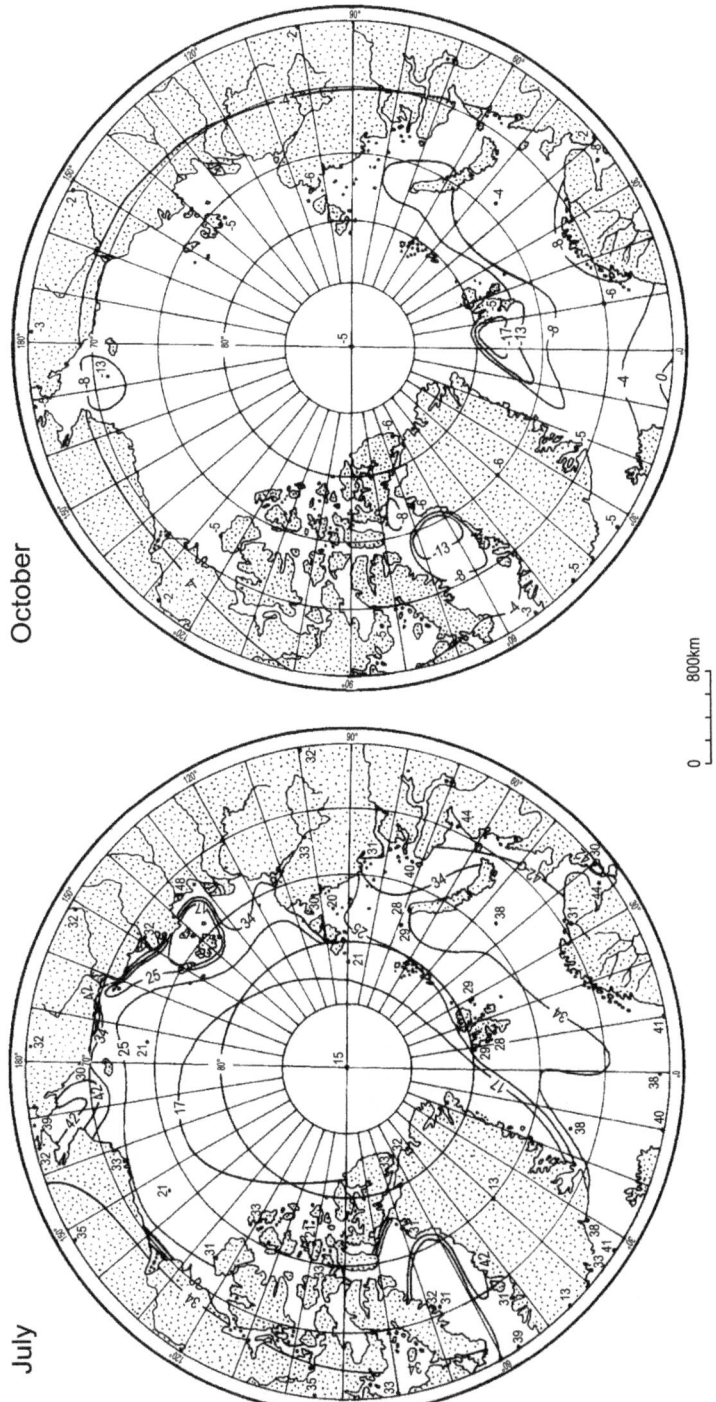

Figure 3.9. Average monthly (July and October) totals (in kJ/cm²) of the net radiation in the Arctic (after Khrol 1992).

3.5.2 Sensible Heat and Latent Heat

The net radiation balance presented in the previous section is the most important component of the heat balance of the surface. Yet, as we know, the energy is not only transported by radiation. It can also be transported from the surface to the atmosphere by evaporation and sensible heat and from the atmosphere to the surface by condensation and sensible heat. However, our knowledge concerning these two fluxes is still limited. As may be seen from the previous section, radiation balance computations can be compared with observations. This permits us to check the correctness of the formulas used for the radiation balance computations. Such a possibility does not exist in the case of the sensible and latent heat fluxes because no accurate direct measurement techniques exist. Thus, our knowledge about this part of the heat balance comes only from computations, which use for this purpose both different climatic data (mainly air and sea/land temperature, air humidity, and wind speed) and different characteristics of land and sea surface. Shuleykin (1935), Budyko (1956), Untersteiner (1964), Vowinckel and Taylor (1965), Ariel et al. (1973), Khrol (1976), and Murashova (1986) developed methods of computing these fluxes. Calculating geographical distributions of the elements of the heat balance of the Arctic surface is a difficult and time-consuming task. Therefore, the existing literature is very meagre. Vowinckel and Taylor (1965) computed evaporation and sensible heat fluxes separately for the following areas: the central Polar Ocean, Kara-Laptev Sea, East Siberian Sea, Beaufort Sea, and the 5° latitude belts in the Norwegian-Barents Sea. For more details see this paper or Vowinckel and Orvig (1970). Only Russian climatologists have presented results of the distribution of the heat balance elements in the Arctic in the form of maps (Budyko 1963; Gorshkov 1980; *Atlas Arktiki* 1985; Khrol 1992). Budyko's maps concern the whole Earth and therefore they include only the southernmost parts of the Arctic. The maps presented here come from Khrol (1992). The values of sensible heat and latent heat fluxes were calculating using methodology proposed by Ariel et al. (1973), Khrol (1976) and Murashova (1986).

3.5.2.1 Sensible Heat

In January (Figure 3.10), the sensible heat flux is positive over almost the entire Arctic, except the open water areas in the Greenland, Norwegian, and Barents seas, and in the Denmark Strait and Baffin Bay (including the polynya known as North Water). The highest positive values occur in the central Arctic, Greenland, the northern continental parts of the Russian Arctic, and the northern part of the Canadian Arctic Archipelago (4–5 kJ/cm²). The greatest loss of energy (up to –60 kJ/cm²), may be observed in the areas occupied by warm sea currents (West Spitsbergen Current, Norwegian Current, Murmansk Current and West Greenland Current).

In April (Figure 3.10), the pattern of distribution of sensible heat is very similar to that in January. The main observable differences concern the magnitude of the fluxes. In April both positive and negative sensible heat fluxes are lower. The highest values oscillate between 2 kJ/cm² and 3 kJ/cm², while the lowest are between –25 kJ/cm² and –34 kJ/cm². The average decrease of the sensible heat in comparison with January is equal to 1–2 kJ/cm² in most of the Arctic (except in open water, where this decrease is greater).

In July (Figure 3.11), the sensible heat is negative in the central Arctic, with highest values near the North Pole (–2 kJ/cm²). These negative values are spread more to the south (up to 70°N) from the Pacific region side. Greater negative values occur in the interior of Greenland (–8 kJ/cm²) and most of all in the continental part of the Arctic (up to –15 kJ/cm²). Between these two regions with negative values of sensible heat, there is a belt with positive values reaching as high as 15–17 kJ/cm². Even higher values may be noted locally in the south-western part of the Canadian Arctic. This is connected with the advection of warm continental air from the South.

In October (Figure 3.11), the sensible heat fluxes again became positive in the Arctic Ocean (up to 3 kJ/cm²) and in Greenland (up to 5 kJ/cm²). In the Arctic seas covered by sea ice, the fluxes are mainly slightly negative (except the Beaufort Sea and possibly the Laptev Sea) and are influenced by the advection of cold air from the continent. The negative values are stronger (up to –25 kJ/cm²) in the seas with open water (the Norwegian, Greenland, Barents, and Chukchi seas, the Danish Strait, Baffin Bay and the Bering Strait). The northern part of the continental Arctic (including the Canadian Arctic Archipelago) has slightly positive sensible heat (rarely exceeding 2 kJ/cm²).

Annual values of sensible heat (Figure 3.12a) are positive in the Arctic Ocean covered by perennial sea ice (up to 21 kJ/cm²) from the Canadian side, in Greenland, and in the Greenland Sea occupied by the cold East Greenland Current (up to 63 kJ/cm²). Very high positive values are also observed locally in sea water areas between the islands of the Canadian Arctic Archipelago. Moderate negative sensible heat fluxes are noticeable in the continental part of the Arctic (up to about –22 kJ/cm²). On the other hand, very great losses of energy occur mainly in the eastern part of the Greenland Sea, where the warm West Spitsbergen Current reaches the sea-ice edge (–368 kJ/cm²). A significant loss of energy also occurs in the polynyas and leads areas (up to –42 kJ/cm²).

56 The Climate of the Arctic

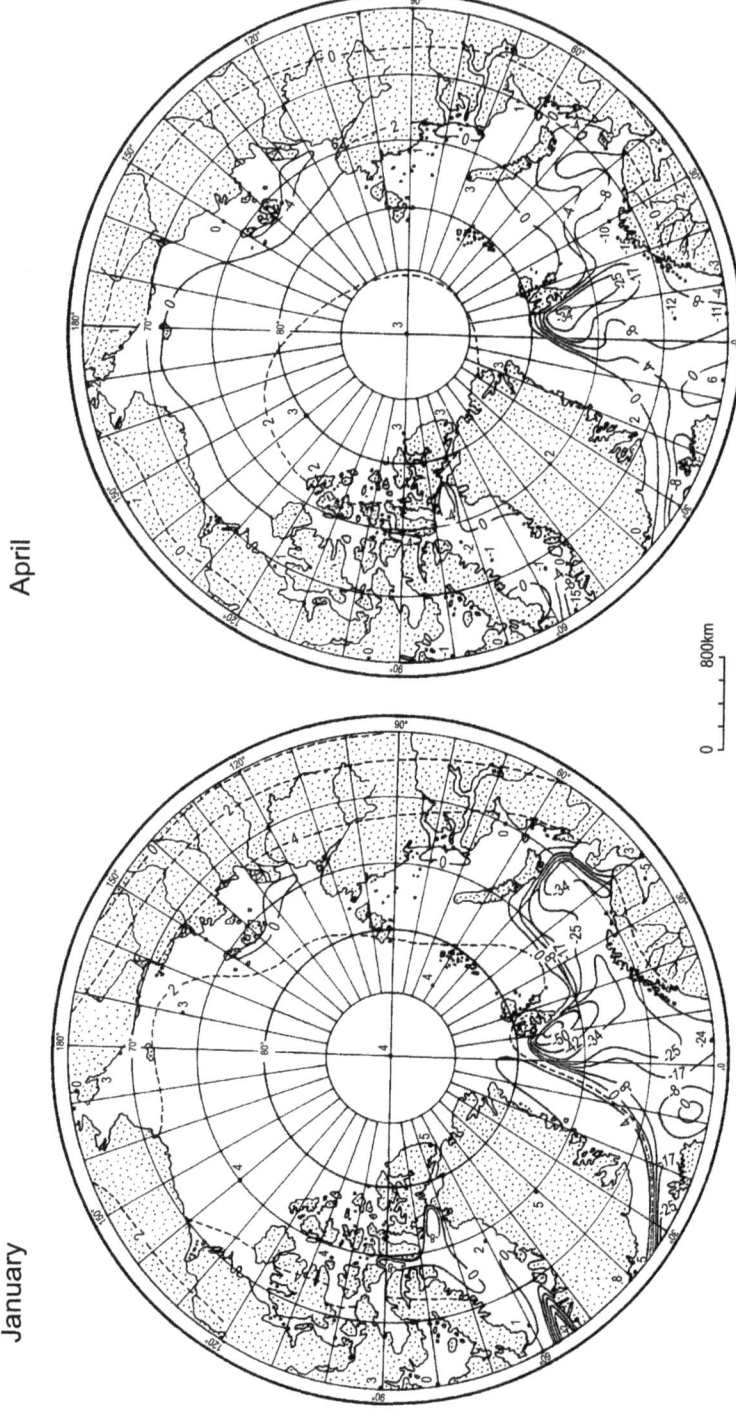

Figure 3.10. Average monthly (January and April) totals (in kJ/cm^2) of the sensible heat in the Arctic (after Khrol 1992).

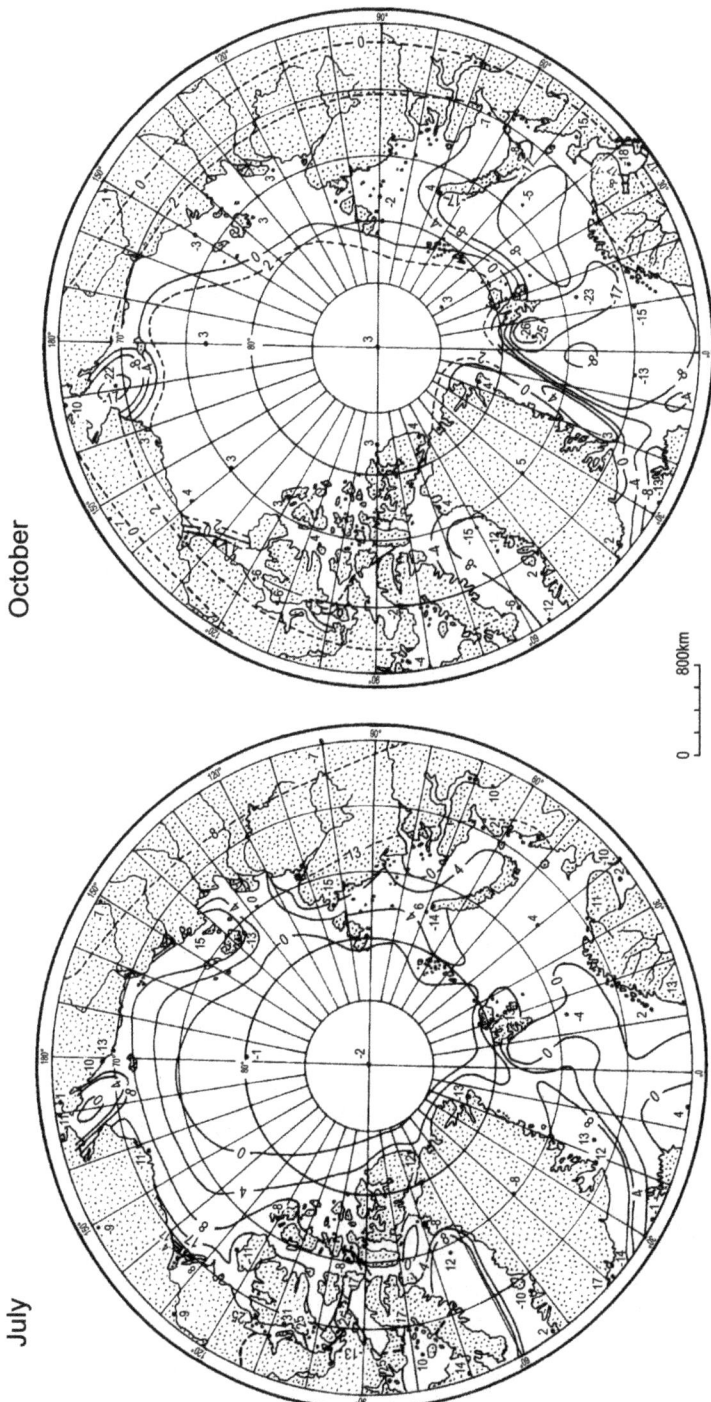

Figure 3.11. Average monthly (July and October) totals (in kJ/cm²) of the sensible heat in the Arctic (after Khrol 1992).

58 *The Climate of the Arctic*

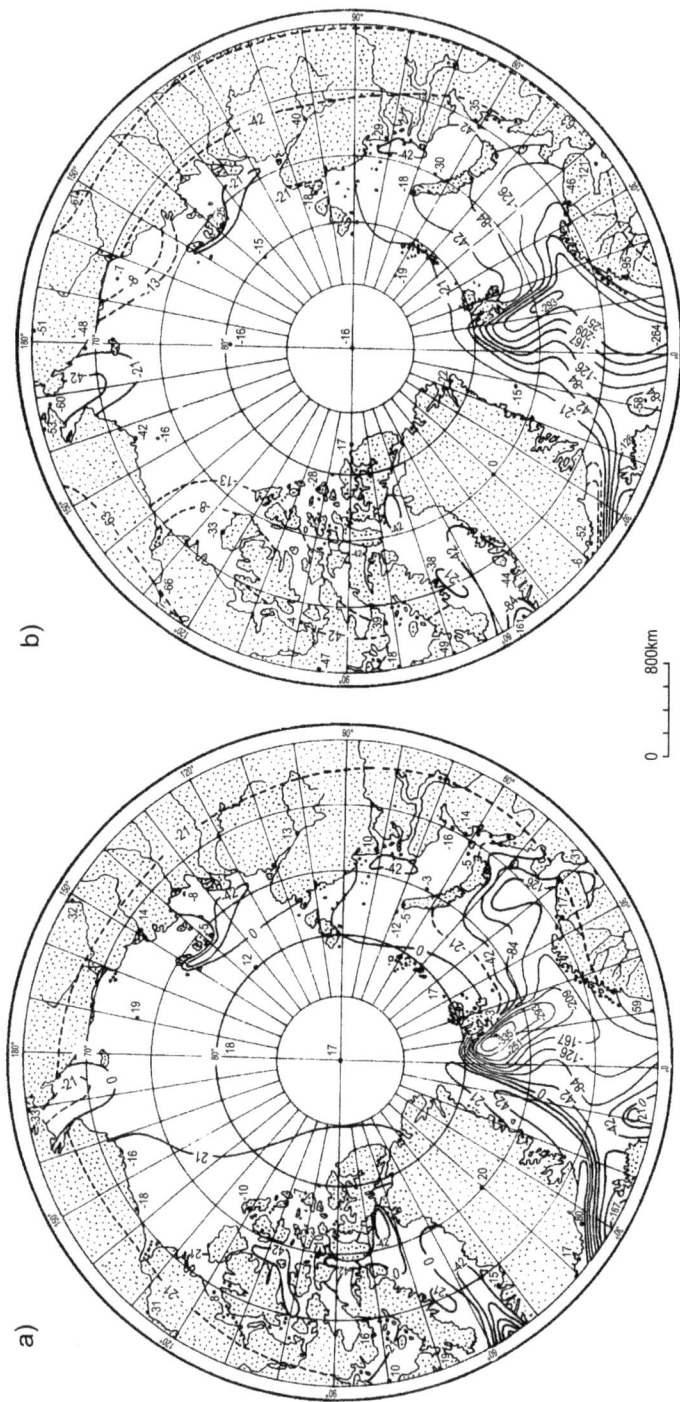

Figure 3.12. Annual totals (in kJ/cm^2) of sensible heat (a) and of latent heat (b) in the Arctic (after Khrol 1992).

3.5.2.2 Latent Heat

Latent heat fluxes in the Arctic are significantly weaker in January than sensible heat fluxes. In this month (Figure 3.13), evaporation is very slight in the Arctic because of low temperature and a surface covered by sea ice and snow. As a result, the latent heat fluxes do not exceed -1 kJ/cm^2 in the central Arctic. Near the sea-ice edge the loss of energy gets higher (-4 kJ/cm^2) and in the open water areas it reaches its maximum (-39 kJ/cm^2). Polynya areas show a loss of energy up to -4 kJ/cm^2.

In April (Figure 3.13), the situation is very similar to that of January. The loss of energy in most of the Arctic is only a little stronger.

In July (Figure 3.14), the Arctic losses significantly more energy via evaporation than in winter. The highest negative fluxes occur in the continental parts of the Arctic and also in the coastal areas of the Arctic islands (-14 kJ/cm^2 to -19 kJ/cm^2). Also negative fluxes (but about 6–7 times smaller) occur in the Arctic Ocean (-2 kJ/cm^2 to -3 kJ/cm^2). Between these two areas, i.e. in the Arctic seas (excluding the Norwegian Sea and the western part of the Barents Sea and the southern part of the Kara and Chukchi seas), positive latent heat fluxes occur (up to 4–8 kJ/cm^2). Polynya areas lose up to -4 kJ/cm^2.

In October (Figure 3.14), the latent heat fluxes in the entire Arctic become negative again. In the Arctic Ocean and seas covered by sea ice, the negative values oscillate from -0.8 kJ/cm^2 (the North Pole) to about -4 kJ/cm^2 and -6 kJ/cm^2 near the sea-ice edge. On the open water areas the loss of energy is significantly greater and in the Norwegian and Greenland seas it reaches a maximum equal to -28 kJ/cm^2 and -24 kJ/cm^2, respectively.

Annual values of latent heat fluxes in the Arctic (Figure 3.12b) are negative in all areas. In the Arctic Ocean these values oscillate between -16 kJ/cm^2 and -19 kJ/cm^2 and their absolute values are only slightly lower than the sensible heat fluxes. Thus, for the Arctic Ocean these two fluxes almost cancel themselves out (compare Figure 3.12a and Figure 3.12b). However, for the Arctic seas they already have mostly the same signs (excluding the Beaufort, Chukchi, and East Siberian seas, as well as the western part of the Greenland Sea). The negative latent heat fluxes are stronger here than the sensible heat fluxes, apart from a small part of the Greenland Sea near the coast of Spitsbergen, where they are very similar. The losses of energy from the continental part of the Arctic are more than twice as large as the sensible heat fluxes and reach almost -50 kJ/cm^2.

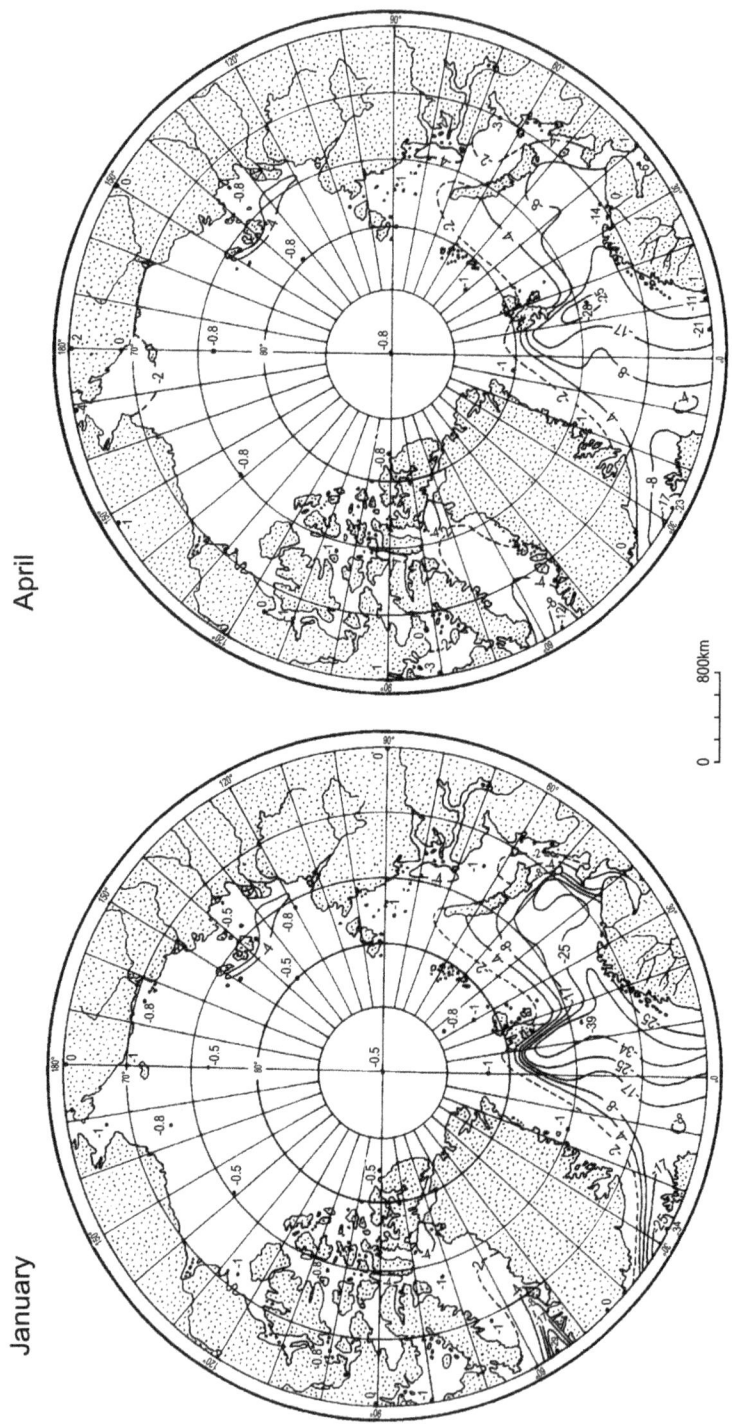

Figure 3.13. Average monthly (January and April) totals (in kJ/cm²) of the latent heat in the Arctic (after Khrol 1992).

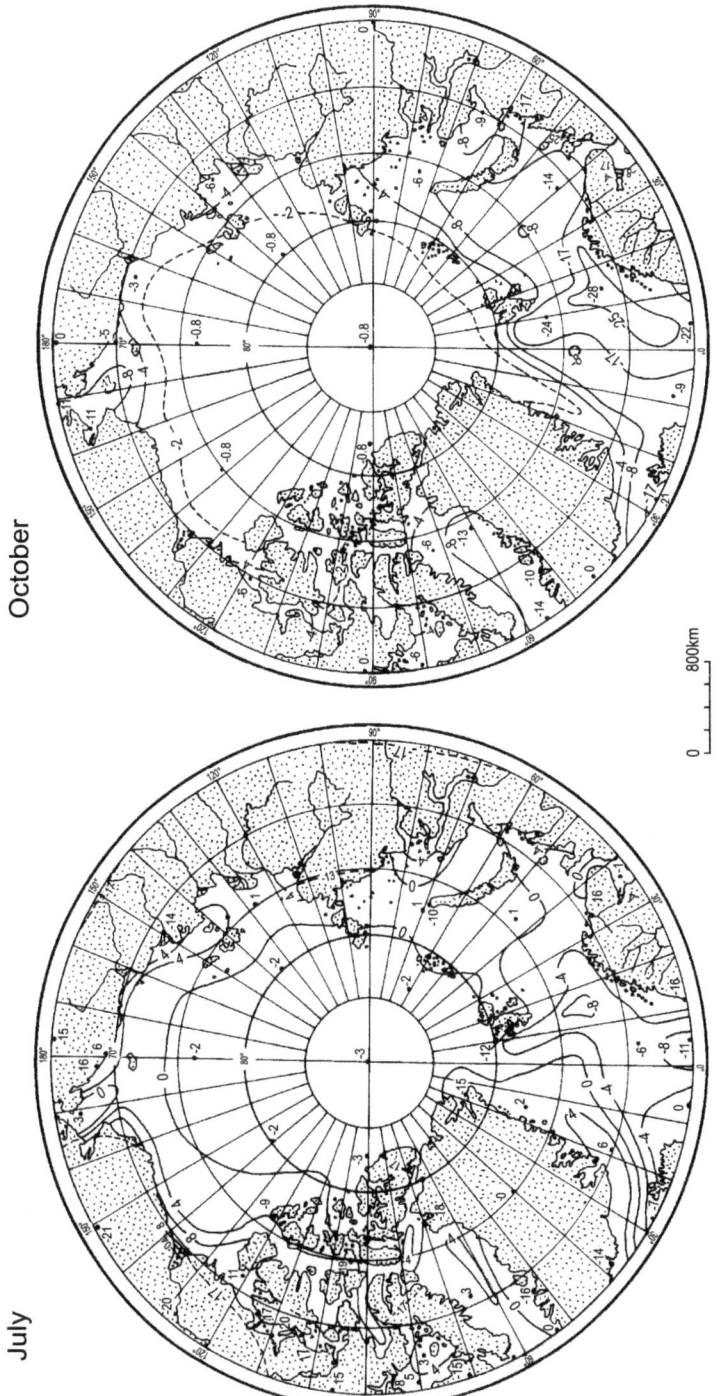

Figure 3.14. Average monthly (July and October) totals (in kJ/cm²) of the latent heat in the Arctic (after Khrol 1992).

Chapter 4

AIR TEMPERATURE

4.1 Mean Monthly, Seasonal, and Annual Air Temperature

Air temperature is the most important, and therefore also most often studied, climatological element. This is as true for the Arctic as it is for everywhere else. For this reason, our knowledge about this element, in comparison with others, is the best, but is still not sufficient for some parts of the Arctic (e.g. the central Arctic and the inner part of Greenland).

The instrumental records of Arctic temperature are brief and geographically sparse. Only five records (Upernavik: commenced 1874; Jakobshavn: 1874; Godthåb: 1876; Ivigtut: 1880 and Angmagssalik: 1895) extend back to the second half of the 19th century. As can be seen, all climatic stations operating during the nineteenth century were located in Greenland. Outside of Greenland, the first station was established in Spitsbergen in 1911 (Green Harbour). In the 1920s, the next seven stations came into operation, mainly in the Atlantic region of the Arctic. Following the Second International Polar Year (1932/1933) most of the Russian stations were established, while most of the Canadian stations were founded after World War II. For this reason, both spatial distribution and reliable estimates of air temperature characteristics in the Arctic are only possible for the last 40–50 years.

Besides the stations, extensive meteorological data have also been gathered during the well-known Fram (1893–1896) and Maud (1918–1925) expeditions. Later on, since 1937 Soviet drifting stations have supplied a large stream of different kinds of meteorological data for the central part of the Arctic Ocean. Unfortunately, this long-term project ended in 1991. Luckily, however, the Polar Science Center at the University of Washington ran a new research project "The Arctic Ocean Buoy Program" in 1979. Early that year an array of automatic data buoys was established in the Arctic Ocean. The main objectives of this programme are to provide measurement of surface atmospheric pressure over the Arctic Ocean and to define the large-scale field of motion of the sea ice. The temperature sensors are installed inside the buoys and therefore give only rough information about this element. Since 1992, however, they have been mounted outside, so the quality of data is significantly greater. Satellites also constitute new and extremely powerful sources of information about temperature in the Arctic.

Although, as has been seen, the first reliable climatological estimation of the spatial air temperature distribution in the Arctic was practically possible only in the second half of 1950s, there were also some earlier attempts to do this. Mohn (1905) was probably first to publish maps showing the spatial distribution of mean air temperatures for all months and for the year. The next proposition was given by Brown (1927), though only for January and July. The maps are, of course, in both cases very schematic. For example, Brown (1927) drew isotherms every 30°F. In the central Arctic, temperatures in January and July were estimated to be about −30°F (−34.4°C) and 30°F (−1.1°C), respectively. More accurate maps (with isotherms every 5°C) were published by Mecking (1928). In his maps the lowest temperatures in the Arctic occur near the North Pole and on the Greenland Ice Sheet, where they reach about −40°C (January) and 0°C (July). Sverdrup (1935) presented a significantly more detailed spatial distribution of air temperature (isotherms every 2°C) for almost the whole Arctic and for every second month (starting from January). The coldest month was January with the mean temperature below −36°C, including the area spreading from the North Pole to the northern parts of Greenland and the Canadian Arctic Archipelago. In comparison with Mecking's map, Sverdrup assumed that the mean temperature in July (near 0°C) is not only present around the North Pole, but over the whole Arctic Ocean, including large parts of the Arctic shelf seas.

Reviewing the climatological literature after World War II, we find only a few more propositions (besides those presented above) showing the spatial distribution of the temperature in the Arctic (Petterssen *et al.* 1956; Prik 1959; Baird 1964; Central Intelligence Agency 1978; Herman 1986; Parkinson *et al.* 1987 (based on data from Crutcher and Meserve 1970) and Przybylak 1996a, b, 2000a). Most other sources (e.g. Prik 1960; Sater 1969; Vowinckel and Orvig 1970; Donina 1971; Sater *et al.* 1971; Barry and Hare 1974; Sugden 1982; Martyn 1985) mainly reprint some of Prik's (1959) maps. In turn, maps published in well-known Russian atlases (Gorshkov 1980 and *Atlas Arktiki* 1985) are also updated versions of maps which had been authored and edited by Prik (1959). The majority of the sources cited, as Przybylak (1996a) has noted, present only the spatial distribution of temperature for January and July. Only Przybylak (1996a) has published maps for the four meteorological seasons popularly used in climatology (Dec.–Feb., March–May etc.) and for the year as a whole. About 40 homogeneous continuous series of temperature from the period 1951–1990 have been used to draw these maps. Out of the regional research, one should mention such works as those by Rae (1951), Donina (1971), Maxwell (1980), Barry and Kiladis (1982), Ohmura (1987), Calanca *et al.* (2000).

In this monograph, the maps published by Przybylak (1996a) are presented because: 1) they show actual mean seasonal and annual temperature conditions in the Arctic, 2) Prik's maps are easily available in the existing literature, and 3) Ohmura (1987) showed that the distribution of temperature over Greenland in Prik's maps contains "serious climatological errors". I must add that these errors are eliminated in the maps presented in *Atlas Arktiki* (1985), which is, however, a less available source. Przybylak (1996a) does not give the temperature for Greenland. Therefore, the temperature maps for Greenland after Ohmura (1987) and Calanca (personal communication) are also presented.

4.1.1 Annual and Daily Cycles of Temperature

The annual cycle of air temperature is a result of
1. changes in the amount of energy received from the sun, which depends on the geographical latitude and season of the year,
2. changes in atmospheric circulation,
3. changes in the physical properties of the underlying surface.

Petterssen *et al.* (1956) have distinguished three well-defined types in the annual cycle of temperatures in the Arctic: 1) maritime, 2) coastal, and 3) continental. These types can be seen in Figure 4.1. Jan Mayen has a maritime type, Malye Karmakuly and Egedesminde have a coastal type, and the rest of the stations have a continental type. From the map presenting the thermic continentality of the Arctic climate (Figure 4.2), we can see that Jan Mayen has a continentality below 20%, Malye Karmakuly and Egedesminde about 40–50%, and the rest of the stations above 60%. Thus, we can assume that regions of the Arctic with a degree of continentality lower than 30% probably have a maritime type, areas with a continentality ranging from 30% to about 55% have a coastal type, and regions with a continentality > 55% have a continental type. From Figure 4.2 it is evident that the continental type is the most common type, occurring in almost 80% of the Arctic, excluding mainly the Atlantic region, the southern part of Baffin Bay, and probably also the Pacific region. The second most common variety is the maritime type. What are the main features of these types?

Maritime type (Jan Mayen). Very small annual range of temperature (i.e. difference between the highest and the lowest mean monthly temperature) slightly exceeding 10°C. A curve presenting the mean temperatures of the summer months (June–August) shows a small variation only between 4–5°C. There is a similar situation in the winter months (from December to March) when the temperature oscillates between −5°C and −6°C. The maximum temperature is shifted to August, and the minimum temperature to March.

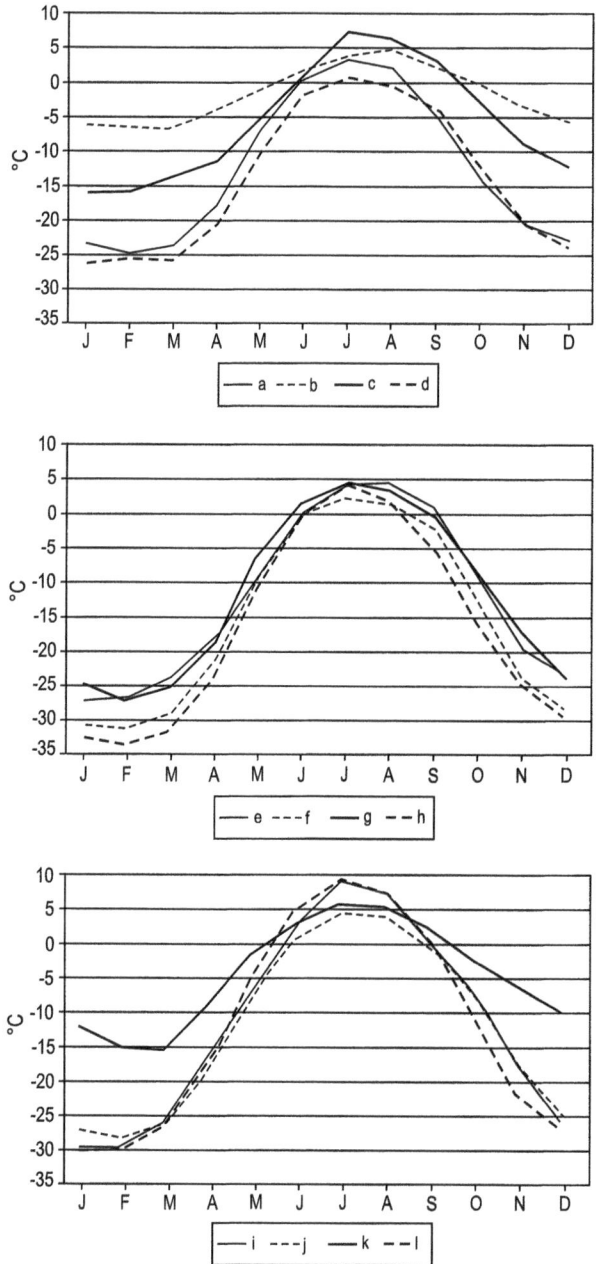

Figure 4.1. Annual course of air temperature (according to monthly means) in the selected Arctic stations, 1961–1990 (after Przybylak 1996b). a – Danmarkshavn, b – Jan Mayen, c – Malye Karmakuly, d – Polar GMO E.T. Krenkelya, e – Ostrov Dikson, f – Ostrov Kotelny, g – Mys Shmidta, h – Resolute A, i – Coral Harbour A, j – Clyde A, k – Egedesminde, l – average temperature for the 65–85°N zone for the period 1961–1986 (after Alekseev and Svyashchennikov 1991).

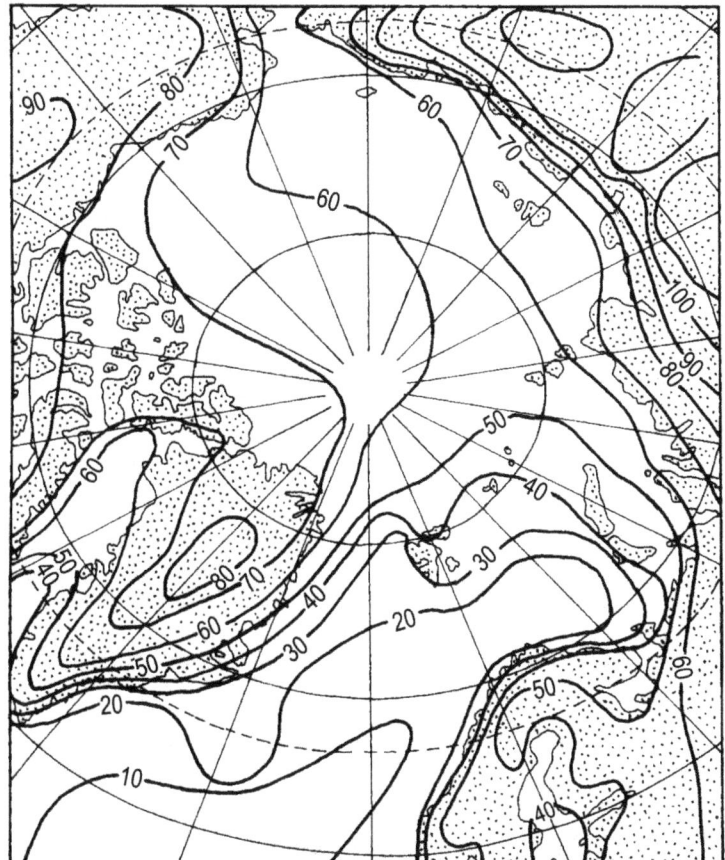

Figure 4.2. Thermic continentality of the climate in the Arctic (after Ewert 1997).

Coastal type (Malye Karmakuly, Egedesminde). This type, in principle, can be called "the transitional type" between the first (maritime) and the third (continental) types. The annual range of temperature (20°C) here is about twice that of the first type and half that of the continental type. Air temperature in winter is markedly lower than in the maritime type and also significantly greater than in the continental type. The minimum temperature is often shifted to February and sometimes, like in the case of Egedesminde, even to March. In summer, air temperatures are similar or higher than in the maritime type. The maximum can sometimes be delayed as Petterssen *et al.* (1956) note, but in our case such a situation does not occur.

Continental type (rest of stations). This type is characterised by the highest annual range of temperature (about 40°C), the lowest winter temperature (monthly means oscillate between –30°C and –35°C) occurring mostly in January and rarely in February. Summer temperatures are relatively very high, especially in southern parts of the Arctic and can reach values near 10°C. In the high Arctic, however, they are reduced to only 1–3°C (Figures 4.1a, d, g, and h). At the North Pole, the mean temperature of the warmest month (July) is only equal to –0.5°C (see Table 4.1). In the central part of the Greenland Ice Sheet (Eismitte) the mean monthly temperatures oscillate from about –42°C (February) to about –12°C (July) (Donina 1971).

Table 4.1. Monthly and annual mean air temperature (°C) (after Radionov et al. 1997)

Region	Month												Year
	Jan.	Feb.	March	April	May	June	July	Aug.	Sept.	Oct.	Nov.	Dec.	
North Pole	–32.3	–35.4	–33.8	–25.8	–12.1	–2.4	–0.5	–2.2	–9.5	–19	–28.1	–31.5	–19.4
Siberian	–31.5	–31.8	–31.4	–24.9	–10.8	–1.8	–0.1	–1.3	–7.2	–17	–25.3	–30.9	–17.8
Pacific	–30.8	–31.2	–29.7	–22.6	–10.5	–2.2	–0.1	–1	–6.5	–18.3	–25.7	–29	–17.3
Ocean Central	–32.4	–34.4	–32.8	–25.9	–12.1	–2.3	–0.3	–1.7	–8.9	–19.4	–28.3	–31.2	–19.1

According to the presented temperature data from stations, the warmest month is July. Out of 11 stations representing different climatic regions of the Arctic, this was the case in as many as 9 stations. Only in Jan Mayen and Ostrov Dikson was the warmest monthly temperature in August. The coldest month is most often February (6 stations), January (3 stations), or March (2 stations). These results are in line both with recently reported areally-averaged monthly temperatures for the Siberian, Pacific, and Central Ocean regions and with results from the North Pole (see Table 4.1), where clearly the highest temperatures occur in July. The lowest temperatures are noted in all regions in February, but in the Siberian and Pacific regions the temperature differences between February and January are very small (< 0.5°C). The annual cycle of mean temperature from the latitude band 65–85°N shows similar results (Figure 4.1l)

The mean monthly daily courses of air temperature have a clear asymmetric course throughout the year, except during the polar night. The second half of the day is usually warmer. On average, the highest temperature occurs between 13°° and 15°° (Figure 4.3). In the polar night, Przybylak (1992a) distinguished five basic types of daily courses of temperature in Hornsund (Spitsbergen): 1) a "normal" pattern with a maximum temperature in "daytime" hours and minimum values at "night"; 2) a "reverse" pattern with a maximum temperature at "night" and minimum in "daytime" hours; 3) with a tendency towards increasing temperatures throughout the entire 24 hours; 4) with a tendency towards decreasing temperatures throughout the entire 24 hours and 5) with a nearly constant temperature throughout the entire 24 hours. During the four winter months (Nov. – Feb.) of the period 1978–1983, types 4 (25.9%) and 2 & 3 (23.3% each) were most frequent. The "reverse" daily course of temperature occurred mainly in December (25.3%) and January (17.4%). This was most apparent on fine days (Przybylak 1992a).

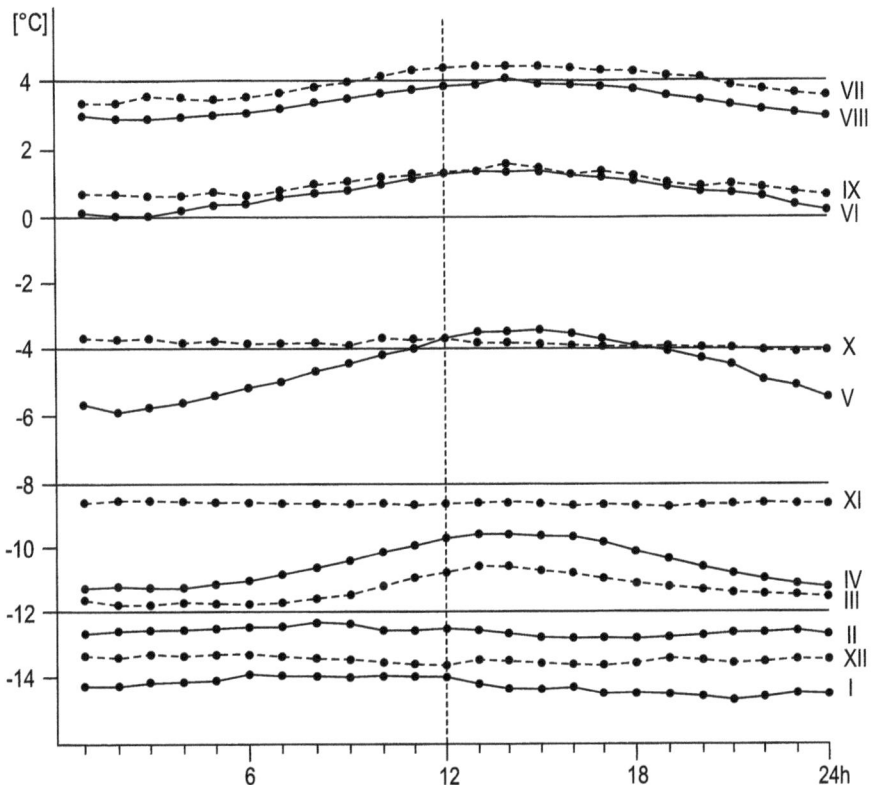

Figure 4.3. Mean monthly daily courses of air temperature in Hornsund (Spitsbergen), 1979–1983 (after Przybylak 1992a). I – January, II – February, III – March etc.

Przybylak (1992a) also found that the occurrence of the "reverse" daily course of temperature during the polar night in Hornsund is usually accidental and mainly connected with nonperiodic changes of temperature resulting from intensive cyclonic activity. Even if we assume that there are other factors favouring the occurrence of the "reverse" daily course of temperature (e.g. daily periodicity of radiation balance and the influence of ozone on it, or daily changes of geomagnetic activity) they may manifest themselves only in some synoptic situations (usually anticyclonic, non-advectional) which, however, occur very rarely during this period.

The clearest daily courses occur on days with little cloudiness. This can be seen in all seasons, except during the polar night, when the differentiation role of the cloudiness is negligible (see Figure 15 in Przybylak 1992a).

4.1.2 Spatial Temperature Patterns

Air temperature in the Arctic shows a great spatial variability in all seasons, but particularly in the cold half-year (see Figures 4.4–4.6). The well-developed atmospheric circulation (and cyclonic activity in particular) during this period is the main factor responsible for this situation. The coldest place in the Arctic in all seasons is the northern part of the Greenland Ice Sheet. The thermal regime of this part of Greenland is shaped by high elevation above sea level, the character of the underlying surface (snow and ice), and the occurrence of quasi-stationary anticyclone circulation.

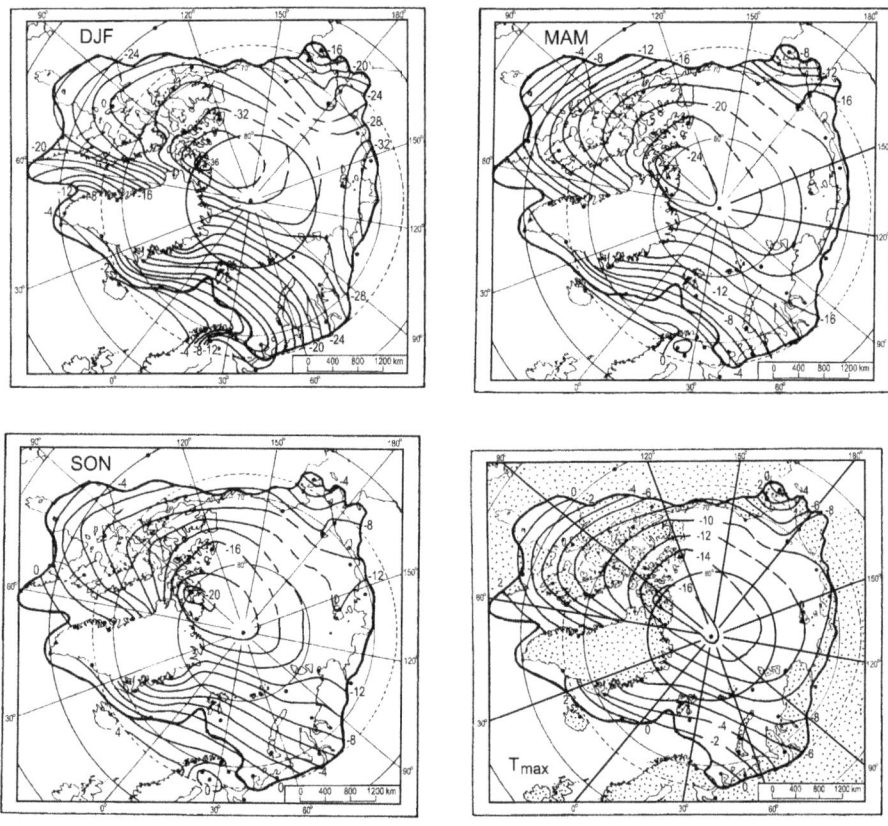

Figure 4.4. Spatial distribution of mean seasonal (DJF, MAM etc.) air temperature (in °C) in the Arctic, 1951–1990 (after Przybylak 1996a, modified for the central Arctic).

In winter the mean temperature in Greenland drops below –40°C (Figure 4.6a), reaching almost –45°C. The secondary minimum temperature is shifted from the North Pole towards the Canadian Arctic Archipelago and Greenland, where the mean winter temperature is around –36°C in the vicinity of the Eureka station (Figure 4.4). A belt of low temperature (< –30°C) spreads from this area through the North Pole towards the central part of the Siberian region. The balance between loss of heat from the snow and ice surface by radiation and the gain in energy conducted by the surface from the water under the ice and transported from the lower latitudes by the atmospheric circulation can explain the existence of this large homogeneously cold area. The highest temperatures in the Arctic occur in the southernmost parts of the Atlantic and Baffin Bay regions, where they oscillate between –2°C and –6°C. These high temperatures are the result of the transport of warm air masses within the very intense cyclonic activity connected with the Icelandic low (see Figure 2.3a). Cyclones which enter the Pacific region are weaker than Atlantic cyclones, but their warming effect is also evident (see the shape of isotherms). Generally, as rightly noted by Radionov *et al.* (1997), the patterns of the isotherms and of the isobars are in good agreement with one another (compare Figure 2.2a with Figure 4.4 or see *Atlas Arktiki* 1985).

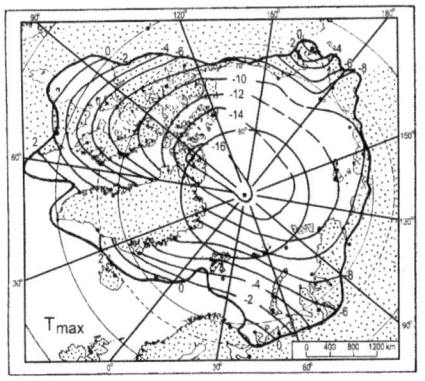
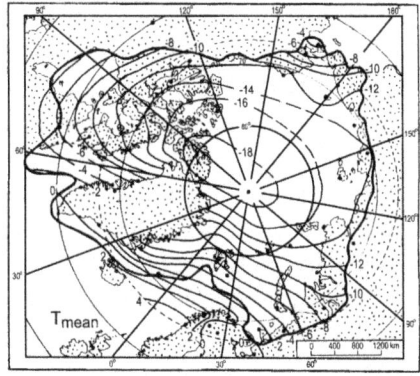
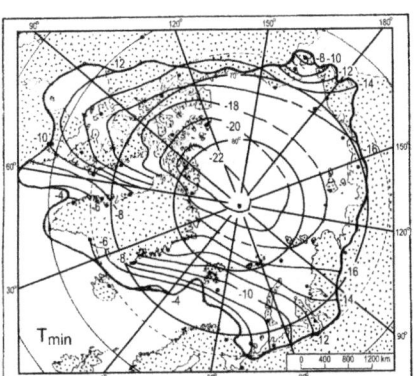

Figure 4.5. Spatial distributions of mean annual maximum (T_{max}), minimum (T_{min}) and average daily (T_{mean}) air temperature (in °C) in the Arctic, 1951–1990 (after Przybylak 1996a).

In spring and autumn, the general patterns of temperature distribution in the Arctic are very similar to those in winter (Figure 4.4). The main difference lies in the magnitude of the temperatures, which, of course, are significantly higher in transitional seasons by about 10–15°C. Comparing, however, temperatures in spring and autumn, one can see that spring temperatures are markedly lower by about 6–8°C. Such a situation is not only observable in the central part of the plateau of the Greenland Ice Sheet, where temperatures are similar or even colder in autumn (compare Figures 4.6b and 4.6d). At the Eismitte station the mean temperatures in spring and autumn, calculated according to data published by Ohmura (1987), are equal to –30.9°C and –31.0°C, respectively.

In summer, the distribution of temperature is clearly more dependent upon the insolation (polar day) than upon the atmospheric circulation. As a result, the courses of the isotherms are more latitudinal. In comparison with other seasons, the smallest spatial air temperature variation is also noted (the smallest horizontal temperature gradients). As mentioned earlier, the lowest mean summer temperature occurs in the northern part of the Greenland Ice Sheet (about –15°C), but, in comparison with other seasons, this temperature may be seen to shift significantly to the South, to the region of the highest elevation in Greenland (Figure 4.6c). In July the temperature drops slightly here below –10°C, and in June and August below –15°C. The secondary temperature minimum occurs in the central Arctic, where the prevailing melting of snow and ice holds the surface temperature slightly below freezing point (Figure 4.4). The highest summer temperature (> 8°C) is observable in the central continental parts of the Canadian and Russian Arctic. Areas where strong cyclonic activity prevails (the Atlantic and Baffin Bay regions) tend to be relatively cold (4–8°C).

Annual mean temperatures depend mainly on temperatures occurring in the cold half-year. Therefore, the patterns of annual, winter, autumn, and spring temperature in the Arctic are very similar (compare Figure 4.5 with Figure 4.4). The lowest temperatures occur in the Greenland Ice Sheet (> 2000 m a.s.l.), where the mean annual ones above 70°N are below –20°C. In the central part, this minimum drops even below –30°C. Outside Greenland, the lowest temperatures (< –18°C) occur in the same region as in winter, spring, and autumn, i.e., in the north-eastern part of the Canadian Arctic Archipelago. The Eureka station has noted the lowest mean annual temperature (–19.7°C). Only a slightly warmer temperature occurs around the North Pole. The mean annual temperature at the North Pole for the period 1954–1991 was equal to –19.4°C (Table 4.1). Slightly warmer conditions occur in the central part of the Arctic (above 85°N from the Atlantic side and above 80°N from the Pacific side), where mean annual temperatures oscillate between –16°C and –18°C. The warmest parts of the Arctic are those where cyclonic activity is greatest, i.e., firstly the regions spreading from Iceland to the Kara Sea (Atlantic region) and then the Baffin Bay and Pacific regions. In all these areas the isotherms are bent to the north.

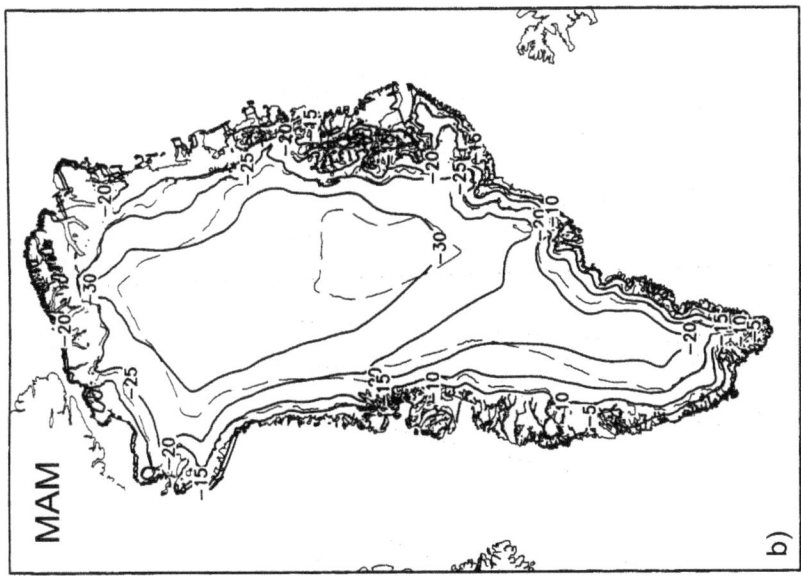

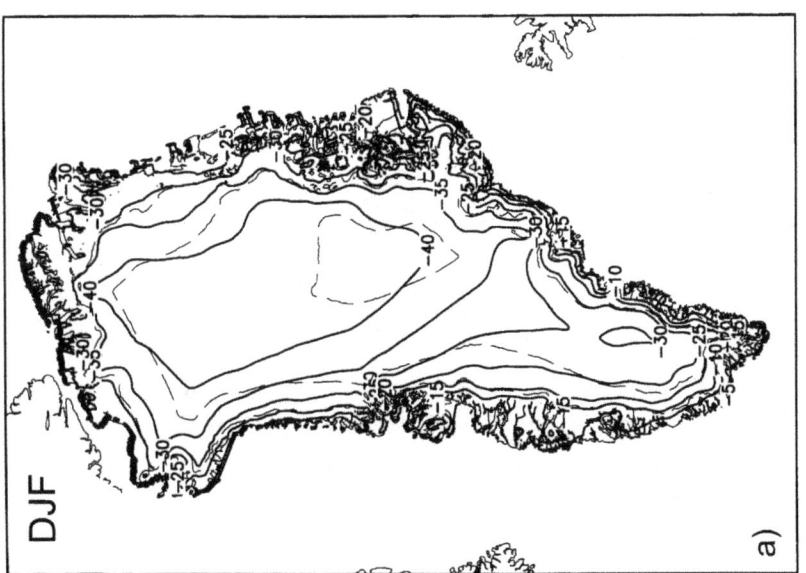

Figure 4.6. Spatial distribution of mean seasonal (a – DJF, b – MAM, c – JJA, d – SON, after Calanca, personal communication) and annual (e, after Ohmura 1987) air temperature (in °C) in Greenland.

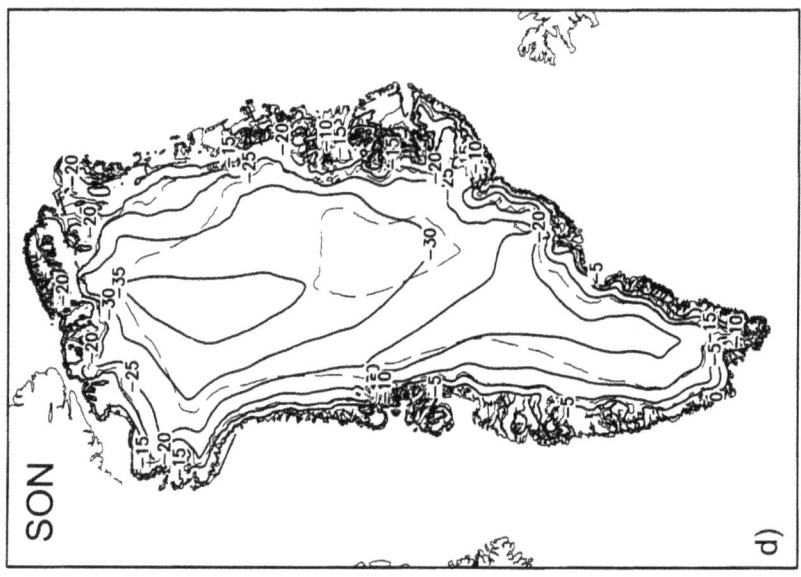

Figure 4.6. cont.

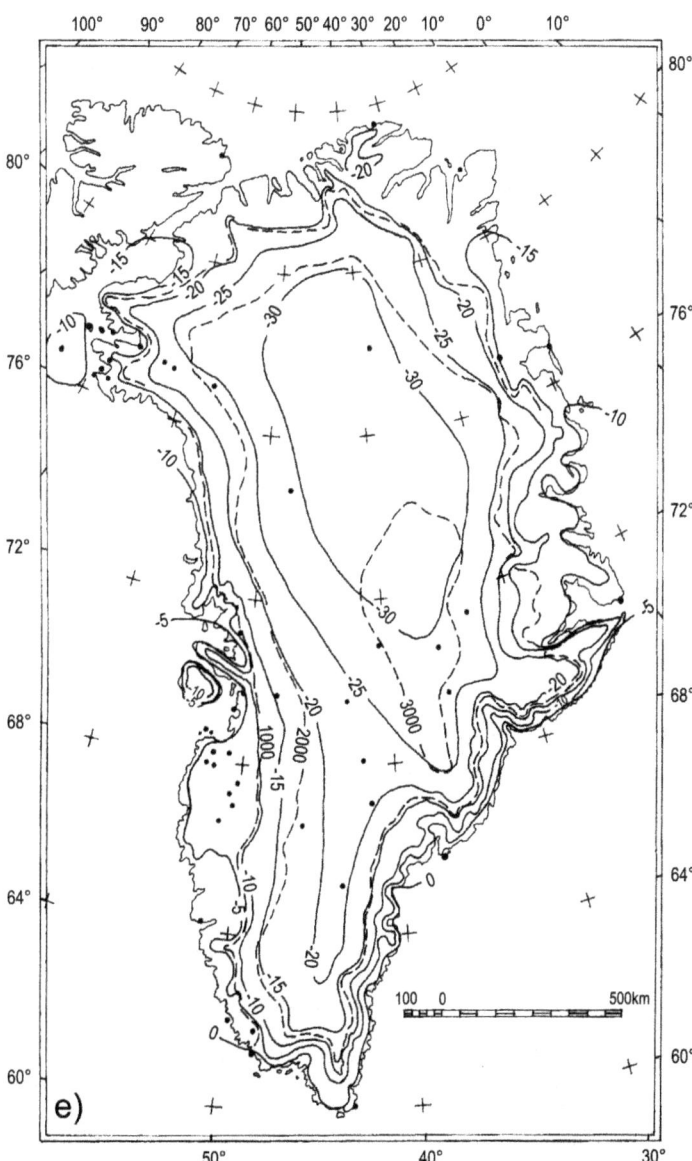

Figure 4.6. cont.

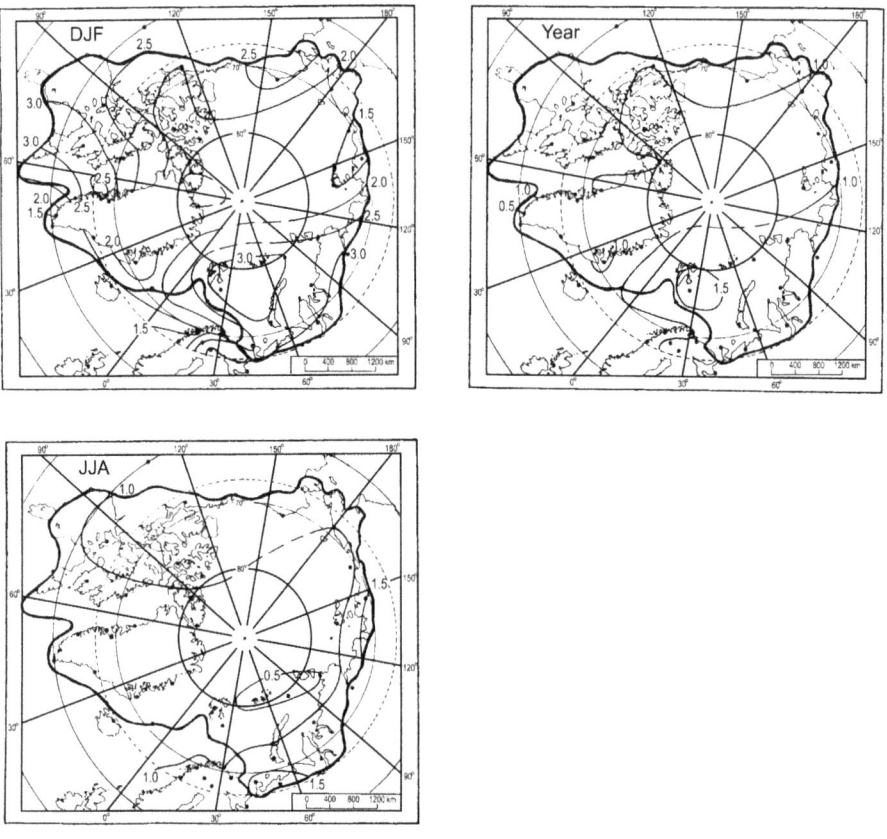

Figure 4.7. Spatial distribution of the standard deviations (in °C) of winter (DJF), summer (JJA), and annual (Year) air temperature in the Arctic, 1951–1990 (after Przybylak 1996a).

The variability of the annual mean temperature (Figure 4.7) is the greatest ($\sigma > 1.5°C$) in the area between Spitsbergen, Zemlya Frantsa Josifa, and Novaya Zemlya, and the smallest ($\sigma < 1.0°C$) in the greater part of the Siberian region, in the north of the Canadian Arctic Archipelago, and probably also in the central Arctic Ocean, particularly from the Pacific side. Significantly, the mean winter temperature has the greatest variability. Przybylak (1996a) has distinguished three regions of the highest variability ($\sigma > 2.5°C$): 1) the central and eastern part of the Atlantic region, 2) the belt encompassing the southern part of the Baffin Bay region and the south-eastern part of the Canadian Arctic, 3) the eastern part of the Pacific region (mainly Alaska). The reason for this high variability here is, without doubt, the vigorous cyclonic activity. The cyclones bring into the Arctic warm air masses from the lower latitudes. The greatest variability, however, does not occur in these areas where the cyclone activity is most common (see Figure 2.3a). This has been noted in the areas where changes of different kinds of air masses most often occur, e.g. air masses of maritime origin (warm) transported by cyclones and of Arctic or polar-continental origin (cold) flowing in from the northern sector. Other Arctic areas which are most often occupied throughout the year by either cold (central Arctic) or warm (seas in contact with the mid-latitudes) air masses, have the lowest variability. Mean summer temperatures have the lowest variability (σ rarely exceeding 1.5°C). This occurs only in some areas of the Russian Arctic (Figure 4.7). The greatest stability of summer temperature may be seen in the region from Spitsbergen to Severnaya Zemlya (σ < 0.5°C). The smallest summer temperature variability can be explained by (Przybylak 1996a): 1) the lowest thermal differentiation of inflowing air masses (Przybylak 1992a), 2) small daily differences in the altitude of the sun (polar day), and 3) the stabilising influence of the melting of snow and sea ice, which is especially strong in the Arctic Ocean.

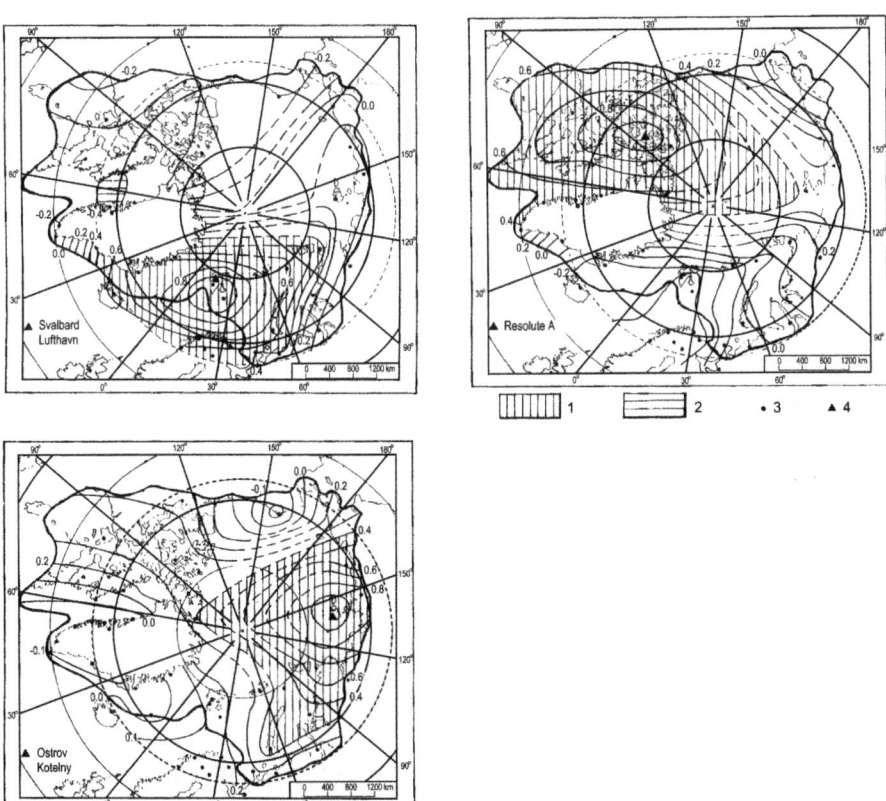

Figure 4.8. Isocorrelates of mean annual air temperature in the Arctic in relation to Svalbard Lufthavn, Ostrov Kotelny, and Resolute A stations, 1951–1990 (after Przybylak 1997b). 1 – positive correlations statistically significant at the level 0.05, 2 – negative correlations statistically significant at the level 0.05, 3 – meteorological stations, 4 – station in relation to which the correlation coefficients of temperature were computed.

Alekseev and Svyashchennikov (1991) and Przybylak (1997b) found that air temperature in the Arctic is most spatially correlated in winter and spring, and least in summer. Mean annual temperatures reveal a slightly stronger correlation than winter temperatures (see Przybylak 1997b). The radius of the extent of statistically significant coefficients of correlation of temperature changes around the stations Svalbard Lufthavn (Atlantic region), Ostrov Kotelny (Siberian region), and Resolute A (Canadian region) is equal to 2500–3000 km for annual values, 2000–2500 km for winter, and 1500–2000 km for summer (Figure 4.8). From Figure 4.8 it can be seen that of the three analysed climatic regions, the highest spatial correlation of temperature occurred in the Canadian region, probably due to the highest stability of atmospheric circulation, especially in the winter and spring (Serreze *et al.* 1993).

The strong correlation of the winter temperature in the Atlantic and Baffin Bay regions seems to be due to very common vigorous cyclonic activity (Baranowski 1977b; Niedźwiedź 1987, 1993; Przybylak 1992a; Serreze *et al.* 1993). This circulation which carries warm and humid air masses from the lower latitudes, diminishes local and even regional features of climatic variations. Cyclones move most often along the Iceland-Kara Sea trough. As a result, the isocorrelates in the eastern Atlantic region have a north-eastern bend. This bend is also present in isocorrelates of the annual temperature (see Figure 4.8). The correlation of winter temperature changes in these regions is also undoubtedly caused by a lack of solar radiation over most of the area. In the other Arctic regions, the strong correlation of temperature change is probably also determined by the predominance of anticyclonic circulation as well as by a high uniformity of the ground. In spring – almost over the whole Arctic – high coefficients of correlation of temperature change are most probably connected with the highest simultaneous homogeneity of the ground (the largest part of the Arctic is covered by snow and sea ice) which, moreover, favours the development and upholding of anticyclones. The low correlation of the summer temperature change is probably caused by: (i) the greatest differentiation of the ground during this season, (ii) weak and evenly distributed cyclonic and anticyclonic circulation (see Figure 2.3), (iii) the influence of local conditions, which are remarkable during this season (it is at this time that the highest values of incoming solar radiation can be noted during a polar day).

Working from the least geometrical distances and using the dendrite method, Przybylak (1997b) delimited 9 groups of stations with the most similar (coherent) mean annual temperature in the Arctic. The schematic distribution of these regions is presented in Figure 7 (Przybylak 1997b). It can be seen that most regions consist of two isolated areas. For example, the south-eastern part of the Canadian Arctic has a similar annual temperature to the area of the south-western Kara Sea and its surroundings in the Atlantic region, and the Pacific region has the same annual temperature as the northeastern part of Greenland.

Przybylak (1997b) found that the coefficients of correlation between seasonal and annual areally averaged Arctic (and also Arctic climatic regions) and Northern Hemisphere (two series: land only and land+sea, after Jones 1994) temperatures are not strong. The average Arctic temperature computed from 27 stations is statistically insignificantly positively correlated with both series of Northern Hemisphere temperature (for annual values $r = 0.18$). Out of the five series of the mean annual regional temperatures, the highest correspondence with the previously mentioned hemispheric series may be noted with the temperature of the Canadian region. The temperature of the Pacific region also has a high correlation, but only for the Northern Hemisphere temperature computed from the land stations ($r = 0.40$). Such relations are typical for almost all seasons. The examined series of temperatures are most strongly correlated in spring and (especially) summer. For the latter season, the statistically significant correlations were computed between the Northern Hemisphere temperature and temperatures of the Atlantic and Canadian regions (see Table 4 in Przybylak 1997b).

4.1.3 Frequency Distribution

In the previous section, the mean seasonal and annual patterns of temperature distribution in the Arctic in the period 1951–1990 have been presented. Also knowledge about the occurrence frequency of different temperature intervals is particularly useful, especially for weather and climate forecasting. Przybylak (1996a) investigated this problem using both data from individual stations and area-average for climatic regions (see Figure 1.2). The results for such data are very similar. Figure 4.9 presents the relative frequency of occurrence of mean winter, summer, and annual temperatures of five analysed climatic regions and for the Arctic as a whole in 1°C intervals. Climatic regions where cyclonic activity dominates (Atlantic, Pacific, and Baffin Bay regions) have the greatest variability of mean winter temperature. Their frequency distributions are characterised by a wide range and a more steady occurrence frequency of individual temperature intervals. Usually the occurrence of three intervals (from –14°C to –17°C) both in the Atlantic (about 70%) and Baffin Bay (about 40%) regions is not accidental. The cyclones, which are connected with the Icelandic low and have the same physical characteristics, are directed simultaneously, as we know, to the Atlantic and Baffin Bay regions. In the Baffin Bay region, however, the dominance of these cyclones and their strength is lower. As a result, the occurrence frequency of these three intervals is almost twice as low as in the Atlantic region. The most normal frequency distribution shows the mean temperature for the Siberian region, where two intervals markedly dominate: –29°C to –30°C (35%) and –30°C to –31°C (about 20%). There is a 50% chance that the mean winter temperature of the whole Arctic will be between –21°C and –22°C. There is also a high frequency (about 30%) that the interval –20°C to –21°C will occur.

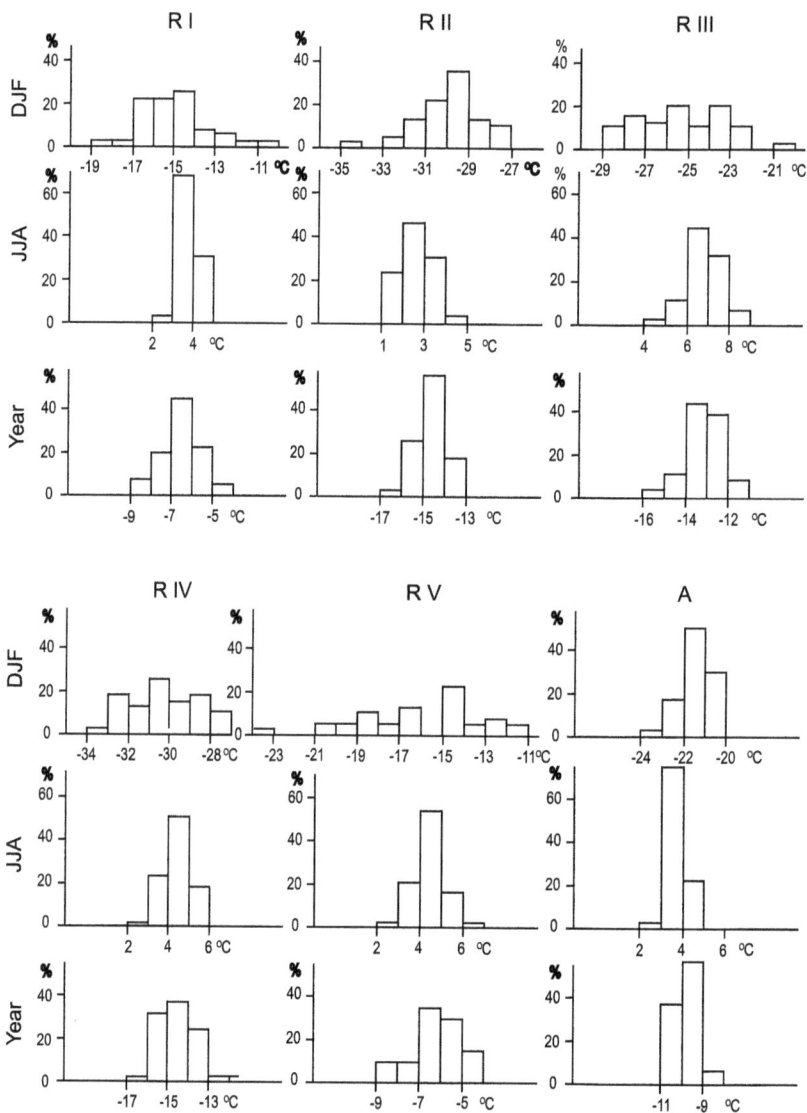

Figure 4.9. Relative frequency (in %) of occurrence of mean winter (DJF), summer (JJA) and annual (Year) temperature of climatic regions of the Arctic (RI–RV) and the Arctic as a whole (A) in 1°C intervals, 1951–1990 (after Przybylak 1996a). RI – Atlantic region, RII – Siberian region, RIII – Pacific region, RIV – Canadian region, RV – Baffin Bay region; see also Figure 1.2.

In summer, the mean temperature of all the climatic regions and of the Arctic as a whole has clear normal frequency distributions. In every region one interval occurs with a frequency of at least 45%. The lowest range of temperature variability occurs in the Atlantic region (only three intervals). Przybylak (1996a) reported that this is caused by the influence of atmospheric circulation, which is still strong in this season. In summer (opposite to winter), the thermal differentiation of air masses incoming here from different directions is markedly lower (Przybylak 1992a). There is as much as a 70% chance that the mean Arctic temperature will range from 3–4°C.

The frequency distribution of the mean annual temperatures, similar to the summer temperature, is nearly normal. The mean annual Arctic temperature in about 95% of cases oscillates between –9°C and –11°C. What is interesting is that such mean annual temperatures do not occur in any climatic region. They are either warmer (Atlantic and Baffin Bay regions) or colder (the other three regions).

4.2 Mean and Absolute Extreme Air Temperatures

As results from investigations conducted by Przybylak (1996a, 1997a), the patterns of spatial distribution of mean 40-year normal (T_{mean}) and extreme (T_{max} and T_{min}) air temperatures in the Arctic are similar in all seasons and for the whole year. The differences, of course, occur only in the magnitudes of temperatures. Usually the mean seasonal T_{max} and T_{min} are warmer and colder respectively than T_{mean} by about 4°C, and for the year as a whole by about 3°C. The only exception to this rule is the Arctic Ocean in summer, mainly in July, when the prevailing melting of snow and sea ice significantly reduces these differences. From the above-mentioned reasons only the mean annual spatial distribution of T_{max} and T_{min} is shown here (Figure 4.5). For Greenland, unfortunately, such maps do not exist, but one can probably assume that the relationships described between the thermal parameters mentioned earlier also occur here. In particular years the highest deviations from the norm of mean extreme air temperatures are noted for winter (3–8°C) and the lowest for summer (1–3°C for T_{min} and 1–4°C for T_{max}). Annual mean temperatures have anomalies of values similar to those of summer temperatures (Przybylak 1996a, 1997a). The spatial distribution of the variability parameter (σ) computed for mean seasonal and annual T_{max} and T_{min} is very similar to that of T_{mean} (Figure 4.7). Przybylak (1996a) found that T_{max} has slightly higher, and T_{min} slightly lower σ than T_{mean}.

The influence of cloudiness on T_{max} is opposite in warm (June – September) and cool (October – May) half-year periods (Table 4.2). In summer the highest T_{max} is connected with clear days and the lowest with cloudy days. Positive anomalies during clear days (see Table 4.2), are especially high (3–7°C) in the most continental part of the Russian and Canadian Arctic. They are much lower in the western and central parts of the Atlantic region (1–2°C). An increase in cloudiness in summer leads to a cooling of the whole Arctic, but especially of the parts of the Arctic which are located near its southern border and are characterised by a high continentality of climate: the stations Naryan-Mar, Chokurdakh, and Coral Harbour A. For these stations the mean differences of T_{max} between clear and cloudy days vary from 5°C to 7°C, while in the Norwegian Arctic (maritime climate) they only differ from 1°C to 2°C (see Table 4.2). In the cool half-year, the influence of cloudiness on T_{max} is opposite to that of summer i.e. an increased cloudiness leads to a warming of the Arctic. The positive anomalies of T_{max} on cloudy days are the highest in winter (above 4°C) almost over the whole Arctic, except for the regions represented by the Jan Mayen and Mys Shmidta stations. It is noteworthy that most of the Arctic (excluding the Siberian region and the western part of the Atlantic region) has higher positive anomalies on cloudy days in spring than in autumn. On clear days, the highest negative anomalies occur in autumn, except for the Jan Mayen and Naryan-Mar stations. In the annual course the lowest differentiated influence of cloudiness on T_{max} occurs at the turn of May/June and September/October, when the described relations between cloudiness and T_{max} change rapidly from one mode to another (see Figure 4 in Przybylak 1999).

Table 4.2. Mean seasonal anomalies of T_{max} (in °C) in the Arctic on clear (1), partly cloudy (2) and cloudy (3) days over the period 1951–1990 (after Przybylak 1999)

Season	Element	DAN[#]	JAN	HOP	NAR*	DIK*	CHO*	SHM*	RES**	COR**	CLY**
1	2	3	4	5	6	7	8	9	10	11	12
DJF	1	−3.1	−5.2	−10.1	−14.1	−8.4	−4.8	−5.9	−3.2	−6.3	−4.3
	2	0.0	−1.1	−1.4	−3.8	−1.7	−0.6	−0.4	0.5	0.4	0.3
	3	4.5	0.9	5.2	4.9	4.8	4.0	2.5	5.8	8.9	4.8
	Mean	−18.5	−2.6	−9.2	−12.4	−21.4	−29.7	−21.2	−27.8	−24.0	−22.6
MAM	1	−2.3	−0.2	−8.9	−7.2	−8.7	−4.6	−6.1	−5.6	−6.0	−4.5
	2	0.1	−1.1	−1.9	−0.9	−3.5	−0.3	−1.0	−1.0	−0.6	−0.3
	3	2.2	0.7	3.8	1.7	4.0	3.6	3.1	7.8	6.3	4.4
	Mean	−11.7	−1.0	−6.7	−1.9	−12.9	−12.7	−12.6	−18.4	−11.5	−13.1
JJA	1	2.0	1.7	1.0	6.1	7.1	4.7	3.9	3.4	2.8	2.6
	2	0.8	0.5	0.2	3.4	2.8	2.9	2.1	1.4	1.3	1.4
	3	−1.7	−0.2	−0.1	−2.8	−0.9	−2.2	−1.3	−1.3	−2.4	−1.8
	Mean	5.3	5.9	3.2	15.2	5.3	12.6	6.7	4.4	10.1	6.3

Table 4.2. cont.

1	2	3	4	5	6	7	8	9	10	11	12
SON	1	-3.7	-3.1	-12.4	-7.1	-15.4	-10.9	-6.6	-10.8	-12.4	-10.3
	2	0.1	-0.7	-2.5	-1.5	-6.2	-3.5	-2.0	-2.6	-0.9	-0.9
	3	3.0	0.3	1.6	1.2	3.6	4.1	1.5	5.8	3.6	2.1
	Mean	**-9.4**	**2.2**	**-1.4**	**1.2**	**-6.3**	**-9.7**	**-5.8**	**-12.0**	**-4.4**	**-5.5**

Key: DAN – Danmarkshavn, JAN – Jan Mayen, HOP – Hopen, NAR – Naryan-Mar, DIK – Ostrov Dikson, CHO – Chokurdakh, SHM – Mys Shmidta, RES – Resolute A, COR – Coral Harbour A, CLY – Clyde A, # – Data for the period 1955–1990; * – Data for the period 1967–1990; ** – Data for the period 1953–1990, 1 – C < 2; 2 – 2 ≤ C ≤ 8; 3 – C > 8; Mean – mean seasonal T_{max}

Generally speaking, the influence of cloudiness on T_{min} is roughly similar to that on T_{max}, but there are also several important differences (compare Tables 4.2 and 4.3). One such difference is the opposite influence of cloudiness on T_{min} and on T_{max} in summer in the Norwegian Arctic and in the southern Canadian Arctic. During this season T_{min} is higher on cloudy days than clear days (see Table 4.3). Moreover, positive anomalies of T_{min} in the rest of the Arctic are significantly (2–3 or more times) lower than anomalies of T_{max} (compare Tables 4.2 and 4.3).

Table 4.3. Mean seasonal anomalies of T_{min} (in °C) in the Arctic on clear (1), partly cloudy (2) and cloudy (3) days over the period 1951–1990 (after Przybylak 1999)

Season	Element	DAN#	JAN	HOP	NAR*	DIK*	CHO*	SHM*	RES**	COR**	CLY**
DJF	1	-3.3	-5.2	-8.4	-12.9	-6.9	-3.9	-6.1	-2.5	-4.9	-3.9
	2	-0.3	-1.3	-1.9	-5.5	-1.8	-1.1	-1.1	0.2	-0.3	-0.3
	3	5.4	1.1	5.2	6.1	4.6	4.1	3.4	5.8	9.2	6.1
	Mean	**-27.1**	**-7.8**	**-15.9**	**-21.3**	**-28.5**	**-36.7**	**-28.2**	**-35.0**	**-32.3**	**-30.4**
MAM	1	-3.7	-1.2	-8.8	-10.7	-9.1	-5.8	-8.9	-6.1	-7.5	-5.8
	2	-0.2	-1.6	-2.4	-2.7	-4.0	-0.9	-2.1	-1.3	-1.4	-1.1
	3	4.1	1.1	4.3	3.6	4.6	5.3	5.2	9.0	9.2	7.0
	Mean	**-19.7**	**-5.5**	**-12.3**	**-10.4**	**-19.8**	**-21.3**	**-20.0**	**-25.5**	**-21.2**	**-22.4**
JJA	1	0.8	-0.7	-0.3	1.8	4.4	1.6	0.9	1.5	-0.2	0.3
	2	0.0	-0.6	-0.3	1.1	1.1	0.9	0.5	0.3	0.0	0.0
	3	-0.3	0.2	0.1	-0.9	-0.4	-0.7	-0.3	-0.3	0.0	-0.1
	Mean	**-0.3**	**2.2**	**0.1**	**6.4**	**0.9**	**3.7**	**0.9**	**-0.5**	**2.1**	**-0.4**
SON	1	-4.2	-2.4	-13.0	-9.6	-15.1	-11.6	-8.9	-11.4	-12.5	-11.5
	2	-0.3	-0.9	-3.1	-3.3	-7.4	-4.6	-4.0	-3.4	-2.2	-2.2
	3	4.2	0.4	1.9	2.4	4.1	4.9	2.6	6.9	5.3	3.6
	Mean	**-15.6**	**-1.6**	**-5.1**	**-5.2**	**-11.3**	**-15.8**	**-11.3**	**-17.8**	**-11.5**	**-11.3**

Key: Mean – mean seasonal T_{min}; other explanations as in *Table 4.2*

In winter, the influence of cloudiness on both T_{max} and T_{min} is very similar, but negative anomalies on clear days are lower in most of the Arctic in the case of T_{min}. In spring, the influence of cloudiness is significantly greater on T_{min} than on T_{max}. Negative (or positive) anomalies of T_{min} on clear (or cloudy) days are clearly greater than anomalies of T_{max}. This means that an increase in cloudiness results in a much greater rise of T_{min} than T_{max} during this season. A similar situation is also present in autumn, although it is expressed slightly weaker than in spring.

Absolute minimum temperatures in the Arctic defined according to *Atlas Arktiki* (1985) occur on the Greenland Ice Sheet. Temperatures in the winter months (December–March) very often drop below –50°C. The lowest measured temperature occurs in the Northice station (–66.1°C, 9th Jan. 1954). Slightly higher temperatures (–64.8°C) were noted in Eismitte (20th March 1931) and Centrale (22nd Feb. 1950). Outside Greenland, the absolute temperatures below –50°C occur over a large area of the Arctic characterised by the greatest degree of continentality (above 60–70%, Figure 4.2). The temperature drops below –50°C in the belt stretching from the central part of the Russian Arctic through the North Pole to the north-eastern part of the Canadian Arctic (see Gorshkov 1980, p. 44 or 45). Maxwell (1980) reported that the absolute lowest temperature in the Canadian Arctic prior to 1975 was recorded at Shepherd Bay (south of Boothia Peninsula) in February 1973 (–57.8°C). However, Sverdrup (1935) reported (see Table on p. K11) that in the area of Lady Franklin-Bay (in the north-eastern part of Ellesmere Island) the lowest noted temperature reached –58.8°C. The exact date when this temperature occurred is unknown (Sverdrup did not give such information), but analysing his Table 7 we can say that this temperature had to be noted during one of the following winters: 1871/72, 1875/76, 1881/82/83, 1905/1906 or 1908/09. The temperature –58.8°C is probably also the lowest temperature noted in the whole "lower" (without Greenland) Arctic. The absolute lowest temperature in the Northern Hemisphere, as we know, occurred in the Subarctic at Oimekon, where it reached –77.8°C (Martyn 1985). In summer, the absolute temperatures < 0°C occur over the entire Arctic. In the central Arctic these temperatures range from –8°C to –16°C (Gorshkov 1980). In Greenland they drop below –20°C and in June and August even below –30°C (see Putnins 1970, his Table XVIII).

The absolute maximum temperatures occur in summer and show a clear dependence on the latitude (see Gorshkov 1980). In the central Arctic they are always below 10°C and in the Arctic islands they rarely exceed 20°C. The highest recorded temperatures occur in the continental southernmost parts of the Arctic (especially in the western parts of the Canadian and Russian Arctic), where they can even exceed 30°C. The highest temperature in the continental part of the Canadian Arctic was recorded at Coppermine (30.6°C) (Rae 1951). In the Canadian Arctic Archipelago, the highest temperature occurred at Cambridge Bay (28.9°C, July 1930). In the Russian Arctic, the absolute maximum temperatures can also exceed 30°C (Gorshkov 1980). In the coastal region, between 45°E and 60°E, the temperature is even higher than 32°C. For example at Naryan-Mar station the highest recorded temperature in 1967–1990 was equal to 33.9°C (10 July 1990). In winter, the highest temperatures never reach freezing point in the most continental part of the Arctic stretching from the central part of the Siberian region, through the central Arctic, to the eastern and northern part of the Canadian Arctic (see Gorshkov 1980). On the Greenland Ice Sheet, they are very rarely higher than –10°C.

4.2.1 Mean Seasonal and Annual Diurnal Temperature Ranges

The highest mean annual Diurnal Temperature Ranges (DTRs) above 8°C occur over the continental parts of the Canadian and Russian Arctic which are located far from Atlantic and Pacific oceans (Figure 4.10). The lowest DTRs (< 5°C) are noted in the Norwegian Arctic, particularly in those areas which are not covered by sea ice. The region spreading from the Norwegian Arctic to Alaska which encompasses almost all islands lying here (from Spitsbergen to Ostrov Vrangelya) has a slightly higher DTR (5–6°C).

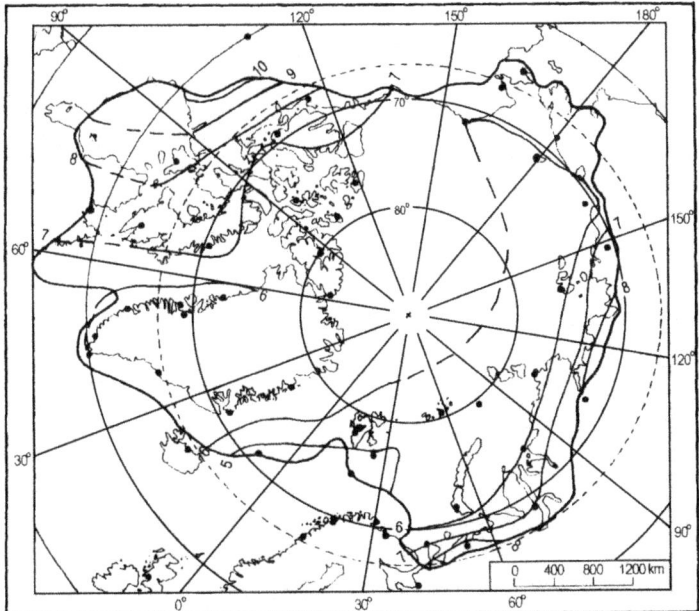

Figure 4.10. Spatial distribution of the mean annual diurnal temperature ranges (DTR, in °C) in the Arctic over the period 1951–1990 (after Przybylak 2000b). Key – dashed lines denote probable course of the isolines.

Probably one of the main factors causing this is the very strong and changeable cyclonic activity occurring here which brings high cloudiness. Its influence in lowering the DTR is noted mainly in the warm half-year. The opposite is true for the cold half-year. As a result, in those regions mentioned, and some others which are also strongly influenced by the atmospheric circulation (the western and northern parts of the Russian Arctic and the western coastal parts of Greenland), the highest DTRs occur in winter (Figure 4.11).

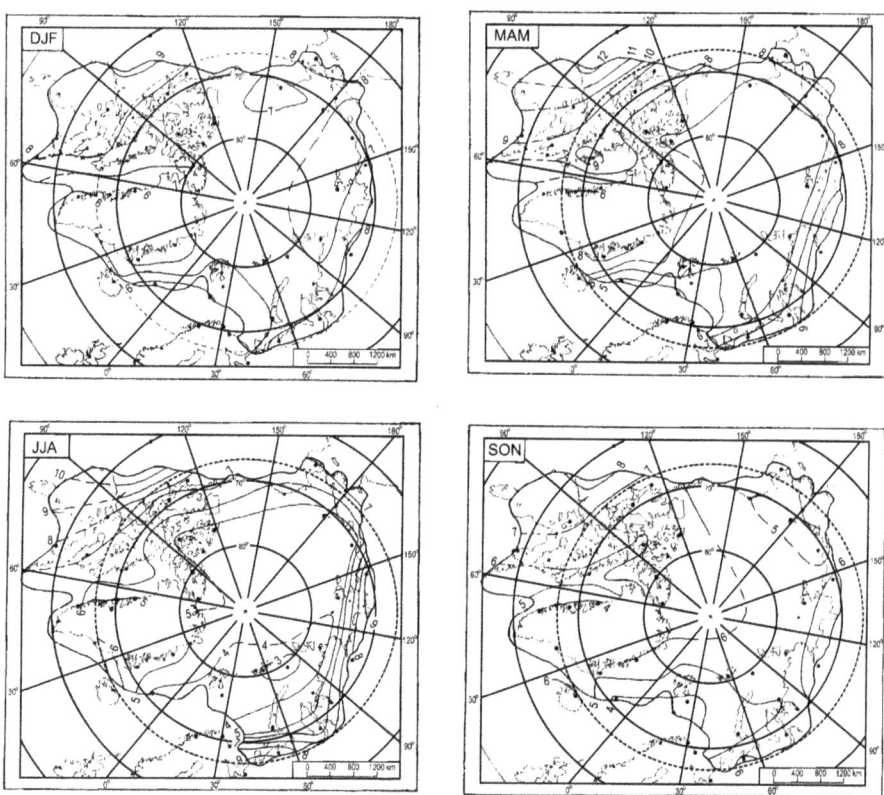

Figure 4.11. Same as Figure 4.10, but for winter (DJF), spring (MAM), summer (JJA) and autumn (SON) (after Przybylak 2000b).

In winter the highest DTRs are noted in the southern continental parts of the Arctic (> 8°C) and the lowest in the southern Norwegian Arctic as well as on the western coast of Greenland (Figure 4.11). In the central Arctic the DTR is equal to about 7°C.

In spring the differentiation of the DTR is significantly higher than in winter and reaches more than 7°C (in winter only 4°C). The highest values (> 9°C) occur in the centre of the southernmost parts of the Canadian and Russian Arctic (characterised by the greatest degree of continentality of climate in the Arctic; see Figure 4.2) and the lowest, similar to those of winter, in the Norwegian Arctic and on the western coast of Greenland (< 6°C).

In summer the mean DTRs are lower than in spring but the differences between the highest (> 10°C) and lowest (< 3°C) values are the same. In this season an exceptionally low DTR is present in the Norwegian Arctic and the north-western part of the Russian Arctic (Figure 4.11) where it drops below 3–4°C.

In autumn the differentiation of the mean DTR in the Arctic decreases (4–5°C). Again, as in other seasons, the highest DTRs are in the centre of the southernmost parts of the Canadian (> 8°C) and Russian (> 6°C) Arctic and the lowest are in the Norwegian Arctic (< 4°C) (Figure 4.11).

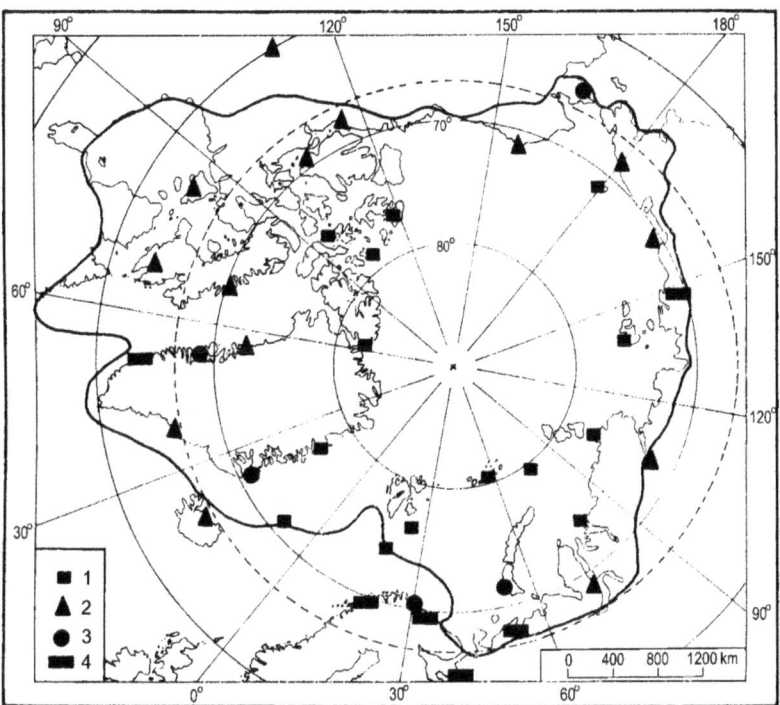

Figure 4.12. Annual courses of the DTR in the Arctic based on their seasonal means from the period 1951–1990 (after Przybylak 2000b).
Key: 1 – maximum of the DTR in winter and minimum in summer, 2 – maximum of the DTR in spring and minimum in autumn, 3 – maximum of the DTR in winter and minimum in autumn, 4 – maximum of the DTR in summer and minimum in autumn.

The annual course of the DTR is particularly interesting. Based on the seasonal means it is possible to distinguish four types of the annual course of this variable in the Arctic (Figure 4.12):
 i) maximum DTR in winter and minimum in summer,
 ii) maximum DTR in spring and minimum in autumn,

iii) maximum DTR in winter and minimum in autumn,
iv) maximum DTR in summer and minimum in autumn.

The first two types clearly dominate in the Arctic. Out of 33 analysed stations, the first type occurred in 14 stations (42.4%) and the second in 11 stations (33.3%). The third and fourth types were present in four (12.1%) and three (9.1%) stations, respectively. Only the DTR in the station Eureka has another pattern. The maximum DTR in winter and minimum in summer occur mostly in the Norwegian Arctic as well as in the western and northern parts of Russian Arctic, and the northern part of the Canadian Arctic. This pattern is probably also present in the central Arctic. It is worth mentioning that these areas are either under the very intense influence of the atmospheric circulation (great frequency of cyclones) or are situated around the North Pole, where the daily contrast of the incoming radiation is the lowest in the Arctic and cyclonic activity, although weaker, is however still present (see Serreze and Barry 1988 or Serreze *et al.* 1993). The second type in the annual course of the DTR (the highest values in spring and the lowest in autumn), which is almost as frequent as the first, occurs in the parts of the Arctic where cyclonic activity is weak and the daily contrast of the solar forcing is the highest (southern parts of the Canadian and Russian Arctic, northern Alaska and in some parts of Greenland). Ohmura (1984) presents a more detailed explanation of the causes of this kind of annual course of the DTR based on investigations of heat balance conducted on Axel Heiberg Island (Canadian Arctic) in the summers of 1969 and 1970. This type in scientific literature is known as the 'Fram' type. The occurrence of the third and fourth types in stations located in different isolated parts of the Arctic can be related to the specific local conditions (radiation and atmospheric circulation).

According to the monthly means of the DTR, the highest values most often occurred in April (63.6% of the stations) or February (18.2%) and the lowest in September (62.1%) or October (16.7%) (Przybylak 2000b).

4.2.1.1 Diurnal Temperature Ranges and Cloudiness

Based on the results presented in Section 4.2, the general pattern of influence of cloudiness on T_{max} and T_{min} seems to be quite similar. However, the existing differences in magnitudes of this influence (expressed by anomalies) are significant during some seasons and should cause appropriate changes of the DTR in the case of increasing or decreasing trends in cloudiness in the Arctic.

The influence of cloudiness on DTR is presented in Table 4.4 and Figure 4.13. These data clearly show that, on an annual basis, increased cloudiness leads to a decrease in DTR. This influence is at its highest in summer, then in spring and also in autumn, except for the region of Jan Mayen. In winter, the situation is much more complicated because the highest positive anomalies of the DTR occur on partly cloudy days in the whole Arctic. A further increase of cloudiness leads to a decrease in DTR. Slightly positive anomalies of the DTR occur also in some parts of the Arctic, both on clear days (Danmarkshavn and Mys Shmidta) and cloudy days (Ostrov Dikson and Chokurdakh). It is noteworthy that in the parts of the Arctic where cyclonic activity prevailed (Atlantic, Pacific, and Baffin Bay regions) the anomalies of the DTR on cloudy days were lower than on clear days (see Table 4.4 and Figure 4.13).

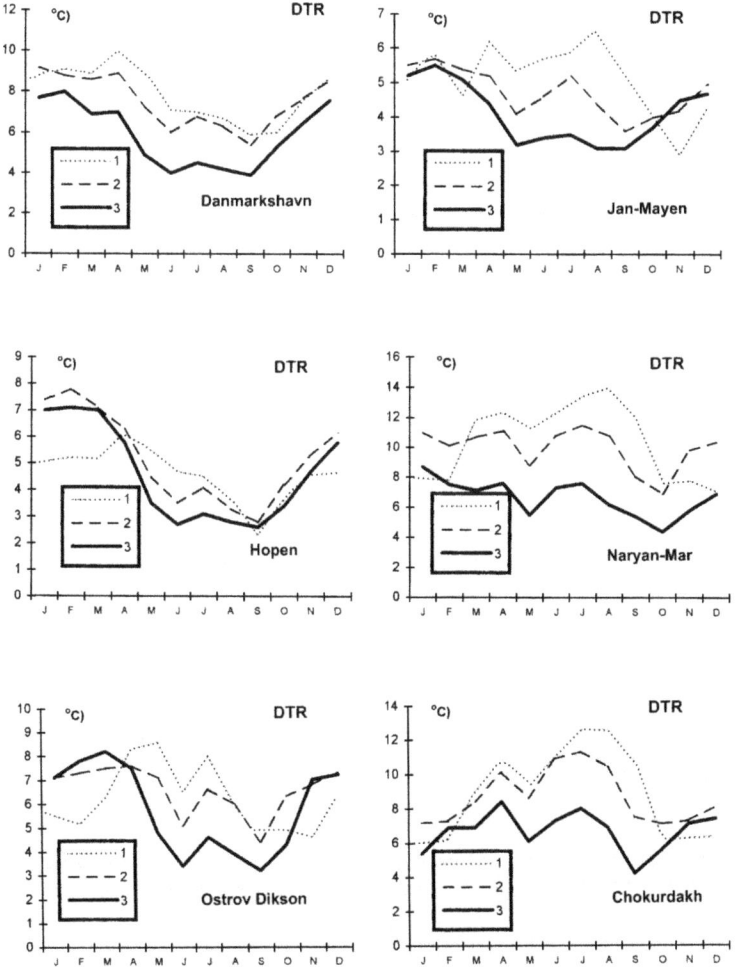

Figure 4.13. Mean annual courses of the DTR on clear (1), partly cloudy (2), and cloudy (3) days at 10 Arctic stations representing the majority of the climatic regions and subregions after *Atlas Arktiki* (1985). After Przybylak (1999).

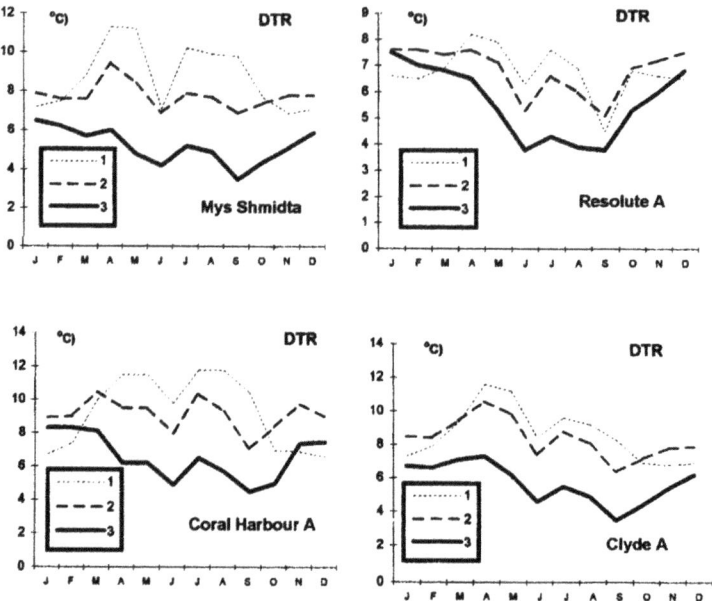

Figure 4.13. cont.

Table 4.4. Mean seasonal anomalies of DTR (in °C) in the Arctic on clear (1), partly cloudy (2) and cloudy (3) days over the period 1951–1990 (after Przybylak 1999)

Season	Element	DAN[#]	JAN	HOP	NAR*	DIK*	CHO*	SHM*	RES**	COR**	CLY**
DJF	1	0.2	0.0	−1.7	−1.2	−1.5	−0.8	0.3	−0.7	−1.4	−0.5
	2	0.3	0.2	0.4	1.7	0.2	0.6	0.8	0.3	0.6	0.5
	3	−0.8	−0.2	−0.1	−1.3	0.2	0.1	−0.8	−0.1	−0.3	−1.3
	Mean	**8.6**	**5.2**	**6.7**	**8.9**	**7.1**	**7.0**	**7.0**	**7.2**	**8.3**	**7.8**
MAM	1	1.3	1.0	0.0	3.5	0.3	1.2	2.8	0.4	1.5	1.1
	2	0.3	0.5	0.6	1.8	0.5	0.5	1.1	0.3	0.8	0.7
	3	−1.9	−0.3	−0.5	−1.9	−0.5	−1.7	−2.1	−1.3	−2.9	−2.6
	Mean	**8.0**	**4.5**	**5.6**	**8.5**	**6.9**	**8.6**	**7.4**	**7.1**	**9.7**	**9.3**
JJA	1	1.3	2.3	1.2	4.3	2.8	3.1	3.0	1.9	3.1	2.3
	2	0.8	1.0	0.5	2.3	1.7	2.0	1.6	1.1	1.3	1.4
	3	−1.4	−0.4	−0.2	−1.8	−0.4	−1.5	−1.0	−0.9	−2.4	−1.7
	Mean	**5.6**	**3.7**	**3.1**	**8.8**	**4.4**	**8.9**	**5.8**	**4.9**	**8.0**	**6.7**
SON	1	0.5	−0.7	0.6	2.4	−0.2	0.7	2.1	0.6	0.2	1.2
	2	0.4	0.2	0.6	1.8	1.2	1.1	1.9	0.9	1.3	1.5
	3	−0.8	0.0	−0.3	−1.2	−0.5	−0.7	−1.2	−1.1	−1.7	−1.5
	Mean	**6.2**	**3.8**	**3.7**	**6.4**	**5.0**	**6.1**	**5.5**	**5.8**	**7.1**	**5.8**

Key: Mean – mean seasonal DTR; other explanations as in *Table 4.2*

An opposite relation occurs in the Siberian and Canadian regions (with prevailing anticyclones) where radiation plays a significantly greater role than in the previously mentioned regions. On cloudy days (Tables 4.2–4.3), the anomalies of T_{max} and T_{min} are nearly the same, but on clear days negative anomalies of T_{max} are significantly greater. In the regions where a great number of clouds are transported together with warm and humid air masses from the lower latitudes by a synoptic-scale cyclonic activity, the influence of cloudiness on T_{max} and T_{min} is different, mainly on cloudy days (opposite to the situation in the Canadian and Siberian regions). This is a consequence of the very similar influence of advection of warm air masses on T_{max} and T_{min}. On the other hand, high cloudiness connected with this advection reduces the loss of the long-wave outgoing radiation to space. This radiation flux is relatively more important during the night than during the day, and the resulting positive anomalies of T_{min} are higher than T_{max} (see Tables 4.2–4.3).

4.3 Temperature Inversions

Surface-based temperature inversions in the troposphere are one of the main features of the Arctic climate, particularly in the low-sun (or no-sun) periods. This differs from normal tropospheric conditions, in which temperature decreases with height from the surface. Because of the very high frequency of the temperature inversions in the Arctic in the annual march, the term 'semi-permanent inversion' is often used. Outside the Polar regions, the semi-permanent inversions occur only in the subtropical belt. In the latter areas, however, the inversions are solely dynamic and are separated from the surface by a highly unstable layer. The polar inversions are generally caused by the negative net radiation balance at the surface. The presence of the Arctic temperature inversions is closely related to the snow and ice surfaces dominant in the area studied. However, in the Arctic there are also upper tropospheric inversions, which are a result either of subsidence of the air in anticyclones or of a warm air advection over underlying cold air masses. One can agree with the view expressed by Vowinckel and Orvig (1970) that the Arctic inversions are a significantly more complex phenomenon than the subtropical type. They distinguished three categories of temperature inversions in the Arctic: surface, subsidence, and advection. The latter two can occur at any height and are characterised by significantly smaller gradients.

Investigations of the thermal structure of the troposphere in the Arctic began at the beginning of the 20th century. In 1906 balloon investigation of the upper levels was sponsored by the Prince of Monaco (Hergessell 1906). Belmont (1958) gave an excellent review of the history of investigations of the thermal structure of the troposphere in the Arctic carried out up to about

1957. Thus, there is no need to reiterate this information. Of the more important works which appeared in 1960s and 1970s one should mention Dolgin 1962; Gaigerov 1964; Stepanova 1965; Vowinckel 1965; Billeo 1966; Vowinckel and Orvig 1967, 1970; Dolgin and Gavrilova 1974. In the 1980s and especially in the 1990s investigations of the Arctic were intensified (Maxwell 1982; Kahl 1990; Nagurnyi *et al.* 1991; Timerev and Egorova 1991; Bradley *et al.* 1992; Kahl *et al.* 1992a, b; Serreze *et al.* 1992; Zaitseva *et al.* 1996). These efforts were undertaken in order to establish and describe the climatology of the temperature inversion. Such knowledge allows researchers to ascertain whether there have been changes in recent years when global warming and greenhouse gas build-up is present.

The characteristics of Arctic temperature inversions are presented here mainly according to the results published recently by Zaitseva *et al.* (1996). They analysed radiosonde data on air temperature (from surface to 3–5 km) over the western Arctic Ocean, which were made during the U.S. Air Force Ptarmigan weather reconnaissance missions (1950–1961) and at the Soviet North Pole drifting stations (1950–1954). For more detailed investigation they have chosen two areas representing contrasting conditions: one situated near the North Pole (1079 soundings) and second in the Beaufort Sea (2040 soundings). The results of inversion frequency, height of inversion base, inversion thickness, and temperature difference across inversion are presented in Figure 4.14. It can be seen from this Figure that temperature inversion occurs with a very high frequency (93%). As would be expected, the highest inversion frequency occurs in winter (98–99%). For clarity, seasons are defined normally (Dec.–Feb., March–May etc.), not as has been done by Zaitseva *et al.* (1996) and for the characteristics of mean conditions median values were used. Most winter inversions (about 75%) begin at the surface. In the Western Arctic they are most common in February (100%). Zaitseva *et al.* (1996) did not find any significant regional differences in winter inversion frequency. This is probably connected with similar weather and surface conditions. The change from winter to spring conditions is very abrupt in the April–May period (similar results were obtained by Vowinckel and Orvig 1967). The frequency of surface-based inversions in May decreased to about 45% and 36% near the North Pole and the Beaufort Sea, respectively.

In summer, the lowest total frequency of inversions (88%), but the highest frequency of the upper inversions (52%) is observed (see Figures 4.14a and b). A significant part of surface-based inversion during this period is connected with the loss of energy used for the melting of snow and ice. In some stations, the secondary maximum of surface-based inversions can be seen (see Figure 4 in Bradley *et al.* 1992). The summertime minimum of surface-based inversions (36%) is, according to Zaitseva *et al.* (1996), connected with the highest cyclonic activity at that period, which causes an intense mixing of the lower atmosphere and results in high cloudiness. In the annual march, the lowest frequency of surface-based temperature inversions occurs in August near the North Pole and in September in the Beaufort Sea, i.e., during the periods occurring just after the end of the summer melting. These results are generally in line with those published by Bradley *et al.* (1992).

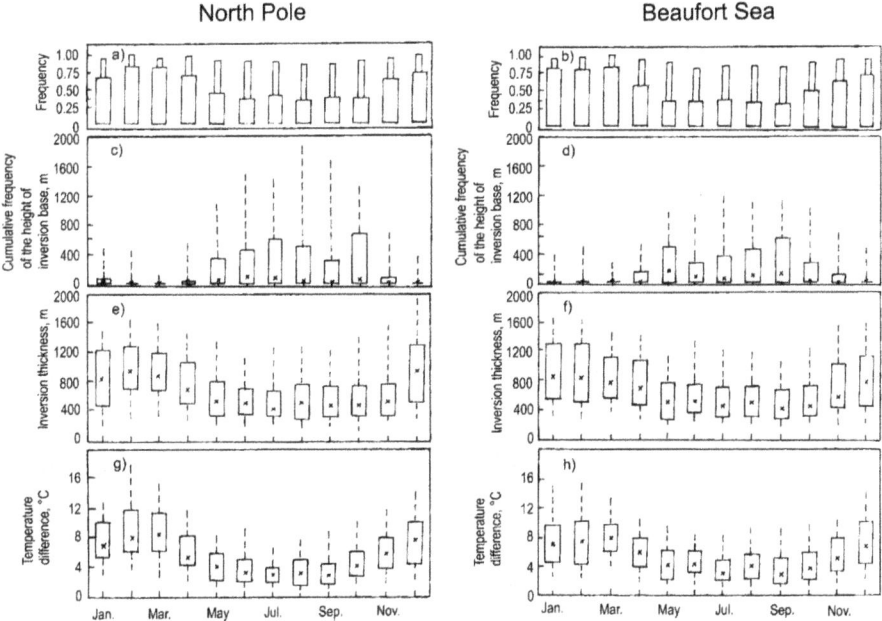

Figure 4.14. Monthly frequencies of (a, b) inversion, (c, d) cumulative frequency distribution of the height of the inversion base, (e, f) its thickness, and (g, h) temperature difference across the inversion for the North Pole (a, c, e, g) and the Beaufort Sea (b, d, f, h) sectors of the Arctic (after Zaitseva *et al.* 1996). Heights of the wide and narrow parts of the bars in Figures a and b denote the frequency of occurrence of surface-based and upper tropospheric inversions, respectively. The combined height of bars is equal to the total inversion frequency.

The height of the inversion base (Figures 4.14c and d) is lowest during the winter (surface-based) and highest in summer near the North Pole (up to about 600 m). In the Beaufort Sea the maxima of the inversion base are observed in May (up to 600 m) and in September (up to 650 m). In summer, due to melting, a significant drop of the inversion base can be seen. It reaches an average level of about 125 m. The inversion base returns to the surface in October (the Beaufort Sea) and in November (near the Pole).

The average thickness of the inversion near the North Pole is about 100 m higher (900 m) than in the Beaufort Sea (about 800 m). In both areas the maximum thickness very often exceeds 1200 m and more rarely 1600 m. On particular days, however, the inversion depth can even reach more than 3000 m (see Table 5 in Bradley *et al.* 1992). The greatest average thickness of inversion is noted near the Pole in February and December, and in the Beaufort Sea, in January. In the latter area, the inversion depth in December and February is only slightly lower. In spring, as a surface mixing layer forms in response to increased solar radiation and extensive cloud cover, the lower parts of inversion layers are destroyed. As a result, a significant decrease of the inversion depth is noted. The mean inversion thickness near the Pole is at its lowest in summer and autumn and slightly exceeds 400 m. In the Beaufort Sea, the situation is very similar when mean seasonal conditions are taken into account, but significantly a lower inversion depth occurs in two autumn months – September and October. Bradley *et al.* (1992) have obtained the same results for Barter Island and Point Barrow stations. Of course, melting processes, which increase the inversion thickness in summer, cause this situation. An abrupt rise of the inversion thickness is noted from November to December (near the Pole) and a month earlier in the Beaufort Sea (see Figures 4.14e and f).

Temperature changes across the inversion layer (called inversion intensity or strength) are highly correlated with inversion thickness and inversely related to surface temperature (see Figs 6 and 7 in Bradley *et al.* 1992). As can be seen from Figures 4.14g and h, the inversions are strongest in winter months when surface temperatures are the lowest. The average temperature difference at this time oscillates between 7°C and 8°C. The highest noted values (> 15°C) occur in both areas in February. On particular days, the temperature across the inversion can significantly exceed 30°C (see Table 5 in Bradley *et al.* 1992). For example, on Barter Island on January 25, 1983 it reached 35.7°C (6.7°C/100 m). Such a situation most often occurs when lower tropospheric warming due to a subsidence (upper inversion) merges with the strong surface-based radiation inversion. The intensity of inversion significantly decreases in April and May. Near the North Pole, the weakest inversions (about 3–4°C) occur in the period June–September. Then a gradual increase of intensity towards the winter maximum is observed. On the other hand, in the Beaufort Sea, a relative flat minimum (3–4°C) occurs from May to October. The mean monthly differences across the inversion in the mentioned periods very rarely exceed 8°C.

It has been shown that the surface-based inversions are most frequent and intensive during clear sky periods in winter. In turn, the upper inversions show an opposite pattern; they are most pronounced in summer and are connected with great amounts of cloud. The intensity and thickness of the upper inversions are considerably less: 1.2°C and 0.5–0.9 km on average (Vowinckel and Orvig 1970). The mean duration of this inversion is also significantly lower in the cold half-year, but greater in summer.

The lowest frequency of surface-based inversions in the Arctic occur in the southernmost part of the Atlantic region, oscillating between 20% and

30% in January and between 30% and 40% in July (see Figure 30 in Dolgin and Gavrilova 1974). Thus the annual cycle of frequency inversions occurring in other parts of the Arctic is reversed in this area. This is connected with the strong cyclonic activity which is common here as well as with the character of the surface (open, relatively warm water). Dolgin and Gavrilova (1974) showed that the distribution of surface-based inversion closely correlates with the mean conditions of atmospheric air pressure and the characteristics of the surface. The highest inversion frequency in the Arctic occurs in the areas where anticyclones and snow and pack-ice dominate. On the other hand, both frequent cyclone occurrence and a surface not covered by snow and sea ice significantly reduce the inversion frequency. However, we must add that the inversions are not only characteristic features in anticyclone systems, but in cyclones as well. The mean annual frequency of inversions in cyclones was found to be 69% (Dolgin 1960; Gaigerov 1962). In the northern sections of the cyclones, the frequency of inversions was greater than in the southern parts.

Chapter 5

CLOUDINESS

Our current knowledge concerning cloudiness in the Arctic, as has already noted been by Raatz (1981), Barry *et al.* (1987) and Serreze and Rehder (1990), is still remarkably poor. Of the existing 15 distinct global cloud climatologies reviewed by Hughes (1984), only two (Scherr *et al.* 1968 and Berlyand and Strokina 1980) provide information about both poles while a further four have information for only one or other of the poles. In the second half of the 1980s and at the beginning of the 1990s four new cloud climatologies became available; one based on surface observations (Warren *et al.* 1986, 1988) and three based on satellite radiation measurements (METEOR – Matveev and Titov 1985; Aristova and Gruza 1987; Mokhov and Schlesinger 1993, 1994; NIMBUS-7 – Stowe *et al.* 1988, 1989; ISCCP – Rossow and Schiffer 1991; Rossow and Garder 1992). Rossow (1992) comments that these climatologies are in excellent agreement on the geographic and seasonal cloud amount variations, and even on total cloud amounts (except for one), everywhere except in the Polar regions. Here large differences occur both in the average geographical distribution of cloud amounts and in their annual march. Moreover, McGuffie *et al.* (1988) report that "none of the existing global cloud climatologies provides comprehensive information for the Polar regions". Moreover, usually the scale and projection of maps do not allow results to be presented in sufficient detail.

In the 1980s and 1990s quite a large number of papers relating to cloudiness in the Arctic were published. The majority of them use satellite-derived radiation measurements to estimate cloudiness. Two groups of works can be distinguished: 1) those presenting possible analysis methods (manual methods and those using computer-based automatic algorithms) of cloud analyses (Key and Barry 1989; Dutton *et al.* 1991; Carsey 1992; Curry and Ebert 1992; Key and Haefliger 1992; Robinson *et al.* 1992; Serreze *et al.* 1992; Francis 1994; Hahn *et al.* 1995), and 2) those presenting satellite-based cloud climatologies and their comparisons with surface climatologies (e.g. Barry *et al.* 1987; Kukla and Robinson 1988; McGuffie *et al.* 1988; Serreze and Rehder 1990; Rossow 1992, 1995; Schweiger and Key 1992). One should add here that the satellite-derived cloud climatologies which are presently available have many weaknesses: they are incomplete, particularly in seasonal coverage, they are based on short time periods, and their reliability is often questionable as a result of cloud-detection problems.

Intercomparative studies of satellite-derived and surface-based cloud climatologies have shown significant differences in the results obtained (e.g.,

McGuffie *et al.* 1988; Rossow 1992; Schweiger and Key 1992). Schweiger and Key (1992) found that satellite-derived estimates of cloud amounts are generally 5–35% lower than had been indicated by surface observations over the entire Arctic, and regional differences may reach up to 45%. In the annual march these differences are 2–3 times greater from May to October than they are in winter. These authors concluded that at present it is not possible to determine which cloud climatology is 'correct'. Certainly, such a situation is very unfavourable and requires extensive further investigation in the near future, because:

— Polar regions are considered to be of great importance for the global climate (see e.g., *The Polar Group* 1980 or *Arctic Climate System Study* 1994),
— cloud cover is a major component of the Arctic climate system through its influence on both energy and moisture exchange between the elements of the system, i.e. atmosphere, ocean, cryosphere, biosphere, and litosphere.

All hope lies in improvements to the analysis methods of satellite-derived radiation measurements, which in many cases reduce the quality of current analyses of available data sets (Rossow 1995).

From the above review, it seems that the old climatologies, mainly based on surface observations, are still the best source of information (Vowinckel 1962; Huschke 1969; Vowinckel and Orvig 1970; Gorshkov 1980). They show a broad agreement over much of the Arctic in regard to the seasonal cycle of total cloud amount. However, there are some disagreements in the geographical distribution of cloud cover, particularly in the case of low cloud in winter (McGuffie *et al.* 1988). Crane and Barry (1984) also found significant differences in the mean values of cloudiness presented in the above climatologies. These differences are probably connected with the different data sets used (a denser or sparser network of stations, longer or shorter periods of observations). For example, a short period of observations can significantly obscure results, especially those of winter cloud climatologies because of difficulties observing clouds with little sky illumination (see e.g., Schneider *et al.* 1989; Hahn *et al.* 1995).

5.1 The Annual Cycle

Variations of cloud amount in the annual march in different parts of the Arctic are very similar and rather straightforward. Analysing them allows us to distinguish three states: winter, summer, and transitional (spring and autumn) (see Figure 5.1). In cold half-year (from November to April) the mean total cloud amounts are clearly at their lowest and oscillate between 40% and

60%. In May an abrupt increase in cloudiness is observed, especially outside the Canadian Arctic. The highest cloudiness in the Arctic occurs from June to October (about 80%–90%). The autumn decrease of cloudiness is even faster than the spring increase. As can be seen from Figure 5.2, the spring transition and high summer cloudiness are entirely accounted for by low clouds in all regions. Middle cloudiness does not show great changes in the annual cycle, while high cloudiness is clearly at a minimum in the summer. Interesting maxima of middle and high cloudiness occur in October, which according to Huschke (1969) corresponds very well to the high degree of cyclonic activity over the Arctic during that month. Middle cloudiness also shows a small maximum in spring.

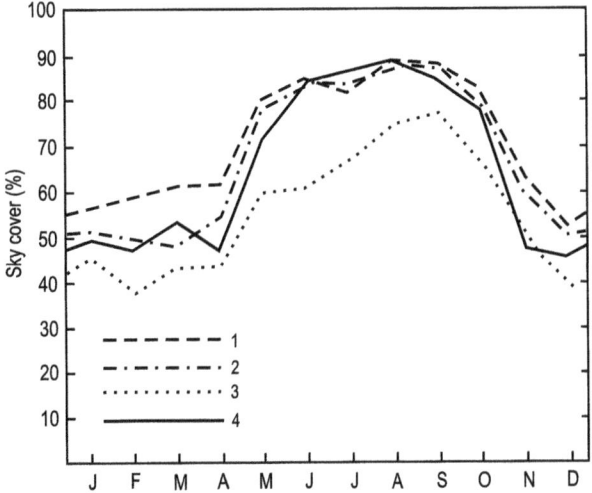

Figure 5.1. Mean monthly total cloud amounts (after Huschke 1969). 1 – West Eurasian Arctic, 2 – East Eurasian Arctic, 3 – Canadian Arctic, 4 – Central Arctic.

The mean annual marches of cloudiness presented here after Huschke (1969) do not indicate the existence of the three types distinguished by Vowinckel (1962). However, when we take into account data from individual stations rather than areally-averaged data, we find at least two of Vowinckel's types. The third type (East Siberian) is characteristic of eastern Siberia, which lies outside the area defined as the Arctic in the present work. The present author would propose a change in Vowinckel's terminology (Norwegian-Sea and Polar-Ocean) to oceanic and continental, respectively. The oceanic type (Figure 5.3) occurs mainly in the Arctic with the lowest degree of climate continentality (mainly the southern and central parts of the Atlantic region and the southern part of the Baffin Bay region). The continental type is most common in the Arctic and probably therefore Huschke's (1969) areally-aver-

100   *The Climate of the Arctic*

age annual courses from different regions, described earlier, can be classified as this type. The oceanic type is characterised by high cloudiness throughout the year, with a slight maximum in summer (Figure 5.3). It is worth noting that even in this area, which is so clearly governed by moving cyclones from the Southwest, maximum cloudiness is observed in summer when the cyclone frequency is lower than in autumn and winter.

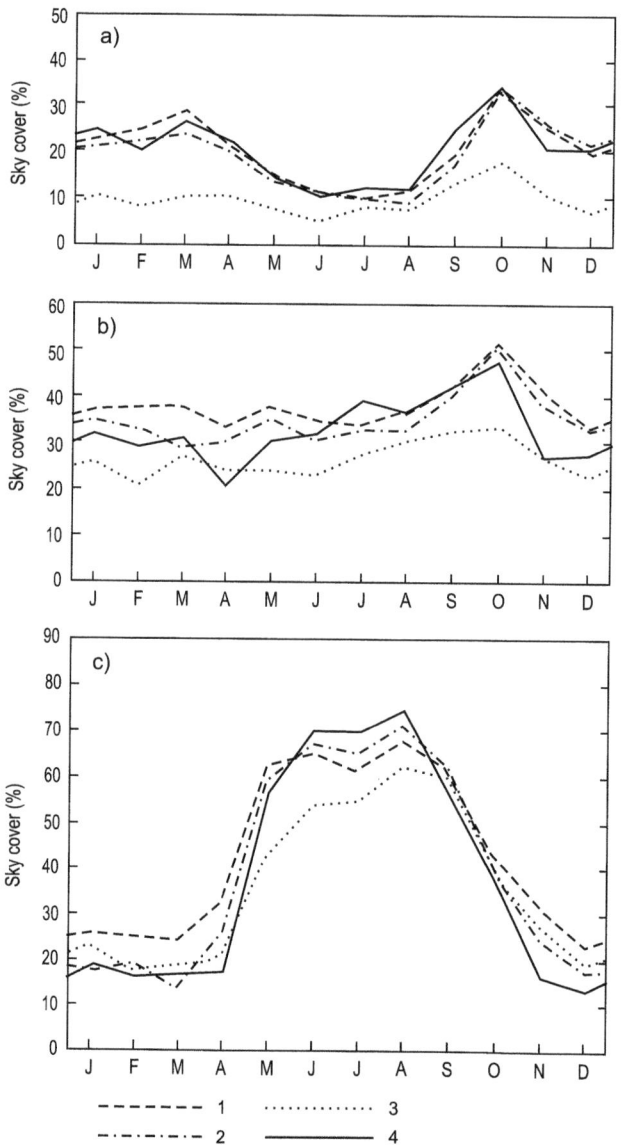

Figure 5.2. Mean monthly high-cloud (a), middle-cloud (b) and low-cloud (c) amounts in selected regions of the Arctic (after Huschke 1969). Key as in Figure 5.1.

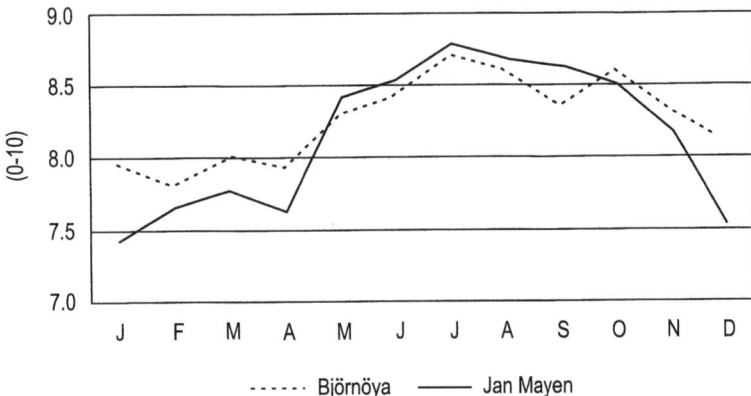

Figure 5.3. Mean monthly total cloud amounts in the Atlantic region of the Arctic, 1951–1998.

5.2 Spatial Patterns

In January, representing winter conditions, the spatial distribution of cloudiness shows greater variation than in summer. The zone with highest cloudiness (> 80%) spreads from the Norwegian Sea to Novaya Zemlya, covering a large part of the Barents Sea and even the southern part of Spitsbergen (Figure 5.4). Cloudiness above 60% occurs in the whole Atlantic region and in the south-eastern part of the Baffin Bay region. The lowest cloudiness (< 40%) includes the belt spreading from the central part of the Siberian region through the North Pole to Greenland and the eastern part of the Canadian Arctic. In this area the absolute minimum occurs over the plateau of the Greenland Ice Sheet and in the Arctic Ocean from the Siberian side (Figure 5.4). In Arctic areas with the highest winter cloudiness on the one hand, as well as in the Arctic Ocean from the Canadian and Greenland side on the other hand, the low clouds dominate. In the rest of the Arctic the middle and high clouds are more common (see Vowinckel 1962 or Vowinckel and Orvig 1970).

From January to July, the most dramatic change in cloudiness (from 35% to more than 90%) occurs in the central part of the Arctic Ocean (Figure 5.4). Very high cloudiness (even greater than in winter) is still observed in the Atlantic region. The cloudiness between the northern part of Norway and Greenland, similar to that of the central Arctic, exceeds 90%. A likely reason for this may be the fact that a slight reduction in cloudiness connected with only slightly lower cyclonic activity can be fully compensated for by more cloudy air masses inflowing from the Arctic Ocean in this season than in winter. The lowest cloudiness (< 60%) occurs on the Greenland Ice Sheet, with a minimum in its north-eastern part (< 50%).

102   *The Climate of the Arctic*

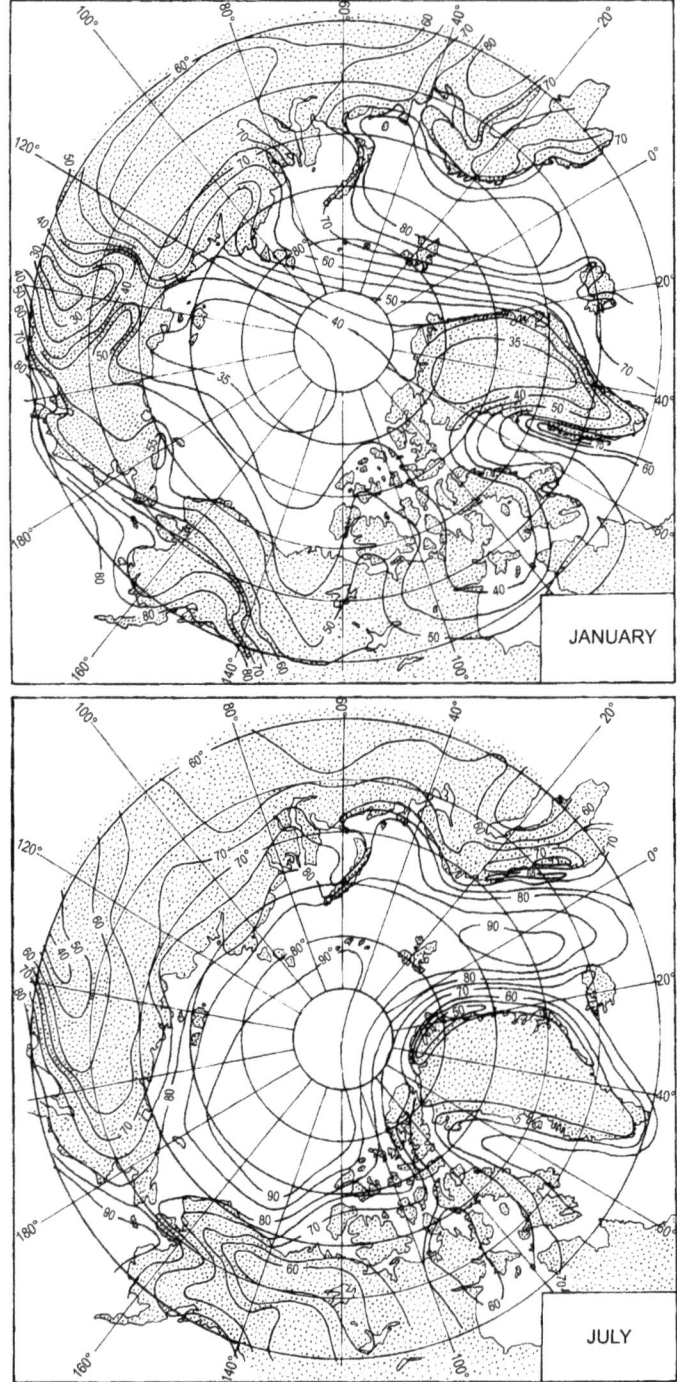

Figure 5.4. Spatial distributions of mean monthly (January and July) total cloud amounts (in %) in the Arctic (after Vowinckel 1962).

The numbers of clear (0–3 degree in 0–10 scale) and cloudy (7–10 degree) days in the sea area of the Arctic for February and August are shown in Figure 5.5. There are a significantly greater number of clear days in wintertime. In February there is a clear day every other day over most of the Arctic Ocean (Figure 5.5c). The largest number of clear days occurs in the southwestern part of the Canadian Arctic Archipelago (> 60%). The clear days are a very rare phenomenon (< 10–20%) in most parts of the Atlantic region and in the southern part of the Baffin Bay region, i.e., in the areas strongly influenced by vigorous cyclonic activity. In summer (August) sunny weather occurs very rarely (Figure 5.5d). In most areas there is a chance of less than 10% frequency. Only on coastal areas of Greenland and Ellesmere Island can clear days occur with a 20–30% frequency.

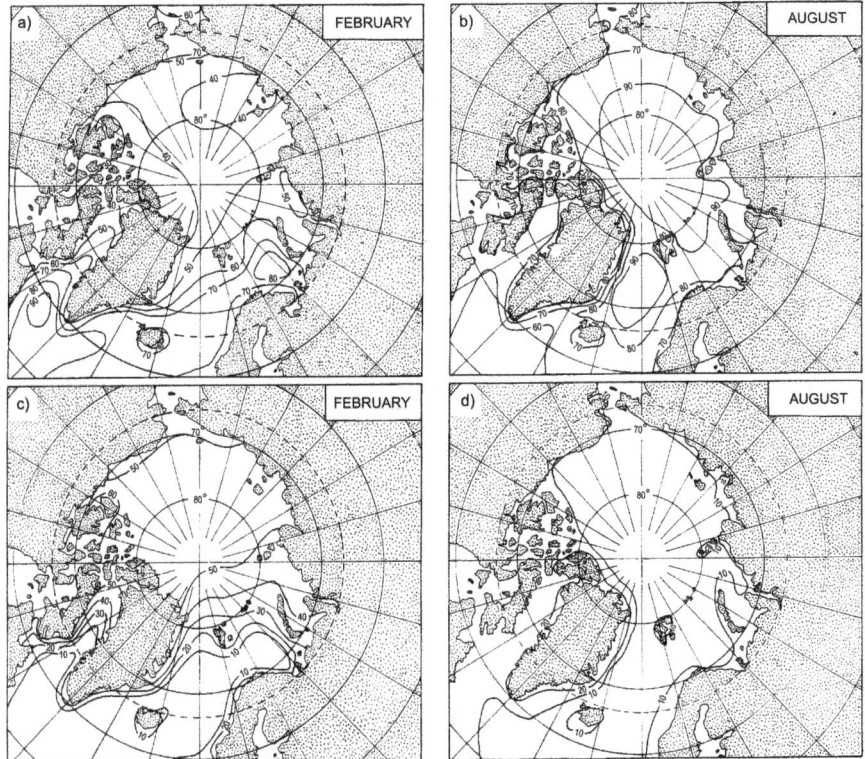

Figure 5.5. Frequency of occurrence (in %) of cloudy sky (7–10 tenths) and clear sky (0–3 tenths) days in February (a and c, respectively) and in August (b and d, respectively) in the Arctic (after Gorshkov 1980).

Cloudy days in summer occur with greater frequency than in winter over almost the entire Arctic (compare Figures 5.5a and 5.5b). The large difference can be seen mainly in the part of the Arctic with a high degree of climate continentality. Cloudy days are almost constant (> 90%) in the central part of the Arctic Ocean and between Spitsbergen, Björnöya, and Jan Mayen. In the rest of the Arctic (excluding the waters near Greenland, Baffin Bay, the whole Canadian Arctic, and the southern parts of the Barents Sea), cloudy days occur with a frequency of 80–90%. The frequency of cloudy days below 60% occurs only near Greenland and Ellesmere Island. In winter, in the entire Arctic (except the Atlantic and the Baffin Bay regions) the frequency of cloudy days is about two times lower. On the other hand, in the Atlantic region and the northern part of the Baffin Bay region, the frequency is generally 10–20% lower. Only in a small area between Greenland and the Labrador Peninsula may a greater amount of cloudiness be noted in winter than in summer.

5.3 Fog

Four types of fog can be distinguished in the Arctic:
— advection fog,
— radiation fog,
— "steam fog" or "Arctic smoke",
— ice fog.

The most common type is advection fog occurring mainly in summer (particularly from June to September) when relatively warm, moist air flows in over a cold surface. The most favourable conditions for the formation of this type of fog are the open waters of the Kara, Laptev, East Siberian, and Chukchi seas. Warm water carried by the northern extension of the Gulf Stream system significantly reduces the frequency of this type of fog over the Norwegian and Barents seas. The frequency of these fogs decreases rapidly from the coastline inland and diminishes less rapidly over the pack ice. Fogs of this kind and other types also do not occur at wind speeds above 10 m/s (Vowinckel and Orvig 1970).

The second type of fog – radiation fog – occurs mainly in winter, when small cloudiness favours large long-wave upward radiation. This radiation cooling is more effective in producing fog over coastal and inland areas because here the flux of subsurface heat is significantly smaller than over the oceanic ice. Radiation fogs are usually shallow and have a light density because they occur in very low temperatures.

The third type of fog – "steam fog" – is not very often observed in the Arctic. This fog can be found mainly over open water during the advection of very cold air. The necessary condition for its occurrence is the great contrast between air temperature and water temperature. Under such conditions the flux of water vapour to the atmosphere is greater than the surface cold air is able to hold. As a result, the excess moisture quickly condenses into fog. This fog occurs most frequently over rivers, unfrozen lakes, open leads, or polynyas. This type of fog does not "live" long; it is very quickly dissipated by wind. Therefore, it is seldom of any great horizontal or vertical extent.

The fourth type of fog – ice fog – is formed when air temperature is low enough (usually below –30°C) to cause direct atmospheric sublimation of moisture in the form of ice crystals. Light ice particles with small full velocity remain suspended in the stagnant air near the surface for a long time. This type of fog occurs mostly near inhabited areas, which are a local source both of moisture and pollutants. Stable air-conditions (little or no wind) during cold spells lead to a large concentration of atmospheric pollutants. Therefore, the ice fog is considered to be one of the types of air pollution (Benson 1969; Maxwell 1982). Its thickness usually oscillates from about 15 m to 150 m. For more details see Berry and Lawford (1977) and Maxwell (1982).

During the course of the year, fogs occur with the highest frequency in the summer months. For example, in the Arctic Ocean their frequency is very high and ranges mainly from 65% to 80%. On the other hand, in winter fogs are observed very rarely (5–10%). However, in some areas of the Arctic a maximum frequency can also occur in every month. Inland areas, where radiation type fogs dominate, show a maximum during one of the cooler months and particularly during autumn (Petterssen et al. 1956). For example, in the Eismitte station (central Greenland) maximums occurred in October (26 days) and December (20 days) (Georgi et al. 1935). Moreover, Prik (1960) revealed that the occurrence of fog depends on the sea-ice concentration (Figure 5.6). The highest frequency of fogs occurs over the sea areas which have a 70–90% sea-ice cover. If there is less or more ice, a decrease in fog frequency is observed. The duration of fogs is different. Usually they do not occur for very long (< 6–12 h), but sometimes their duration can reach as much as 76 h (Prik 1960).

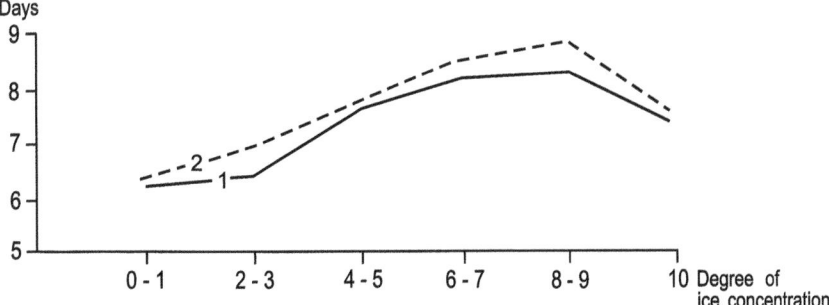

Figure 5.6. Average number of days with fog during 10-day periods in summer months (July and August) for different degrees of ice concentration (after Prik 1960). 1 – Ostrov Ruskiy, 2 – Ostrov Uedinenya.

Information about geographical distribution and also about other characteristics of fogs in the Arctic is somewhat limited. Most existing publications analyse this atmospheric phenomenon for individual points or small areas (e.g. Loewe 1935; Bedél 1956; Kanevskiy and Davidovitch 1968; Krenke and Markin 1973a, b; Pietroń 1987). Among the works giving more a comprehensive and general insight into the problem one can mention Rae 1951, Petterssen *et al.* 1956, Prik 1960, Andersson 1969, Ukhanova 1971, and Maxwell 1982. Surprisingly, such well-known studies as those of Vowinckel and Orvig (1970) and Putnins (1970), give very short notes not exceeding half a page. Only Prik (1960) presents maps with the frequency of fogs in the whole Arctic (for July) and Ukhanova (1971) for the non-Soviet Arctic (for July and for the year as a whole). The isolines must be treated, however, as a rough approximation because, for example:

1) the network of stations was too sparse,
2) different definitions of fogs were used,
3) observations were made with different frequencies in the course of a single day.

The accuracy of fog observations is greatest in the warm-half year. In winter, particularly during the polar night, the reliability of observations is lower. In this season, and also during the night hours in the transitional seasons, a spurious increase of fog frequency was noted (Sverdrup 1933).

In January, in the non-Soviet Arctic, the highest mean frequency of fogs occurs in Alaska, where it can exceed 20%, as in the case of the Barrow station (21%). In the rest of the area studied (excluding the inland region of Greenland) the frequency is lower than 10% (i.e. fewer than three days with fog). In the majority of the Canadian Arctic Archipelago and Greenland stations there is less than one day of fog in January.

In July (Figure 5.7), as was mentioned earlier, the frequency is at its highest almost everywhere. In the central part of the Arctic about 20–25 days with fog have been noted. A similar frequency also occurs in the northern part of the Barents and Chukchi seas, and probably in the East Siberian and Laptev seas (see Prik 1960). Over land areas, the frequency significantly decreases by up to 5–10 days. A smaller number of fogs are also observed in the southern part of the Atlantic region (10–20 days).

The annual number of days with fogs is greatest in the central part of the Arctic (> 140 days). Over almost the entire Arctic Ocean there are more than 100 days with fog. Frequent fogs resulting from the cold East-Greenland Current are also noted between Greenland and Spitsbergen. The northern parts of all the Arctic seas also probably have more than 100 days with fog (except the Norwegian Sea and Baffin Bay). In the central part of Greenland, during the Wegener expedition in the years 1931/1932, as many as 133 days with fog were recorded. However, two thirds of these fogs were due to the presence of clouds at the ice surface, and the remaining third were radiation fogs (Putnins 1970). Moreover, information from just one year is also rather unreliable. For example, on the French station Centrale situated not very far from the Eismitte station in 1949/1950, only 56 days with fog were observed. On coastal stations, the frequency of fogs is significantly lower and most often oscillates between 30 and 60 days, only rarely exceeding 100 days in some stations like Barrow (Ukhanova 1971), Ostrov Hejsa (Krenke and Markin 1973a, b). Krenke and Markin (1973a, b) showed that occurrence of fogs depends significantly on local conditions. More fogs are usually observed in more open coastal stations.

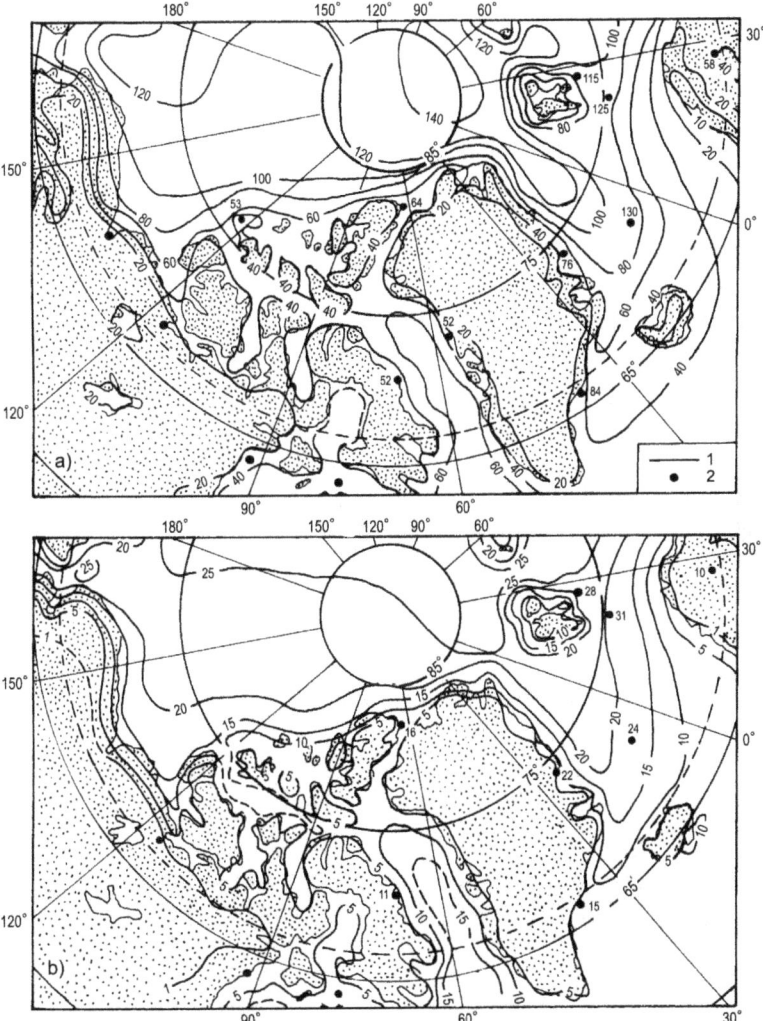

Figure 5.7. Average number of days with fog for the year (a) and July (b) (after Ukhanova 1971). 1 – average number of days with fog, 2 – maximum number of days with fog.

Chapter 6

AIR HUMIDITY

Water vapour is a very important meteorological element because it is a crucial link in water circulation on the globe. Air humidity is most often characterised in meteorology using the following characteristics: actual water vapour pressure, relative humidity, and saturation deficit. It should be mentioned here that, when relative humidity is used to describe the humidity conditions in the Arctic, a distinction should be made between the expression of relative humidity in terms of percentage of saturation with respect to ice, and its expression in terms of saturation with respect to water. Measurements of air humidity in low temperatures (particularly below –10°C) using both psychrometers and hair hygrometers are highly inaccurate. More details can be found in studies such as those by Koch and Wegener (1930), Loewe (1935), Sverdrup (1935), Gol'cman (1939, 1948), Ratzki (1962), and Prik (1969).

The above difficulties of humidity measurement in the Arctic mean that the quality of the obtained data is often low. Probably this is the most important reason for the small number of publications in which the air humidity in the Arctic is described. In some geographical monographs or even in climatological works this element is totally neglected (Prik 1960; Steffensen 1969, 1982; Barry and Hare 1974; Maxwell 1980, 1982; Sugden 1982) or is only treated very cursorily (*Meteorology of the Canadian Arctic* 1944; Rae 1951; Putnins 1970; Vowinckel and Orvig 1970; Sater *et al.* 1971). The only studies we have which are in any way comprehensive are those which have been presented by Zavyalova (1971), Burova (1983) and *Atlas Arktiki* (1985) for the entire Arctic, and Pereyma (1983), and Przybylak (1992a) for Spitsbergen. On the other hand, in recent years quite a considerable number of papers analysing content, distribution, and transport of water vapour in the Arctic troposphere have been published (Drozdov *et al.* 1976; Burova and Gavrilova 1974; Burova 1983; Calanca 1994; Serreze *et al.* 1994a, b, 1995a, b; Burova and Lukyachikova 1996).

6.1 Water Vapour Pressure

Generally speaking, because of the low air temperatures, water vapour content is also low throughout the Arctic. This results both from limited evaporation and the small amount of water vapour which can be held by the cold air. The annual course of the water vapour pressure is therefore very similar to that of air temperature. In winter months, from November to March, and in some parts of the Arctic even to April (e.g. Spitsbergen), the water vapour pressure is the lowest and shows clear uniformity (Figure 6.1), although the day-to-day changes are almost the greatest in the annual course (see Przybylak 1992a). The values of this element in the Arctic must be the lowest over the Greenland Ice Sheet and probably oscillate near 0.0–0.1 hPa. They are also very low (about 0.2 to 0.6 hPa) in the coldest continental parts of the Canadian and Russian Arctic. In the parts of the Arctic characterised by an oceanic climate (Atlantic, Baffin Bay and the southern Pacific regions) the water vapour pressure oscillates between 2–4 hPa (Zavyalova 1971; Przybylak 1992a).

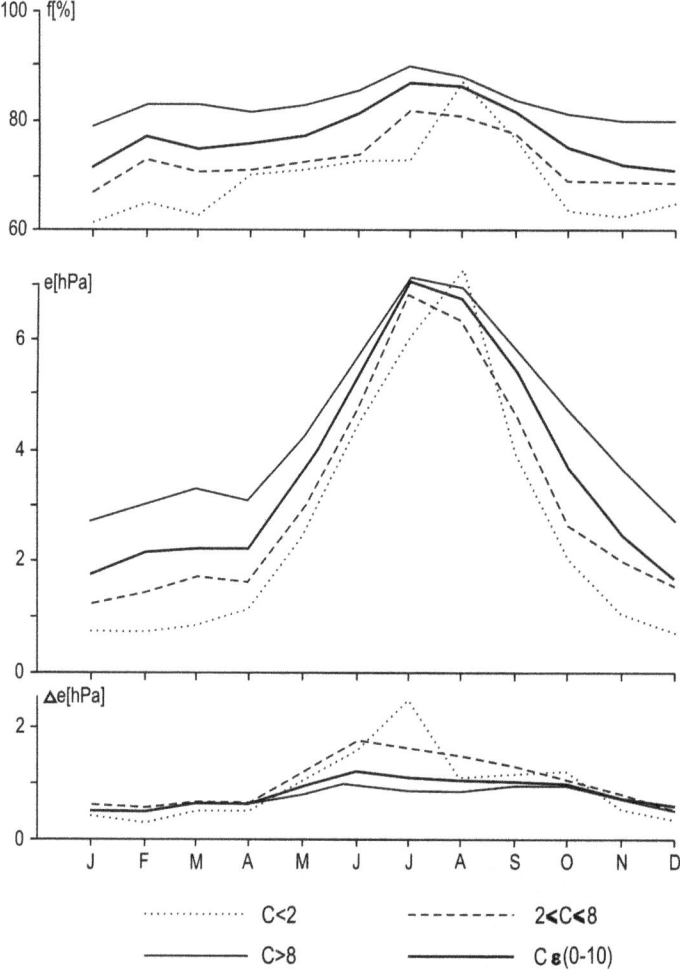

Figure 6.1. Mean annual course of relative humidity (f), water vapour pressure (e), and saturation deficit (Δe) in Hornsund (Spitsbergen) of days of various degree of cloudiness (C) in the period Nov. 1978 – Dec. 1983 (after Przybylak 1992a).

In Spitsbergen, the highest mean daily values rarely exceed 6 hPa and the lowest do not drop below 0.3 hPa. From April (or May) to June a significant increase in water vapour pressure is observed. The annual maximum of this element occurs in July or August. During these months in the southernmost parts of the continental Arctic, the mean water vapour pressure can exceed 9 hPa (e.g., in Coppermine 9.5 hPa and 9.4 hPa in July and August, respectively). In the southernmost maritime areas the water vapour pressure is a little lower because the air temperature is also lower. The highest values occur here mainly in August and reach about 8 hPa (Björnöya – 7.9 hPa, Jan Mayen – 8.0 hPa). In south-western Spitsbergen (Hornsund), the water vapour pressure is lower and the maximum is observed in July (7.1 hPa) (Przybylak 1992a).

The highest mean daily water vapour pressure can reach 12–13 hPa in the southernmost parts of the continental Arctic and 10–11 hPa in the maritime Arctic. In Hornsund the maximum mean daily value for the period 1979–1983 amounts to 9.9 hPa and was connected with long-term (two weeks) inflow of warm and humid air masses from the southern sector. In September and October the greatest decline of water vapour pressure is observed. From Figure 6.1 it can be seen that in Hornsund in all months (except August) the highest water vapour pressure occurs on cloudy days. This results from the fact that cloudy days (except during the high summer months) are warmer than partly cloudy, and particularly, clear days. To explain the relationships between cloudiness and water vapour pressure, one can add that cloudy days are connected with the maritime (warm and humid) air masses coming from the southern sector, while clear days occur when Arctic or polar continental (cold and dry) air masses flow in from northern and eastern sectors. The mean vertically integrated meridional vapour flux over the Arctic for the surface – 700 hPa layer is clearly evident on the maps presented here (Figure 6.2). Poleward moisture transport in all months is the greatest in the Atlantic region where it is associated with cyclonic activity along the primary North Atlantic storm track. On the other hand, the equatorward fluxes dominate over the Canadian Arctic, particularly in July. The result of this is the opposite influence of atmospheric circulation on air humidity in the Atlantic and Canadian regions. In the former it leads to an increase of the absolute content of water vapour, while in the latter it results in a decrease.

The mean monthly daily courses of water vapour pressure in the cold half-year in Spitsbergen are almost uniform, particularly during the polar night, because they depend on non-periodical factors. The highest and lowest values can occur with the same probability in every hour of the day. In the summer months, the differences in the fluxes of solar radiation reaching the Arctic in a daily cycle determine the occurrence of the maximum values in the afternoon and the minimum values of water vapour pressure in the "night" hours. However, their range is very small and does not exceed 0.3 hPa (Przybylak 1992a). The daily cycles are better developed on clear days than on cloudy days.

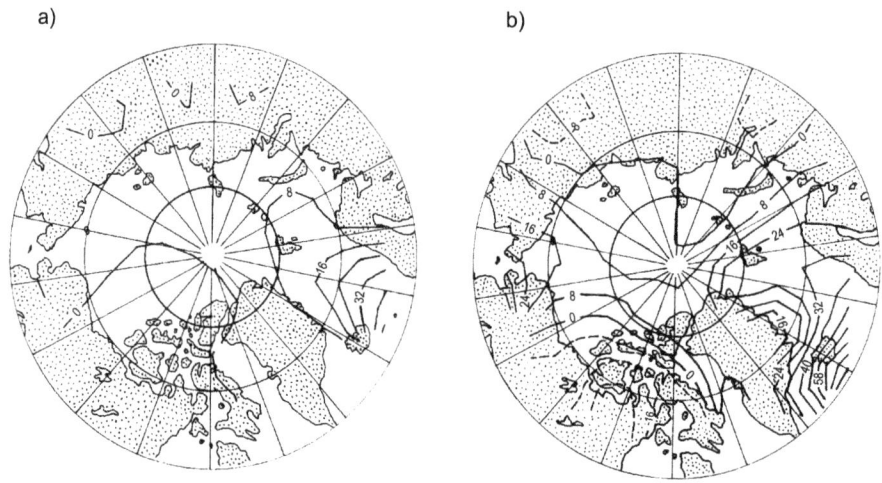

Figure 6.2. Mean vertically integrated meridional vapour flux (kg m^{-1} s^{-1}) over the Arctic Basin for the surface – 700 hPa layer for (a) January and (b) July. Contours over the high (greater than 3000 m) Greenland ice cap have been omitted (after Serreze *et al.* 1995a).

6.2 Relative Humidity

Relative humidity describes the degree of saturation of air by water vapour. This parameter is almost always used to characterise the air humidity in the Arctic. Some of the above cited papers are devoted to studying this parameter, either entirely (*Meteorology of the Canadian Arctic* 1944; Rae 1951; Putnins 1970; Vowinckel and Orvig 1970; Sater *et al.* 1971; Pereyma 1983) or to a great extent (Petterssen *et al.* 1956; Zavyalova 1971). In winter, the relative humidity in the Arctic should be calculated with respect to the saturation vapour pressure over ice, which is lower than it is over water. This permits the air to be slightly supersaturated with respect to ice, while it is not saturated with respect to water. In practise, this means that for winter we receive significantly higher values of relative humidity than when they are calculated in respect to water. From the Figure 6.3 it can be seen that the differences are quite large and exceed 30%. It is well known that the annual course of relative humidity is usually opposite to the course of air temperature. However, in the Arctic such a situation occurs only when the relative humidity in the cold half-year is computed with respect to ice (see Figure 6.3). Supersaturation occurs frequently near the surface in winter as long as no condensation takes place.

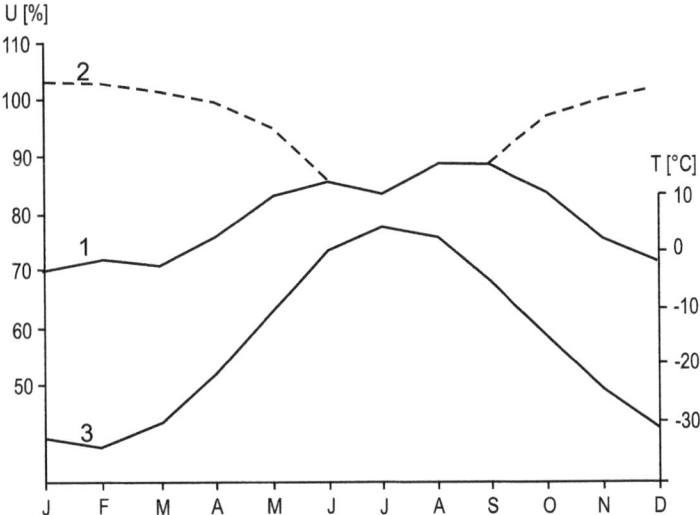

Figure 6.3. Mean annual course of relative humidity (U) and air temperature (T) (after Zavyalova 1971). 1 – relative humidity determined using hygrometer, 2 – relative humidity corrected according to the formula proposed by Malmgren, 3 – air temperature.

The mean monthly relative humidity over ice in the Arctic, computed according to a formula proposed by Malmgren (1926), shows more than 100% in the period from November to April, but on particular days supersaturation can also be observed in October when temperature drops below −25°C (Radionov *et al.* 1997). Radionov *et al.* (1997) found that thermal conditions favorable to the supersaturation of water vapour in the air occur continuously over 75% of the period from December through March, with a maximum in February (89%). In January (Figure 6.4), supersaturation occurs over the entire Arctic, excluding the Atlantic region and the southern and central parts of the Baffin Bay and Pacific regions. This phenomenon also does not occur in the southern coastal part of Greenland. The highest relative humidity (>106%) is observed in the central part of the Greenland Ice Sheet. The secondary maximum (>104%) occurs on Ellesmere Island and in the Arctic Ocean from the Greenland and Canadian Arctic side. The relative humidity computed with respect to water (Figure 6.5) shows almost the opposite pattern and generally 20–30% lower values. In practise, because measurements of relative humidity in low temperatures are still made using hair higrometers, which measure this element with respect to water, the original results of relative humidity measurements and different climatic analyses are presented with respect to water (e.g. Kanevskiy and Davidovitch 1968; Krenke and Markin 1973a, b; Markin 1975; Wójcik 1976; Pereyma 1983; Przybylak 1992a, b).

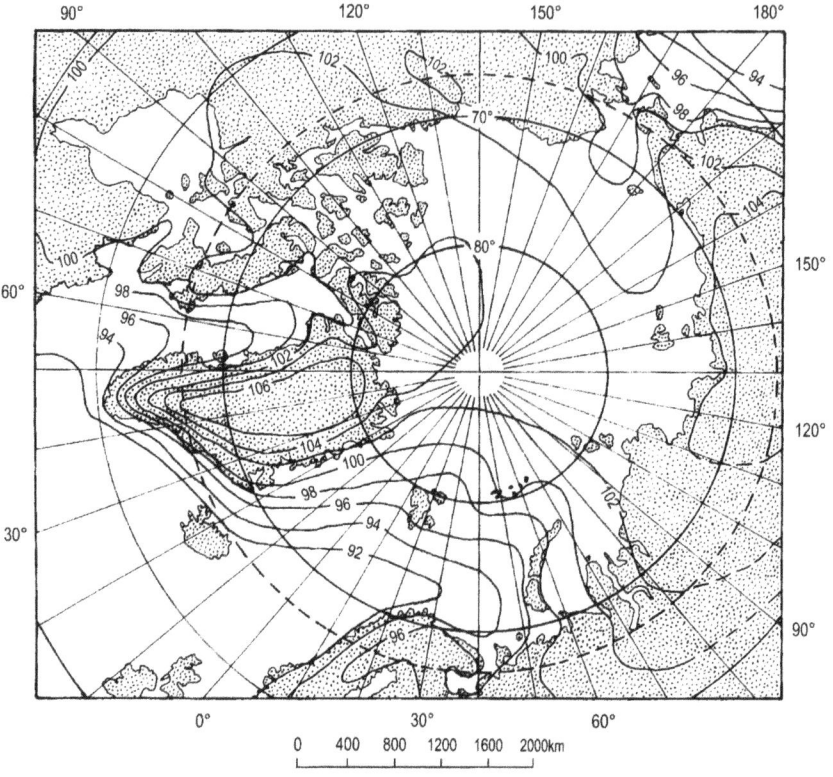

Figure 6.4. Mean relative air humidity in January in the Arctic expressed in terms of percentage of saturation with respect to ice (after *Atlas Arktiki* 1985).

In summer, the accuracy of measurements of air humidity is the highest and, because of mostly positive temperatures in the Arctic, the computed method of the relative humidity with respect to water is correct. The highest relative humidity (> 95%) is noted mainly in the Arctic Ocean and in the northern parts of Arctic seas, excluding the Greenland and the Norwegian seas and Baffin Bay (Figure 6.5). High humidity (above 90%) occurs over a large portion of the Greenland and Norwegian seas and in the other Arctic seas. However, in the Baffin Bay region such high humidity is only observed in its south-western part. The lowest relative humidity is to be found in the southernmost continental parts of the Arctic (about 70–85%) and in the inner part of Greenland, and other large islands.

The relative humidity in Hornsund shows a dependence on cloudiness similar to that of water vapour pressure (Figure 6.1), i.e., significantly higher values (about 15–20%) are observed in all months (except August) on cloudy days than on clear days. The value of the relative humidity in August is computed from only a few days, so its representativeness may be not high.

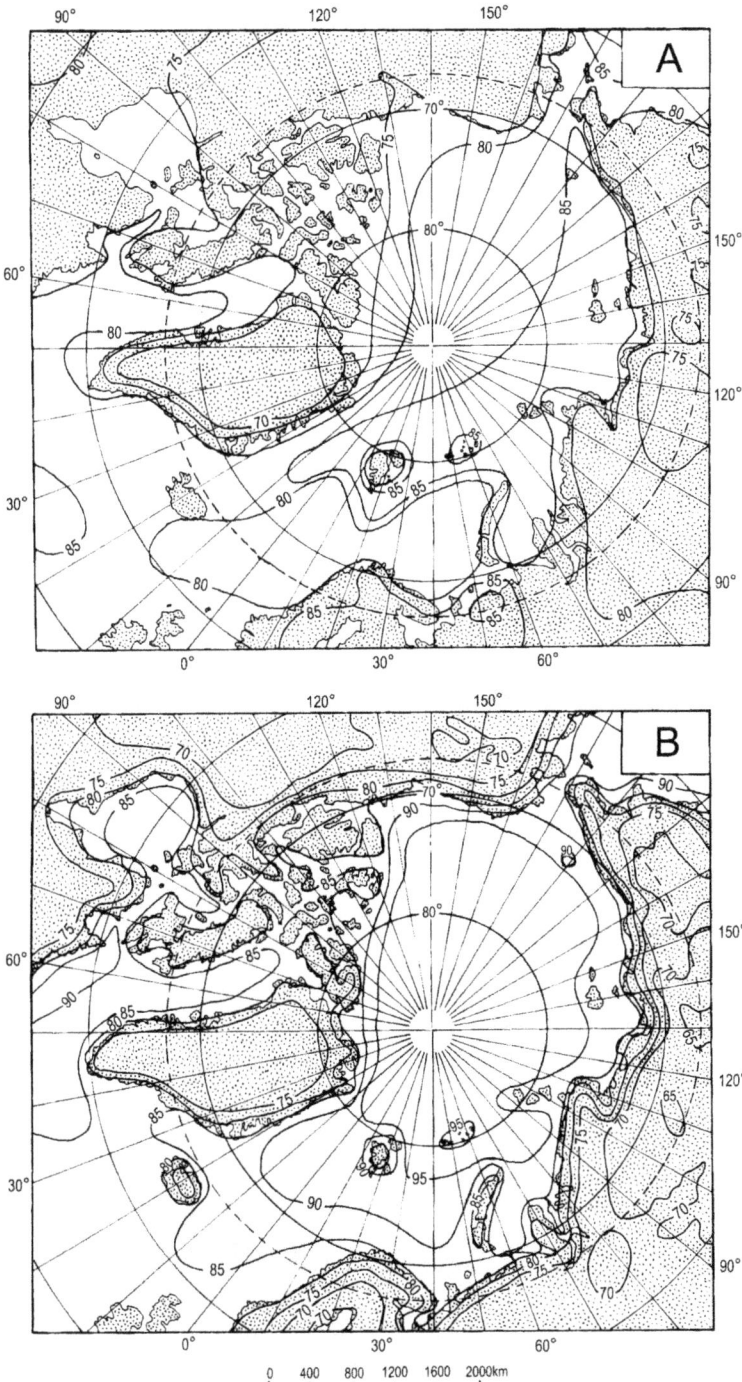

Figure 6.5. Mean relative air humidity in (A) January and (B) July in the Arctic expressed in terms of percentage of saturation with respect to water (after *Atlas Arktiki* 1985).

The mean monthly daily courses of the relative humidity in the Arctic in the cold half-year (October–March) are uniform. From April the daily course becomes clearer and is best developed in the late spring and summer months. For example, in Spitsbergen (Hornsund), the greatest range in the mean monthly daily courses of relative humidity occurs in August (6%) and in May (5%). The lowest values were observed most often at 14°° and the highest in the "night" or early morning hours (Przybylak 1992a). The daily courses of the relative humidity in the warm half-year, similar to those of air temperature, become clearer when the cloudiness decreases.

Chapter 7

ATMOSPHERIC PRECIPITATION AND SNOW COVER

Knowledge about precipitation and its changes in the Arctic is just as important as knowledge about air temperature. This information is needed first of all to correctly estimate the mass balance of the Arctic glaciers and the Greenland Ice Sheet. In turn, the mass balance influences the recession (negative balance) or advance (positive balance) of glaciers. As a result, changes of sea level are observed. Both these processes are very important for natural environment and for human industrial activity. Therefore, during the present period of global warming, due at least partly to human activity, the monitoring of all kinds of ice in the Arctic is crucial. Relying only on temperature investigation, it is difficult to give a credible prediction of the future behaviour of sea ice and land ice. More accurate predictions may be made when tendencies of precipitation are also taken into account.

One can agree with the conclusion of Petterssen *et al.* (1956), that observations of atmospheric precipitation are 'the most unsatisfactory of all Arctic meteorological records'. This is due to the very frequent solid precipitation (snowfall) occurring in the Arctic. In addition, rainfalls in summer have a very low intensity. In winter, the snowfall is most often connected with storm activity and typically takes the form of fine snowflakes. As a result, wind easily lifts and redistributes these snowflakes according to exposure and local topography. Arctic snow usually begins to drift at wind speeds of 7–8 m/s. During the polar night, it is sometimes difficult to distinguish between a period of snowfall and a period of blowing snow. The measurement of rainfall also encounters difficulties because rain in the Arctic occurs mainly as a light steady drizzle. Serious gauge undercatches are caused mainly by wind-induced turbulence over the gauge orifice. Other losses that decrease the gauge catch are related to evaporation from the gauge before the time of reading, and wetting losses due to moisture that adheres to the walls and funnel of the gauge. Legates and Willmott (1990) have estimated that these undercatches could reach about 40%. On the other hand, sometimes overcatches also occur. Prik (1965) found that precipitation in Ostrov Dikson was at its highest on days with strong winds, due to snow blown into the recording gauge (particularly large during snowstorms). Bryazgin (1971) also reported that different types of gauges used in the Arctic show a different sensitivity to factors causing errors in measurement. As a result, it is rather difficult to reliably eliminate these errors from the Arctic precipitation series. Hulme (1992) also came to the same conclusion.

Until now, only Bryazgin, using his own method (Bryazgin 1976a), has undertaken attempts to make adjustments of the precipitation series from the Arctic (Gorshkov 1980; *Atlas Arktiki* 1985). However, it is difficult to say how reliable these results are. For example, comparing the January, July, and annual totals of precipitation for the Canadian Arctic presented in the *Atlas Arktiki* with those published by Maxwell (1980), we notice significant differences. The highest differences are for January precipitation. The map in the *Atlas Arktiki* shows 2–5 times greater precipitation than the map in Maxwell's work. It seems to me that such large differences cannot be explained by measurement errors. The results for July and the annual totals display fewer discrepancies, and are only higher by about 120–200%. More details about the quality of the precipitation series may be found in the following selection of papers: Prik 1965; Bogdanova 1966; Bryazgin 1969, 1976b; Bradley and England 1978; Sevruk 1982, 1986; Bradley and Jones 1985; Folland 1988; Legates and Willmott 1990; Hulme 1992; Metcalfe and Goodison 1993; Peck 1993; Marsz 1994; Hanssen-Bauer *et al.* 1996; Ohmura *et al.* 1999. More recently, Groisman *et al.* (1997) have provided a good brief summary of this problem. All the above limitations of the measurements of precipitation in the Arctic should be kept in mind, particularly when the data for water balance computations is used.

As has already been mentioned, precipitation measurement problems are at their greatest on the Greenland Ice Sheet. Therefore, to estimate precipitation in Greenland, two other methods are proposed (Bromwich and Robasky 1993; Bromwich *et al.* 1998). The first method, the older of the two, was used initially by Diamond (1958, 1960). Precipitation amounts on the Greenland Ice Sheet are computed, taking into account the accumulation of snow over one, or more typically several, years. This net build-up of snow on the surface is the end result of almost exclusively solid precipitation (due to falls of snow and/or ice crystals) minus the net runoff of meltwater, the net flux of water vapour to the surface due to frost formation and condensation minus sublimation and evaporation, and the deposition minus the erosion of snow by drifting (Bromwich and Robasky 1993). This method is also commonly used to estimate precipitation in other glaciated areas. The second method uses indirect meteorological approaches to calculate the precipitation. One technique computes the atmospheric moisture balance. Precipitation is found as the residual from the budgeting of the fluxes of water vapour into and out of an atmospheric volume. The weakness of this method is the fact that the precipitation can be computed only for seasonal and longer time scales and for regions of at least 1 million km^2 that are monitored by a good synoptic network of radiosonde stations (e.g. Rasmusson 1977; Bromwich 1988). Another approach is to add up precipitation amounts calculated from estimates of synoptic-scale vertical motion (Bromwich *et al.* 1993).

A review of the literature shows that most of the geographical and climatological monographs of the Arctic give surprisingly little information about precipitation (e.g., Prik 1960; Sater 1969; Putnins 1970; Vowinckel and Orvig 1970; Sater *et al.* 1971; Barry and Hare 1974; Sugden 1982; Barry 1989). More information can be found in Rae (1951) and Petterssen *et al.* (1956), but particularly in Bryazgin (1971) for the non-Soviet Arctic, in Maxwell (1980) for the Canadian Arctic, and in Przybylak (1996a, b) for the Arctic as a whole. The precipitation has only been presented in cartographic form by Bryazgin, based on the periods 1916–1973 (Gorshkov 1980) and 1930–1965 (*Atlas Arktiki* 1985), by Przybylak (1996a, b) based on the period 1951–1990, as well as by Bogdanova (1997) for solid precipitation. Bryazgin presents results for some months and for the year as a whole. On the other hand, Przybylak additionally provides the results for all seasons. For some parts of the Arctic, maps have also been presented by Bryazgin (1971) for the non-Soviet Arctic, Maxwell (1980) for the Canadian Arctic, and Ohmura and Reeh (1991) and Ohmura *et al.* (1999) for Greenland. In the case of Greenland, there are also earlier attempts to chart the distribution of the annual accumulation (Diamond 1958, 1960; Bader 1961; Benson 1962; Mock 1967 and Barry and Kiladis 1982).

The role of snow in shaping the climate was recognised as early as at the end of the 19th century (see e.g. Voieikov 1889, Brückner 1893, or Süring 1895) and its mechanism is presented in brief in the Introduction to the present volume. The snowfall creates a snow cover when air temperature is ≤ 0°C. A detailed consideration of the snow cover (in terms of depth and density) has often been omitted from climatological studies as being less important. It has, however, been investigated by hydrologists, who need this information for water budget computations. At present, two types of data about the snow cover are available: 1) in situ (standard observations in meteorological stations), and 2) satellite remote sensing. The first type of data gives the best information about physical characteristics of the snow cover but its main weakness is its low spatial resolution (the network of stations is sparse). On the other hand, the satellite data give very good time and spatial resolution of the extent of the snow cover, but limited (if any) information about snow depth and density. One should also add that snow-cover mapping from satellite-derived imagery (with weekly resolution) was begun in November 1966 by the National Oceanic and Atmospheric Administration (Matson 1991). Until 1972 only visible satellite charts were used, and thus there was no information from the dark season (the polar night). Moreover a major problem with snow cover mapping at this time was the difficulty of distinguishing between snow and clouds. Since 1972, however, when passive microwave sensors were introduced, mapping has been possible both in the presence of clouds and darkness (Barry 1985).

There are quite a large number of attempts to deal with the problem of snow cover in the Arctic, in the Northern Hemisphere, and in the world as a whole. A significant increase in the number of published papers has been observed since the start of the satellite era. Of most important works analysing snow cover on the global, hemispheric, and continental scales, one should mention Kotlyakov (1968), Kopanev (1978), Dewey (1987), Dudley and Davy (1989), Cess *et al.* (1991), Robinson (1991), Ropelewski (1991), and Kotlyakov *et al.* (1997). The most recent synthesis presenting different aspects connected with the snow cover in the Arctic Ocean has been provided by Radionov *et al.* (1997). Of the other positions, the following should also be mentioned: Bryazgin (1971), Dolgin *et al.* (1975), Maxwell (1980), Romanov (1991), *Ice Thickness Climatology 1961–1990 Normals* (1992), Brown and Braaten (1998), Warren *et al.* (1999).

7.1 Atmospheric Precipitation

7.1.1 Annual Cycle of Precipitation

The amount of precipitation over any area depends on the moisture content of the air, the pattern of synoptic scale weather systems affecting the area, and the topography and the character of the underlying surface. The moisture content of the air can be described using the concept of precipitable water. Precipitable water is defined as the "depth to which liquid water would stand if all the water vapour in a vertical column of uniform cross-section, extending from the earth's surface to the top of the atmosphere, were condensed" (Maxwell 1980). Serreze *et al.* (1994b) published maps presenting the fields of precipitable water above the Arctic for the surface to 300-hPa layer for January and July (Figures 7.1a and b, respectively).

Precipitable water vapour in January is highest (4–6 mm) in the southernmost part of the Atlantic region, dropping to about 2 mm over the central Arctic Ocean. Serreze *et al.* (1994b) connected this with the spatial distribution of troposphere temperatures. In July, precipitable water reaches its annual maximum, with values ranging from about 12 mm over the northern part of Baffin Bay and 12–13 mm over the central Arctic Ocean to 15–19 mm in the southernmost parts of the Arctic. A clear zonal distribution may be observed in this month, which reflects the zonal tropospheric temperature pattern in summer. Other factors influencing the precipitation in the Arctic have been described in Chapters 1, 2, and 6.

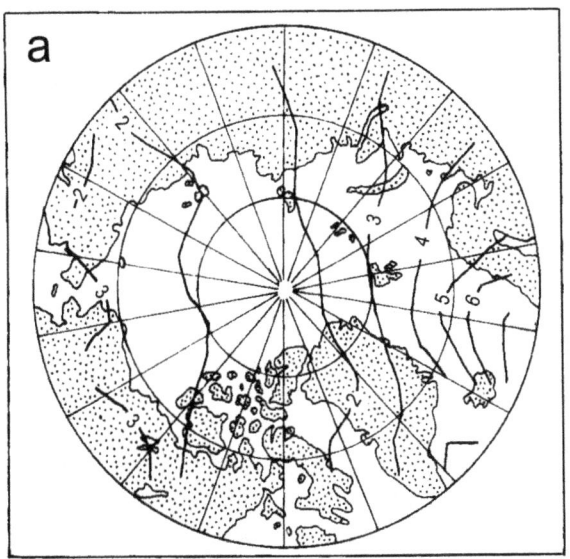

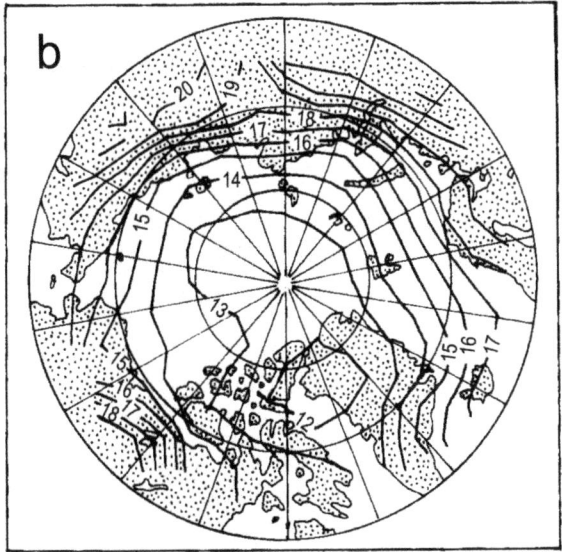

Figure 7.1. Fields of precipitable water for the surface to 300-hPa layer (mm) for January (a) and July (b) (after Serreze *et al.* 1994b).

In most climatological handbooks describing the annual cycle of precipitation in the Arctic, it is usually maintained that the precipitation is at its highest in summer and lowest in winter. This statement is in agreement with what has already been said about the moisture content in the atmosphere. However, it does not take into account other factors, which in some areas of the Arctic can significantly change the "picture". The most important of these is, of course, the synoptic scale atmospheric circulation. The inspection of the monthly totals of atmospheric precipitation in selected stations representing all climatic regions in the Arctic (Figure 7.2) reveals the existence of at least two main types of the annual courses. The first type is characterised by the highest precipitation occurring mainly in the autumn months, when the temperature is still relatively warm (particularly in September and October) and cyclonic activity is only slightly lower than in winter, and the lowest is in spring when the anticyclonic activity is greatest (Serreze et al. 1993). Such annual cycles of precipitation occur in the Arctic areas, where atmospheric circulation is strongest (Atlantic, Pacific and Baffin Bay region). This is particularly evident in the data from Jan Mayen (Figure 7.2b), Mys Shmidta (Figure 7.2g), and Clyde A (Figure 7.2j). The parts of the Arctic with the most continental climate (the Canadian and Siberian regions) show a maximum of precipitation in summer and minimum in winter. In these areas, precipitation depends mainly on air temperature, which determines the magnitude of evaporation and the upper limit of the air's water vapour capacity. This type of annual cycle is clearest in the stations Ostrov Kotelny (Figure 7.2f), Resolute A (Figure 7.2h), and Coral Harbour A (Figure 7.2i). In the central Arctic Ocean the highest precipitation occurs in summer, but the lowest is in spring (see Table 14 in Radionov et al. 1997). The amount of precipitation in the entire Arctic is significantly higher in the second half of the year (60–70% of the annual total).

7.1.2 Spatial Patterns

The precipitation amounts presented for the Arctic in its entirety (except for the inner part of Greenland) come from meteorological stations located on the seacoast below 200-m a.s.l. As has already been shown from measurements carried out on Spitsbergen (e.g., Kosiba 1960; Baranowski 1968; Markin 1975; Marciniak and Przybylak 1985), summer precipitation on the glaciers (200–400 m a.s.l.) is two to three times greater than that measured on the tundra. The mean summer and annual vertical gradients were estimated to be about 35–40 mm/100 m and 80 mm/100 m, respectively (Markin 1975). For this reason, the snow accumulation measurements on the glaciers or on ice caps cannot be used to correct measurements of precipitation on the coastal stations. Some authors compare the results of snow accumulation on glaciated areas with the neighbouring meteorological stations to estimate the magnitude of the impact of wind upon gauge collection of precipitation (see e.g. Bromwich and Robasky 1993).

Atmospheric Precipitation and Snow Cover 123

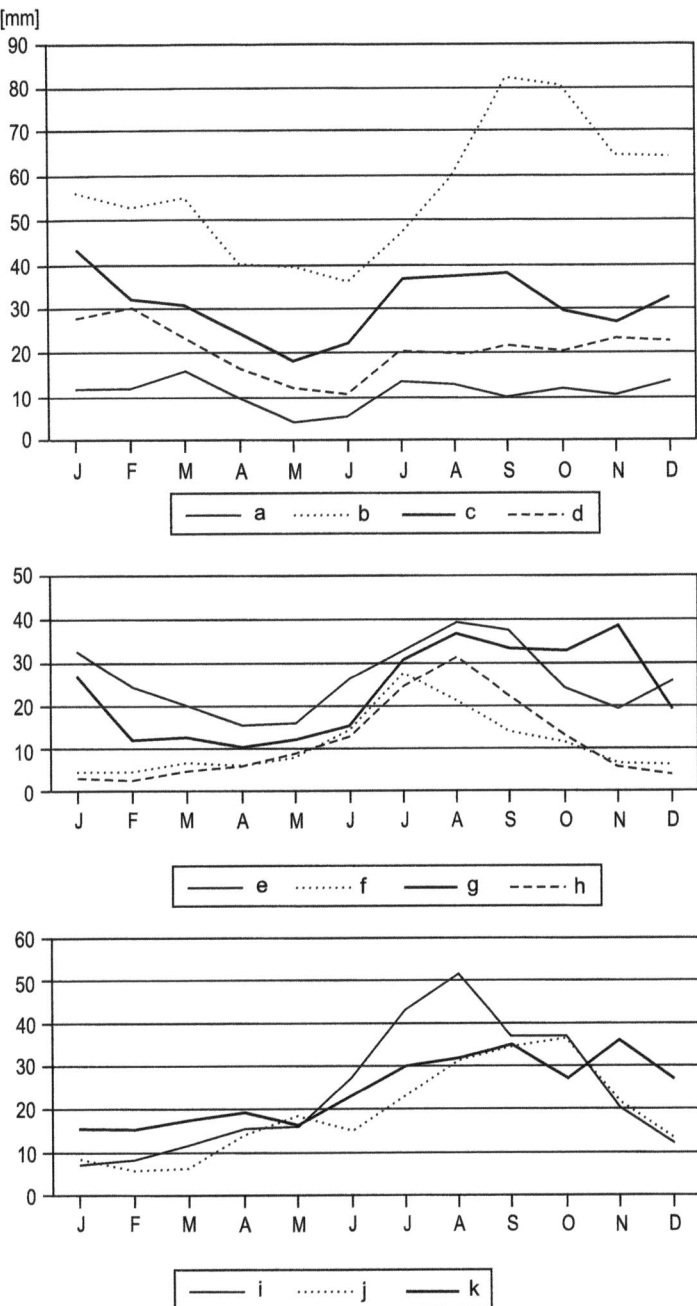

Figure 7.2. Annual course of atmospheric precipitation (according to mean monthly totals) in selected Arctic stations, 1961–1990 (after Przybylak 1996b). a – Danmarkshavn, b – Jan Mayen, c – Malye Karmakuly, d – Polar GMO E.T. Krenkelya, e – Ostrov Dikson, f – Ostrov Kotelny, g – Mys Shmidta, h – Resolute A, i – Coral Harbour A, j – Clyde A, k – Egedesminde.

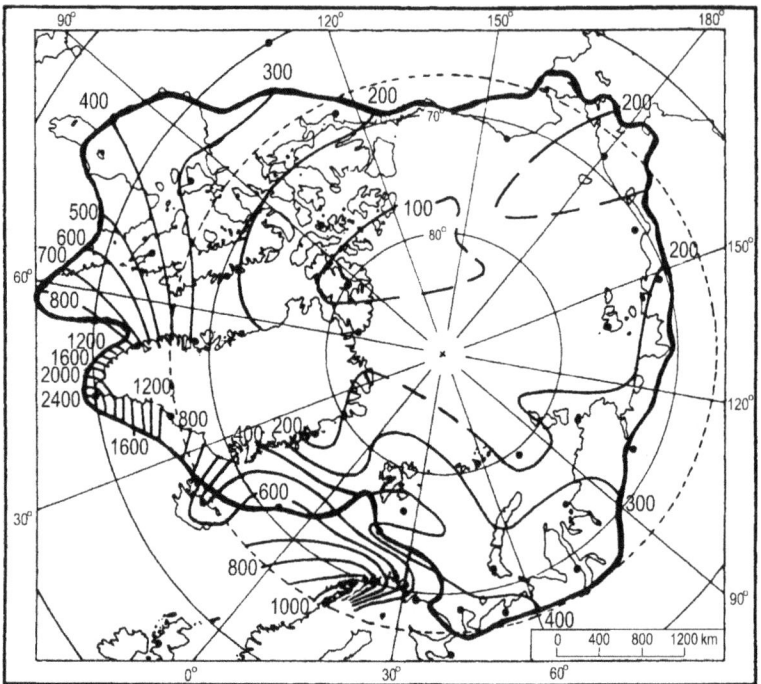

Figure 7.3. Spatial distribution of the annual totals of atmospheric precipitation (in mm) in the Arctic, 1951–1990 (after Przybylak 1996a).

Low air temperature and low atmospheric moisture content in the Arctic significantly limit the values of precipitation. Mean annual totals of precipitation from the period 1951–1990 (Figure 7.3) over almost the entire Arctic (except the southernmost fragments of the Atlantic and Baffin Bay regions) do not exceed 400 mm. The lowest precipitation amounts occur in the coldest part of the Arctic (the northern part of the Canadian Arctic Archipelago above 77°N and the Arctic Ocean from the Canadian side), where they are less than 100 mm. The rest of the Arctic Ocean, the central part of the Siberian region, and the northern part of the Canadian Arctic (70–77°N) also have low precipitation (< 200 mm). Anticyclonic activity prevails over the above areas throughout the entire year (Serreze *et al.* 1993). The highest annual precipitation totals (> 500 mm), on the other hand, occur in the warmest areas of the Arctic, which are also characterised by the most intense cyclonic activity (the southernmost fragments of the Atlantic and Baffin Bay regions, and southern coastal parts of Greenland). Particularly high totals (> 2000 mm) are observed on the south headland of Greenland in the vicinity of the Prins Christian Sund station. The mean annual precipitation sum in this station computed from the period 1951–1980 is 2451.4 mm (Przybylak 1996a). Details about the great precipitation in this region have been omitted from very well known Russian atlases (Gorshkov 1980; *Atlas Arktiki* 1985), which give the annual sum as about 1200 mm. Only Ohmura and Reeh (1991) and Ohmura *et al.* (1999) show in detail on their maps this area of high precipitation (see Figure 7.4). The reason for such great sums of precipitation is twofold: 1) the high frequency of cyclonic activity, and 2) the orographically forced ascent of air masses crossing the Greenland Ice Sheet, which reach an altitude above 2000 m a.s.l. very near the coast (170–180 km from Prins Christian Sund) (Przybylak 1996a). Ohmura and Reeh (1991) note that the south-eastern coast of Greenland is directly hit by the onshore flow from the northern part of the Icelandic low, with a relatively high water-vapour content of 2.1 g/m^3.

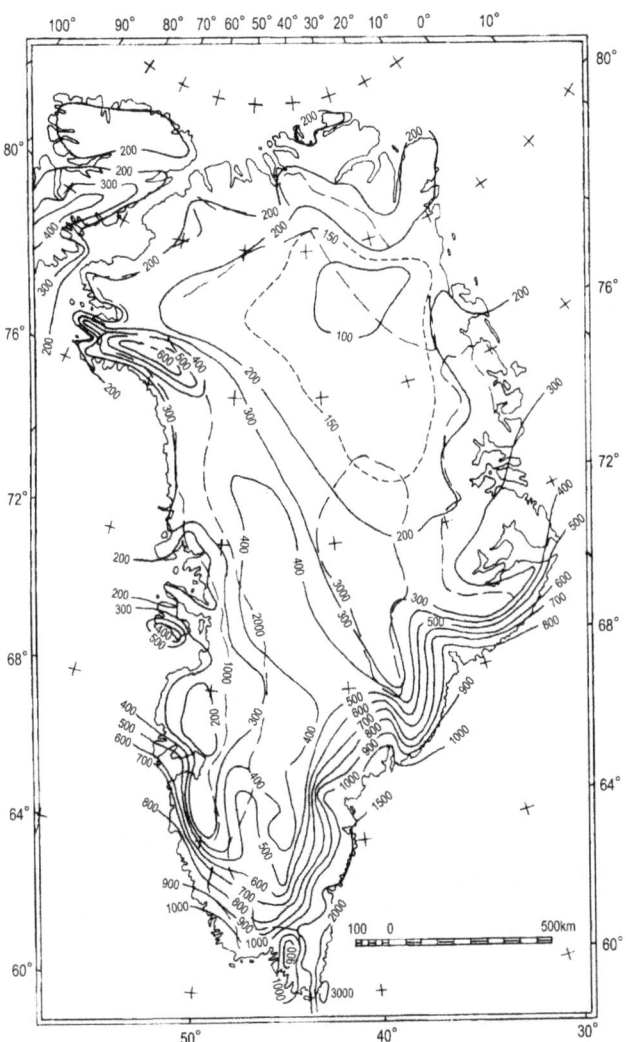

Figure 7.4. Annual total precipitation (in mm) for Greenland (after Ohmura *et al.* 1999).

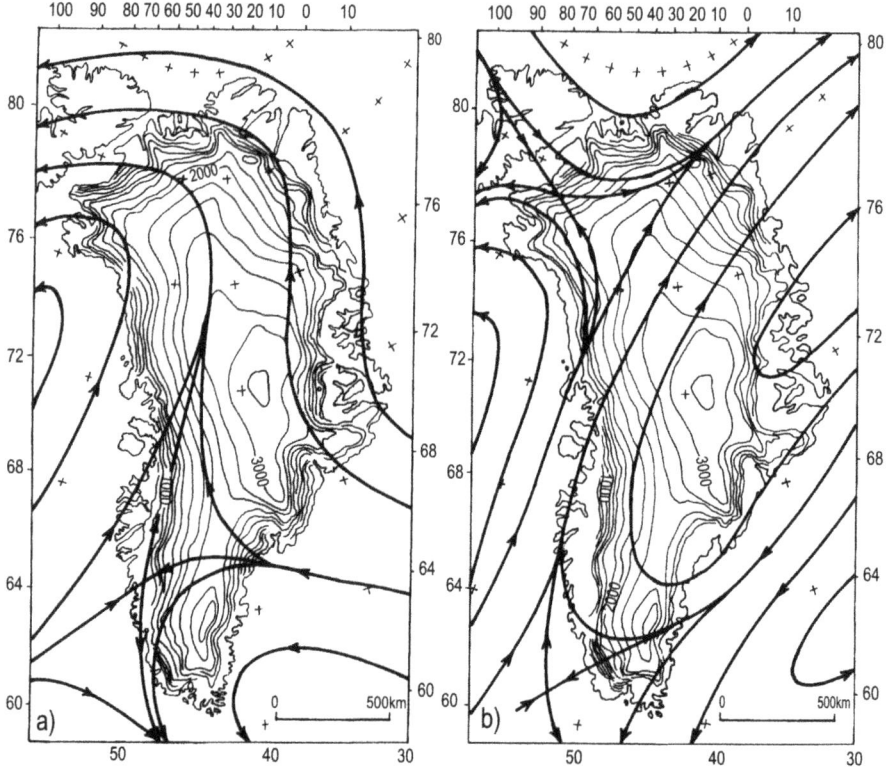

Figure 7.5. Monthly resultant wind streamlines at 850 hPa for (a) January and (b) July (after Ohmura and Reeh 1991).

In comparison with earlier maps of precipitation in Greenland (Diamond 1958, 1960; Bader 1961; Benson 1962; Mock 1967; Barry and Kiladis 1982), more recent maps (Ohmura and Reeh 1991, Ohmura *et al.* 1999) were constructed using not only glaciological data, but also meteorological data. The merging of these two data sets significantly improved our knowledge about precipitation in this part of the world. There is good agreement between Ohmura and Reeh's, and Przybylak's results on precipitation in the coastal parts of Greenland (compare Figure 7.3 and Figure 7.4). The lowest annual totals of precipitation in Greenland occur in its north-eastern part (< 100 mm), which topographically represents the north-eastern slope of the ice-sheet. Computing the monthly resultant wind for January and July for the level of 850 hPa over Greenland (Figure 7.5), Ohmura and Reeh (1991) found that the area in question in both seasons remains in a precipitation shadow, both with respect to the southwesterlies and the westerlies. Generally, there is higher precipitation (> 500 mm) on the slopes of the Greenland Ice Sheet, which are well exposed to the main wind streamlines (south-eastern, southern, and south-western). As was previously mentioned, the greatest annual totals (> 2000 mm) are observed in the south headland of Greenland (Figure 7.4).

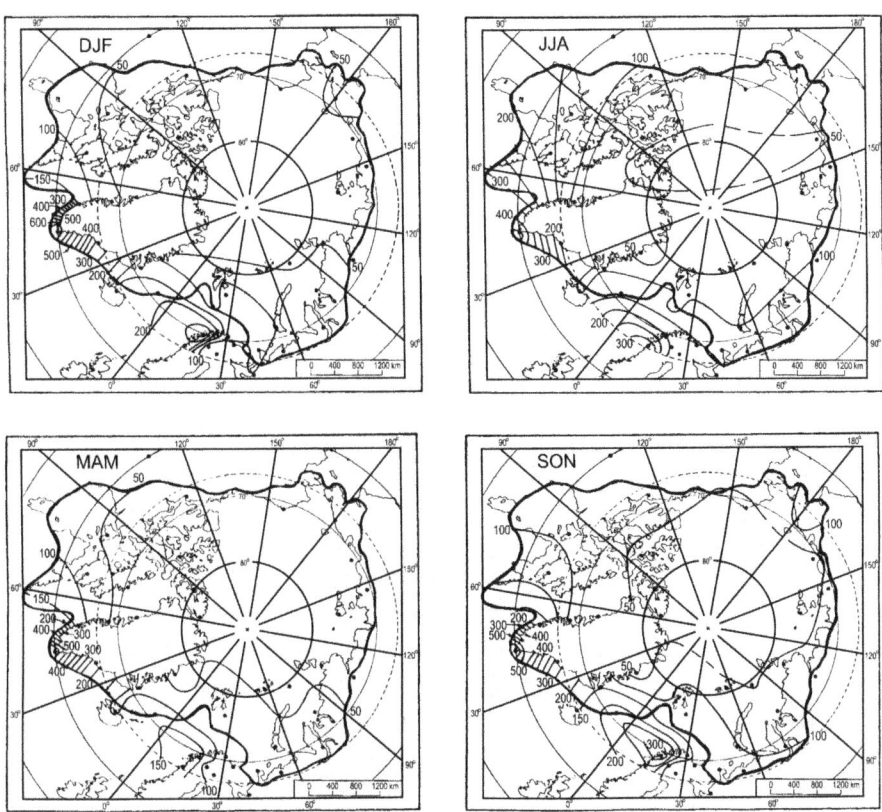

Figure 7.6. Spatial distribution of the winter (DJF), spring (MAM), summer (JJA), and autumn (SON) atmospheric precipitation (in mm) in the Arctic, 1951–1990 (after Przybylak 1996a).

The amounts of seasonal precipitation are presented in Figure 7.6. The lowest sums of precipitation occur in spring. This minimum should be rather connected with the maximum frequency of anticyclones occurring clearly in this season than with the air temperature, the lowest values of which are reached in winter. In addition, the fact that sea ice in this season is near its maximum extent is also important. As a result, the available moisture is at a minimum. Winter precipitation sums are slightly higher than those of spring, but their spatial patterns are very similar. Spring totals of lower than 50 mm occur in about 70% of the Arctic (the Arctic Ocean, the Siberian region, almost the entire Pacific region, and the northern part of the Canadian Arctic). Precipitation of above 100 mm falls only in the south-western part of the Atlantic region and in the southern part of the Baffin Bay region. The highest seasonal mean totals of precipitation occur in summer (excluding the western and southern fragments of the Atlantic region). This maximum one can relate to the highest temperature, water vapour content and cloudiness in this season. Summer totals of precipitation below 50 mm are observed only in the area spreading from the central part of the Siberian region to the north-eastern part of the Canadian Arctic, and on the north-eastern coast of Greenland and the surrounding Greenland Sea. Amounts of precipitation exceeding 100 mm fall only in southernmost parts of the Atlantic and Canadian regions, with highest values (> 400 mm) occurring on the southern headland of Greenland (Figure 7.6). For the inner part of Greenland, no maps exist which present seasonal precipitation totals. However, data for January and July (*Atlas Arktiki* 1985) are available. The spatial patterns of precipitation in both months are similar to those based on the annual totals (Figure 7.4), i.e., the lowest values occur in the north-eastern part, and the highest in the southern and western parts of Greenland. In January, the amounts of precipitation are generally lower than in July in the northern and central parts, and are higher in the southern part of Greenland. As a result, spatial differentiation in January is greater and ranges from 5 mm (north-eastern part) to about 100 mm (southern part), while in July it ranges from 10 mm to only 75 mm, respectively.

Generally speaking, the spatial patterns of precipitation in the Arctic, both seasonal and annual, have a zonal course, i.e. a poleward decrease in precipitation is noted. The greatest disagreements with this rule occur in the areas where climate is mainly formed by very intense atmospheric circulation.

Przybylak (1996a) found, analysing the highest and lowest seasonal and annual totals of precipitation during the period 1951–1990, that in the majority of stations (64%) the maximal annual sums occurred in the coldest decades (1961–1970 and 1971–1980). In the case of the seasonal totals the results were the same, but only for the occurrence of the lowest seasonal sums of precipitation. The highest ones (except for autumn) with a similar frequency occurred in all four decades. Only the highest totals of autumn precipitation occurred more often during the warmest decades. It is a rather surprising result in the context of the predictions presented by the climatic models (see Section 11.2). The highest annual total of precipitation in the Arctic occurred in Prins Christian Sund in 1965 (3299 mm). On the other hand, the lowest were noted in Ostrov Chetyrekhstolbovoy (25 mm, 1988) and Eureka (31 mm, 1956). The lowest seasonal sums of precipitation during the period studied ranged from 1 mm (Resolute A, winter) to 4 mm (Danmarkshavn, autumn).

Figure 7.7 presents 6 regions of coherent annual totals of precipitation in the Arctic. The Wrocław dendrite method was used to delimit these regions. One can see that regions 1, 2, and 3 consist of two separated parts.

This means that teleconnections of precipitation occur in the Arctic. Przybylak (1997c) also found that these are more common than teleconnections of air temperature.

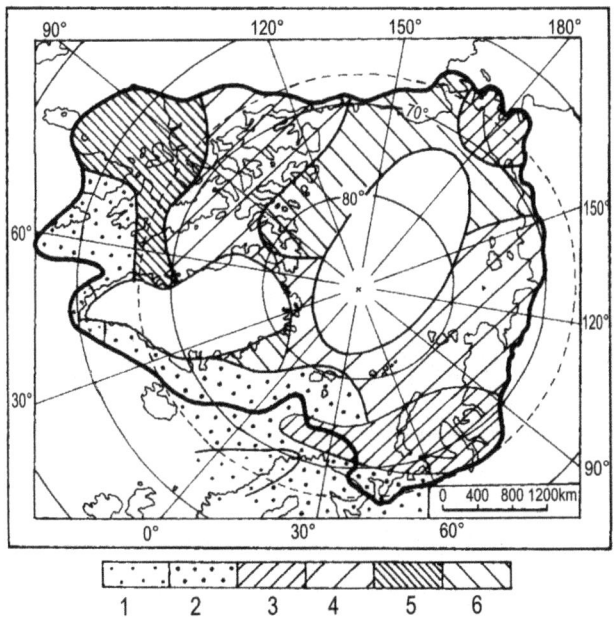

Figure 7.7. Regions of coherent annual totals of precipitation in the Arctic, 1951–1990 (after Przybylak 1997c).

The range of variability of both seasonal and annual totals of precipitation is very high in the Arctic. The ratio of the highest to the lowest sums of precipitation is greatest in the coldest areas of the Arctic, where anticyclones dominate (the northern part of the Canadian Arctic and Siberian region). Przybylak (1996a) computed the variability coefficient (v) for all Arctic stations from the formula $v = \sigma / m$, where σ is a standard deviation and m is a mean value. These coefficients were computed for the annual and seasonal sums of precipitation over the period 1951–1990 (Figures 7.8 and 7.9). The highest variability of the annual sums (> 30%) occurs in the Arctic Ocean from the Pacific side, in the Pacific region, the eastern part of the Siberian region, the northern part of the Baffin Bay region, and the north-eastern coast of Greenland. A large variability (about 30%) is also characteristic of the area lying between Zemlya Frantsa Josifa and Severnaya Zemlya. The above-mentioned areas have the lowest amounts of precipitation in the Arctic. The annual totals of precipitation have the lowest variability (< 20%) in the southernmost parts of the Atlantic region, where the highest precipitation and the greatest occurrence of cyclones are noted.

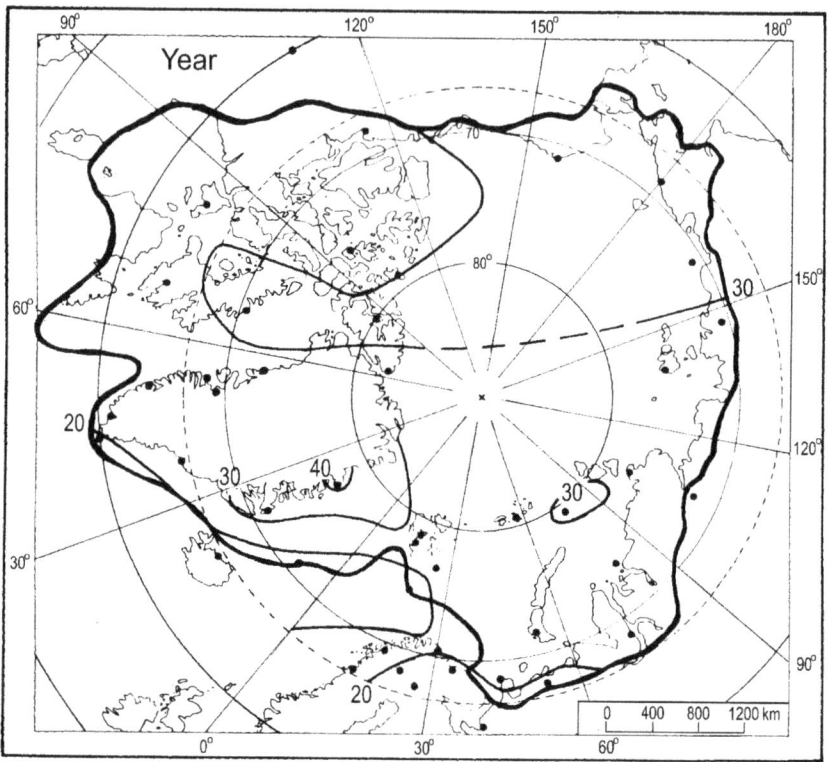

Figure 7.8. Spatial distribution of the coefficient of variability in annual precipitation in the Arctic, 1951–1990 (after Przybylak 1996a).

The dispersion of seasonal precipitation sums is significantly greater than in the case of the annual ones. The highest coefficients of variability occur in the seasons with the lowest precipitation, i.e. in winter and spring. Their values exceed 50% in almost the entire Arctic (Figure 7.9). The spatial differentiation of variability is higher in spring, ranging from about 100% in Alaska to 30–35% in the southernmost areas of the Atlantic region. The highest values of v of winter precipitation totals do not exceed 70%, and occurred in isolated areas located in different parts of the Arctic having a continental climate. On the other hand, their lowest values are observed in the same areas as in the case of spring, but are slightly higher. The variability of summer and autumn precipitation totals is significantly lower than in previous seasons (Figure 7.9), the v ranging mainly between 30 and 50%. Similar to the case of the annual sums, the variability of seasonal precipitation is highest in the areas with the lowest precipitation. This means that areas with low amounts of precipitation are more sensitive to changes in factors determining the precipitation.

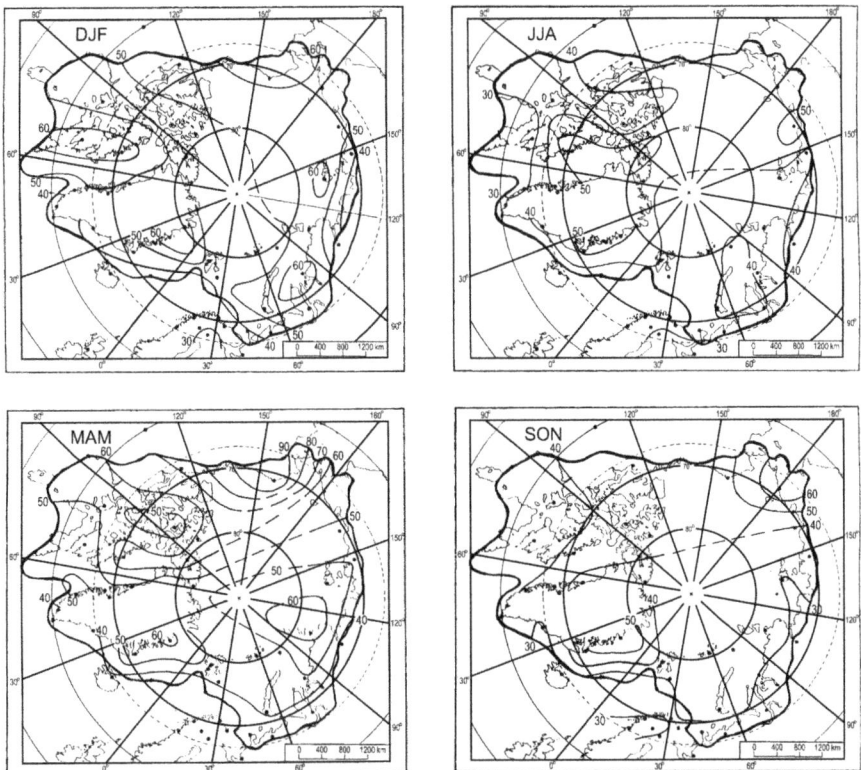

Figure 7.9. Spatial distribution of the coefficient of variability in winter (DJF), spring (MAM), summer (JJA), and autumn (SON) precipitation in the Arctic, 1951–1990 (after Przybylak 1996a).

7.1.3 Frequency Distribution

In climatology, the frequency occurrence of the given element within the arbitrary chosen intervals often supplements the information obtained from the analysis of the mean values. The relative frequency of occurrence of winter, summer, and annual precipitation sums in selected stations representing all distinguished climatic regions in the Arctic, is presented in Figure 7.10. Histograms of the frequency have been drawn for 25 mm and 50 mm intervals for seasonal and annual totals, respectively.

In winter, the precipitation totals in the majority of stations, except the southernmost stations located in the Atlantic region (Jan Mayen and Mys Kamenny), lie in the intervals 0–25 mm or 25–50 mm. In the Canadian Arctic this happens in about 100% of the stations. Jan Mayen is characterised by the greatest range of change of winter precipitation totals (50–300 mm). Precipitation occurs with the highest frequency (only 20%) in the interval 200–225 mm.

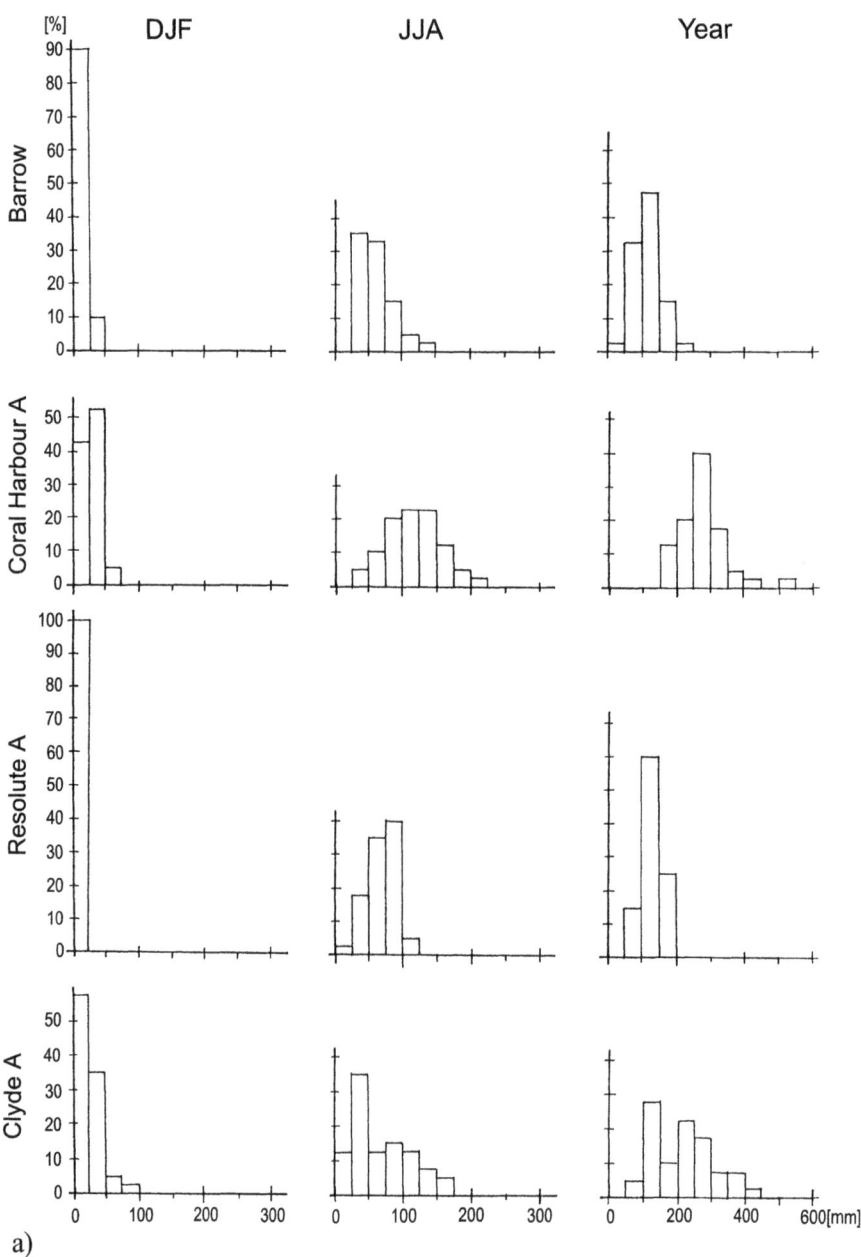

Figure 7.10. Relative frequency (in %) of occurrence of winter (DJF), summer (JJA) and annual (Year) precipitation in selected stations representing particular climatic regions of the Arctic, 1951–1990 (after Przybylak 1996a):
a) Danmarkshavn, Jan Mayen, Ostrov Vize, Mys Kamenny, and Ostrov Kotelny
b) Barrow, Coral Harbour A, Resolute A, and Clyde A.

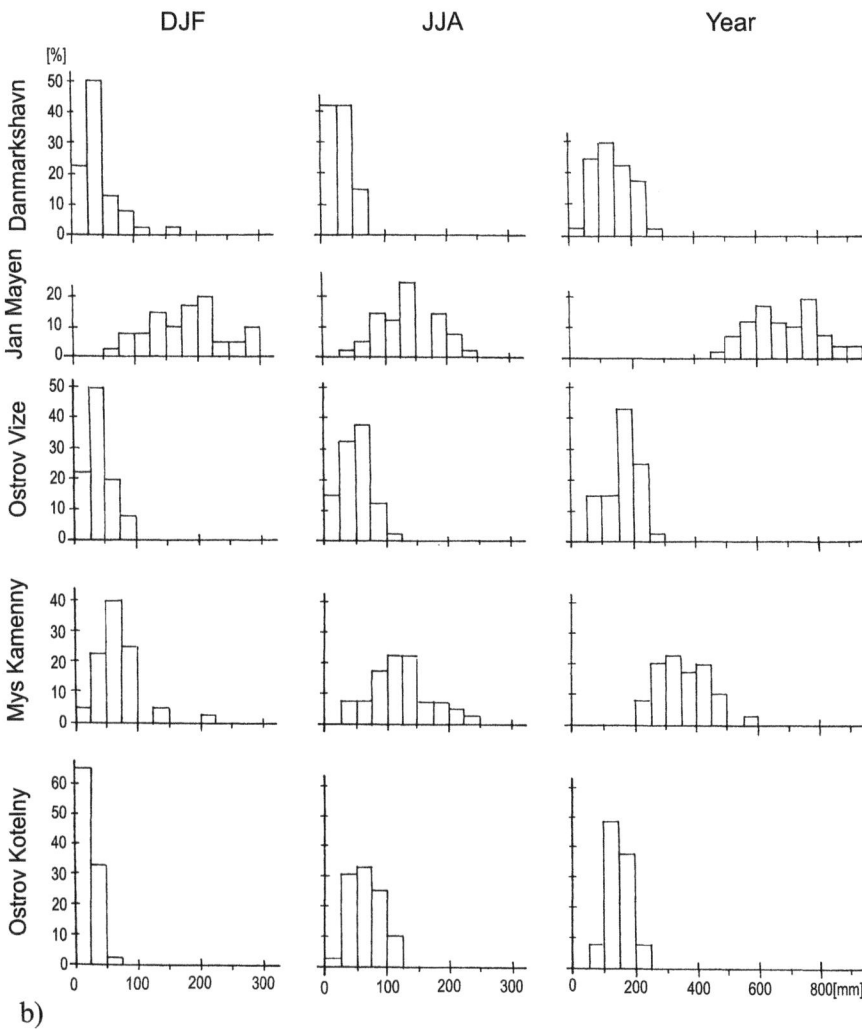

b)

Figure 7.10. cont.

Frequency distribution of summer precipitation is clearly more flat and also shows a more normal distribution than winter precipitation. The intervals which most often occur rarely exceed 40%. The greatest range of annual sums of precipitation (500–1000 mm) is characteristic of the warmest parts of the Arctic (Jan Mayen), and the lowest (20–200 mm) for the coldest parts (Resolute A). The frequency distribution on the areas with the highest cyclonic activity (Jan Mayen, Clyde A) is bimodal. The frequency of occurrence of annual precipitation sums within the analysed intervals exceeds 50% only in the northern part of the Canadian Arctic (Resolute A).

7.2 Number of Days with Precipitation

The number of days with precipitation is a very important characteristic of the precipitation regime. Three categories of days with precipitation are mostly used: ≥ 0.1; ≥ 1.0, and ≥ 10.0 mm. In the Arctic, as was mentioned in the previous section, light (< 1.0 mm) precipitation prevails. More intensive amounts of precipitation occur with a significantly lower frequency. Days with precipitation greater than 10.0 mm are particularly rare. Most authors presenting results for large parts of the Arctic give only information about the number of days with measurable precipitation, i.e., ≥ 0.1 mm (e.g., Bryazgin 1971; Maxwell 1980; *Atlas Arktiki* 1985). The maximum number of days with precipitation in the Arctic throughout the year occurs mainly in one of the months from the second half of the year. On the other hand, the minimum number is observed most often in spring.

In January (Figure 7.11), the greatest number of days with precipitation (> 18) occurs in the parts of the Arctic with the highest cyclonic activity (the south-western part of the Atlantic region and the southern part of the Baffin Bay region). A high frequency (12–18 days) may also be observed in the central part of the Arctic Ocean, the Kara Sea, the eastern part of the Barents Sea, and the continental areas neighbouring these seas, as well as in the central part of the Baffin Bay region. Areas with the most continental climate (the Siberian and Canadian Arctic) have fewer than 9 days with precipitation. However, the minimum (< 6 days) occurs in the central part of Greenland.

In July (Figure 7.11), generally speaking, an increase of days with precipitation is observed in the entire Arctic, except the Atlantic and Baffin Bay regions, where even a decrease is noted, particularly in the area stretching from the southern part of the Baffin Bay region through Iceland to Spitsbergen. The greatest number of days with precipitation (> 15) occurs in the vicinity of the North Pole, in the south-eastern Canadian Arctic, and locally in the western and eastern parts of the Russian Arctic. The lowest number (< 6 days), similar to January, is observed in the central part of Greenland, but the area is larger.

The annual number of days with precipitation is, unfortunately, not presented in the *Atlas Arktiki* (1985). Only Bryazgin (1971) for the non-Soviet Arctic and Maxwell (1980) for the Canadian Arctic give such information. According to Bryazgin's map, the greatest number of days with precipitation occurs in the southernmost parts of the Atlantic region (> 240). A high frequency (> 200 days) is also observed in the Atlantic region southward from Spitsbergen and probably in the southern Baffin Bay region. Quite a large number of days with precipitation (180–200) occur above 85°N. However, Radionov *et al.* (1997) for the Arctic Basin give a significantly lower number (only 152 days).

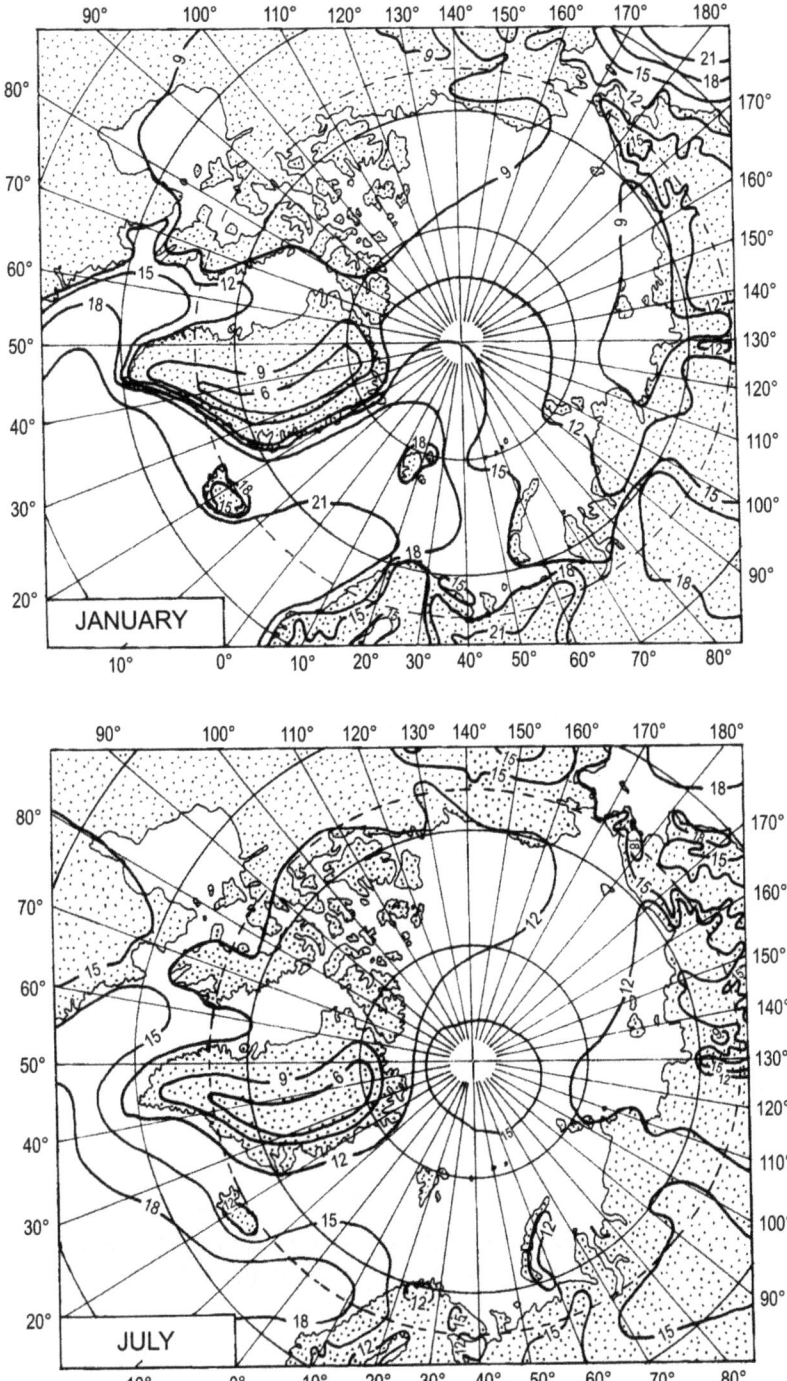

Figure 7.11. Number of days with precipitation ≥ 0.1 mm in the Arctic in January and July (after *Atlas Arktiki* 1985).

The Canadian Arctic, according to Bryazgin's data, has about 120 days with precipitation in the northern part and more than 140 days in the southern part. A significantly lower number of days with precipitation is noted by Maxwell (1980). Generally, on his map, coastal regions in the entire Canadian Arctic have 75–100 days with precipitation. A greater frequency (up to 150 days) is observed only in some mountainous regions neighbouring Baffin Bay and Davis Strait.

7.3 Snow Cover

The snow cover becomes established earliest in the central part of the Arctic in late August (Figure 7.12). According to Radionov *et al.* (1997), stable snow cover at the geographic Pole forms on 20 August. At the height of the Arctic islands (Zemlya Frantsa Josifa, Severnaya Zemlya, Novosibirskiye Ostrova) the autumn formation of snow cover occurs on around 11 September. Ten days later it becomes established near the north of the Svalbard islands, Taymyr Peninsula, and in the Laptev and New Siberian seas. On 1 October snow cover is present over the entire Arctic, excluding the southern part of the Baffin Bay region, and the western and southern parts of the Atlantic region. In the Canadian Arctic, according to the map published by Maxwell (1980), the snow cover forms in the northern part (on 1 Sep.), in the central part (on 15 Sep.), and in the southern part (on 1 Oct.). The decay of the snow cover begins in the south. At the Eurasian Arctic coast melting starts during the first ten days of June (Figure 7.12). In the Canadian Arctic the decay begins about half a month later (on 15 June). In July snow cover still exists only above 80°N. This is also true of the Canadian Arctic. In the central Arctic the decay of snow cover is delayed until the mid-July. At the North Pole, on average, the snow cover decays around 18 July (Radionov *et al.* 1997).

From this information it may be concluded that the number of days with snow cover is greatest in the vicinity of the Pole (more than 350 days) (Figure 7.13). A period of more than 300 days occurs northward from the Arctic islands (above 77°N in the Russian and Canadian Arctic and above 83°N in the Atlantic region). The coastal part of the Russian Arctic has about 260–280 days with snow cover, while in the north-eastern part of the Canadian Arctic the period is longer by about 20 days (280–300 days). In the Atlantic region the number of days with snow cover is lowest and varies from 200–240 days. Of course, this characteristic concerns the formation of snow cover on the coastal parts of land and on the sea ice. Mountainous areas have a longer period of snow cover. For example, in the Canadian Arctic the duration of snow cover can reach 320 days (Maxwell 1980).

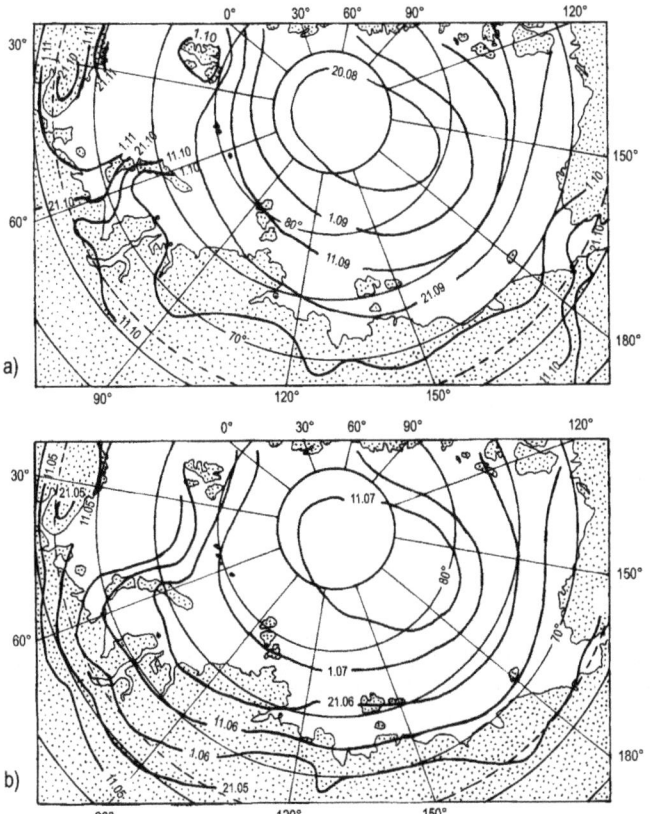

Figure 7.12. Average dates of stable snow-cover formation (a) and its decay (b) in the Arctic, 1954–1991 (after Radionov *et al.* 1997).

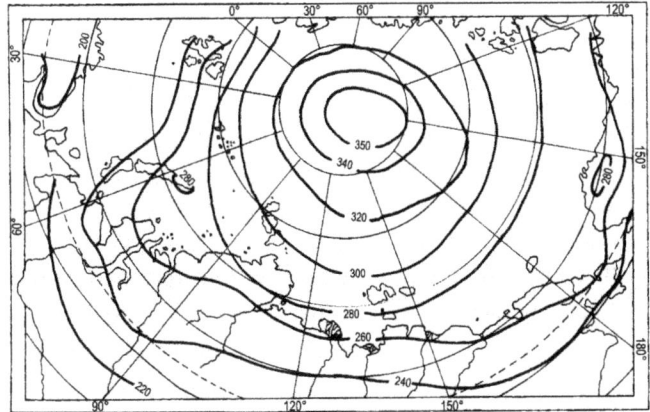

Figure 7.13. Mean annual number of days with the snow cover in the Arctic, 1954–1991 (after Radionov *et al.* 1997).

Besides the duration of the snow cover, a very important characteristic is also its thickness. The maximum snow cover thickness is generally observed from April to May, except in the Canadian Arctic, where it is observed most often about one month earlier (Maxwell 1980). The highest rates of snow accumulation in autumn are observed in the Siberian region (Radionov *et al.* 1997), where from September to November the snow-cover thickness increases, on average, by 14–16 cm. In the central Arctic Ocean and in its part from the Pacific side, the monthly increase in snow cover thickness is smaller and oscillates on average by about 5 cm. During the following months, the rates of snow accumulation over the entire Arctic mostly decrease. The long-term (1954–1991) mean monthly snow cover thickness for May, when the snow accumulation is maximal, is greatest in the central part of the Arctic Ocean (about 40 cm) (Figure 7.14). According to more recent calculations (Warren *et al.* 1999, their Figure 9), this maximum is located rather in the area between the North Pole and the northern parts of Greenland and the Canadian Arctic Archipelago. Moreover, in this area the mean snow depth is greatest in June (40–46 cm). In the Arctic islands the mean snow cover thickness in May is equal to about 30 cm. On the coastal parts of the Russian Arctic, the northern and western parts of the Canadian Arctic, and Alaska, it varies from 20 to 35 cm. In the eastern part of the Canadian Arctic (the eastern coast of Baffin Island and Ellesmere islands), the snow cover is higher than it is near the North Pole and, according to the data presented by Maxwell (1980, his Figure 3.136), oscillates from 50 to 70 cm on April 30 (mean from 1955–1972). Similar values also occur in Spitsbergen (Pereyma 1983, Leszkiewicz and Pulina 1996). Thus, probably here, particularly in the southeastern part of Baffin Island, southern Spitsbergen or somewhere in the southern part of Greenland (there is no data), the snow cover reaches the highest thickness in the Arctic. Very frequent cyclonic activity bringing warm and humid air masses from the Icelandic low and favourable orographic conditions (mountainous ridges and slopes well exposed to the main air streamlines) are the most important factors responsible for this situation. The snow cover thickness described above concerns only the accumulation of snow on the areas covered by sea ice and tundra. In the mountainous regions the snow cover thickness is significantly greater (up to 2–3 times and more) (see e.g. Pereyma 1983; Grześ and Sobota 1999).

The least studied characteristic of the snow cover, but one that is also very important, particularly for the computation of the water budget, is its density. Throughout the year, the snow density increases significantly from September to December (from about 0.2 g/cm^3 to 0.3 g/cm^3), then from December to April the rate of increase is significantly lower. The second largest increase occurs from May to the end of June (from 0.3 g/cm^3 to 0.4 g/cm^3) due to snow melting (Loshchilov 1964; Radionov *et al.* 1997; see also Figure 7.15). The long-term mean snow density for the whole accumulation period changed little across the Arctic Basin, from 0.31 to 0.33 g/cm^3. The spatial distribution of the mean snow density for the month of May presented recently by Warren et al. (1999) confirms entirely the correctness of earlier investigations (Figure 7.16). Non-average densities of snow cover show a significantly greater change from only 0.05–0.09 g/cm^3 for fresh snow to 0.50–0.55 g/cm^3 for melting snow.

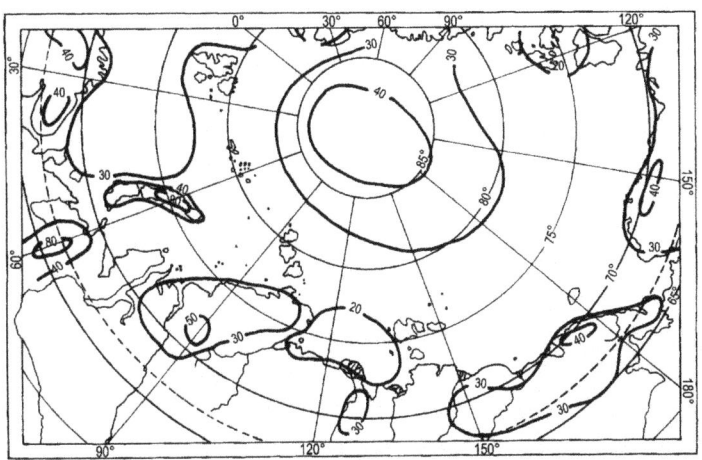

Figure 7.14. Distribution of the mean snow-cover depth (in cm) in the Arctic during May, 1954–1991 (after Radionov et al. 1997).

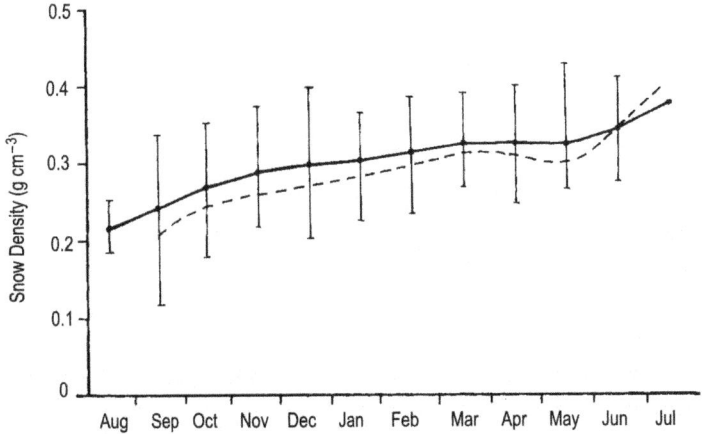

Figure 7.15. Long-term (1954–1991) mean snow density in the Arctic for each month (large solid dots). All available density measurements for each month are used, irrespective of year and geographical location. Error bars indicate one standard deviation. Values of Loshchilov (1964) based on measurements at stations NP.-2 through NP.-9 are shown for comparison as dashed line (after Warren et al. 1999).

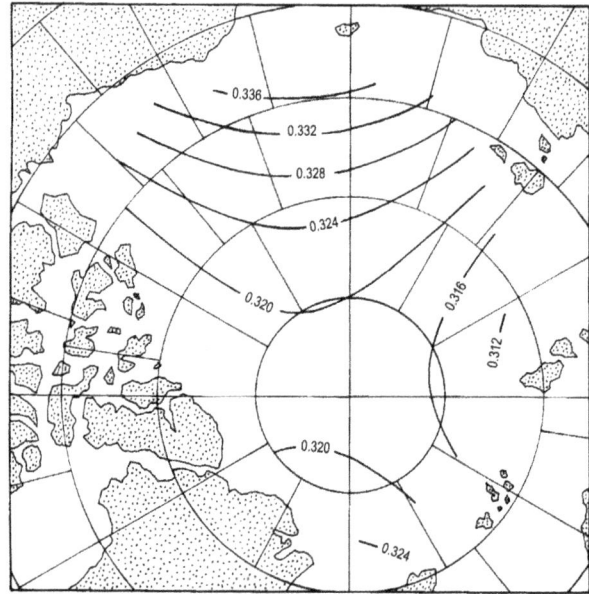

Figure 7.16. Mean snow density for the month of May (in g cm^{-3}) in the Arctic. A two-dimensional quadratic function was fitted to all the available data for May, irrespective of year (1954–1991) (after Warren *et al.* 1999).

Chapter 8

AIR POLLUTION

Until recently, the Arctic environment was treated as a pristine place unspoiled by man. If we take diaries or logbooks of polar explorers from the 19th and early 20th centuries, we will find a large number of phrases underlying the Arctic's cleanliness, its crystal air, and sparkling ice. Opinions about the lack of pollution in the Arctic continued to be held to the beginning of the 1970s, although the first documented report of arctic air pollution (coining the term 'Arctic haze') was published in 1956 (Mitchell 1956). The renewed interest in the nature and origin of the 'Arctic haze' was caused by the growing evidence found during this time that air pollution is not only confined to small areas around urban or industrial sources, but can be transported long distances before being removed to the Earth's surface. This discovery allows us to conclude that the Arctic atmosphere can be polluted, even though it does not have local sources of pollution. Some scientists returned to the observation of 'Arctic haze' made by Mitchell in the early 1950s and the 'ice crystal haze' by Greenaway (a Canadian flight lieutanant) in the late 1940s and pointed out that they were not only just an indication of ice crystals or of wind blown dust, but rather of air pollution originating from the mid-latitudes. From this time, there has been a decline in the view that the Arctic is a place where the original state of the globe exists and can be used as reference to measure the human influence on planet Earth.

The growing awareness that human activity can also destroy the environment of the most remote areas of the world has motivated scientists to begin more detailed studies of this problem. In the last 30 years, the efforts undertaken mainly by scientists of different disciplines from Canada, Denmark, Norway, and the USA, including glaciologists, meteorologists, and atmospheric chemists, led to a significant increase in our knowledge concerning air pollution in the Arctic. During this period hundreds of works were published. In spite of this great scientific interest, particularly in the problem of 'Arctic haze', there are still many questions unsolved (for details see Karlqvist and Heintzenberg 1992). At present, there exist several good reviews of the current state of knowledge, where the reader can find more details than will be given here (e.g., AGASP 1984; Barrie 1986a; Stonehouse 1986; Heintzenberg 1989; Jaworowski 1989; Sturges 1991; Barrie 1992; Shaw 1995).

Jaworowski (1989) has distinguished four types of pollutant sources in the Arctic environment: (1) local natural, (2) local anthropogenic, (3) remote natural, and (4) remote anthropogenic. Most scientists assume that the currently observed levels of pollutants, which are significantly higher than they were before the industrial age, is caused by the fourth type of source and there is some evidence to support this. However, not all scientists share this opinion (see Jaworowski 1989). From a climatological point of view, the most important pollutants in the Arctic are long-lived greenhouse gases (e.g., carbon dioxide, methane, and freons) on the one hand, and short-lived gases building up the 'Arctic haze' on the other. As we know, the greenhouse gases exhibit a much more even geographical distribution and their concentration in the Arctic is at the same level as in the other regions. For this reason, every publication analysing the variation of concentration of these gases (in daily, seasonal, and longer time scales), their sources and sinks, etc. can be taken to characterise greenhouse gases in the Arctic. Here, one should only mention that systematic measurements of carbon dioxide concentration in the near surface Arctic air were started in 1961 near Barrow, Alaska. Such measurements of other radiatively active gases began significantly later (in 1970s or in 1980s). Another important fact is that the increase in the concentration of greenhouse gases will cause significantly greater warming in the Arctic than in the lower latitudes. Such a prediction for CO_2– doubling is given in most of the climatic models (see Chapter 11).

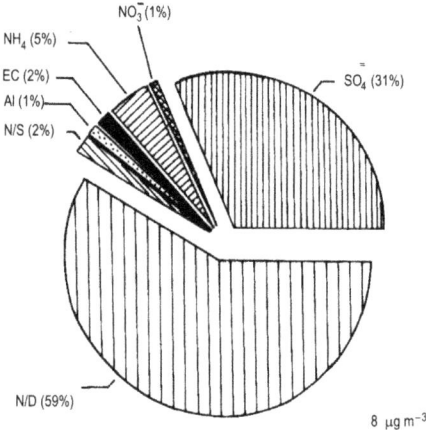

Figure 8.1. The major chemical components in Arctic haze particles < 1 μm radius. Their average total mass is 8 μg m^{-3}. 30%$_m$ were not determined in the chemical analyses; 29%$_m$ are not specified in this graph (after Heintzenberg 1989).

The present chapter, on the other hand, will focus on the problem of the 'Arctic haze' as a well-defined case of air pollution in the Arctic, which influences climate through changes in the radiation balance of the atmosphere. The chemical composition of the 'Arctic haze' was first analysed by Rahn et al. (1977) based on measurements carried out at Barrow between 12 April and 5 May 1976. Its average composition is presented in Figure 8.1. The major components are sulphates (31%), nitrates (6%) and elemental carbon (2%). Still, about 60% of the particles of the 'Arctic haze' are unidentified but are believed to have an organic origin. A considerable fraction of the undetermined part may be water. The size of haze particles oscillates from 0.05 to 1 μm and is a result of the combined effects of sources, transformations, and sink processes (Heintzenberg 1989). During the transport from a distant source (5–10 days) most of the smallest (< 0.05 μm) and largest (> 1 μm) particles are eliminated by coagulation and sedimentation, respectively.

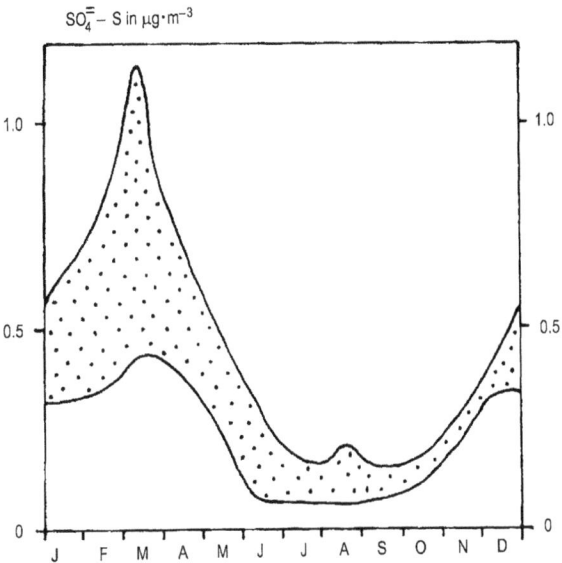

Figure 8.2. Seasonal variations of Arctic haze in terms of sulphate concentrations measured at Ny Ålesund, Spitsbergen during the years 1978–1984. Data from Ottar et al. (1986). The concentrations are given as $SO_4^= - S$ in μg m^{-3}. The measurement values fall within the shaded area (after Heintzenberg 1989).

The strong seasonal variation of haze pollution was first observed at Barrow in the late 1970s. Later on, the measurements carried out in other parts of the Arctic showed that this trend exists in the entire Arctic. The longest chemical time series (1978–1984) of Arctic air pollution comes from Ny Ålesund, Spitsbergen (Ottar et al. 1986). Figure 8.2 presents seasonal variation of sulphate concentration, which is the main component of the 'Arctic haze'. The maximum sulphate concentration was measured at Ny Ålesund in March and the minimum from June to September. Winter concentrations are

10 to 20 times greater than those in summer. In addition, it was recognised that during winter the 'Arctic haze' particles were mostly of man-made origin, while in summer they mainly came from natural sources. To explain the reason for this seasonal difference, an analysis of pollution sources and their transport pathways must be taken into account, along with some meteorological criteria. According to Raatz (1991), industrial source regions covering large areas of the mid-latitudes appear to be the major contributors to arctic air pollution (northeastern USA/southeastern Canada, western and eastern Europe, western USSR, Ukraine, southern Urals, western Siberia, Korea, and Japan). To this list eastern China should also be added, according to some other sources (e.g., Rahn and Shaw 1982, Barrie 1992). The annual emission of SO_2 (the main component of the 'Arctic haze') is highest in these regions (Figure 8.3). From distinct natural sources of 'Arctic haze', Raatz (1991) gives the deserts of northern and western China and those of the southern USSR, as well as the Sahara desert of Africa.

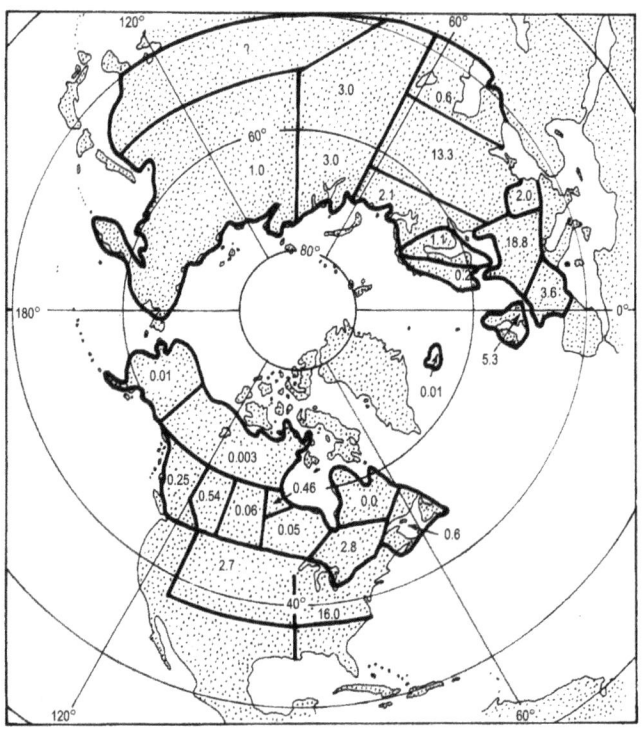

Figure 8.3. Annual emissions of SO_2 (10^6 tonnes) in regions of the Northern Hemisphere that influence the Arctic (after Barrie 1986b).

The importance of each of the above sources in contributing to Arctic air pollution depends on its strength, proximity to the Arctic region (or Arctic masses), and the frequency of synoptic situations favouring a poleward flow. The magnitude of the poleward transport of pollution from mid-latitudes depends significantly on the seasonal variation of the polar front (Rahn and McCaffrey 1980). In winter the polar front is shifted to the south (40–50°N) and the most industrial regions lie to the north of them. In such a situation, the polluted air masses can easily reach the Arctic. The main pathways for the transport of pollution aerosols between the mid-latitudes and the Arctic are shown in Figure 8.4.

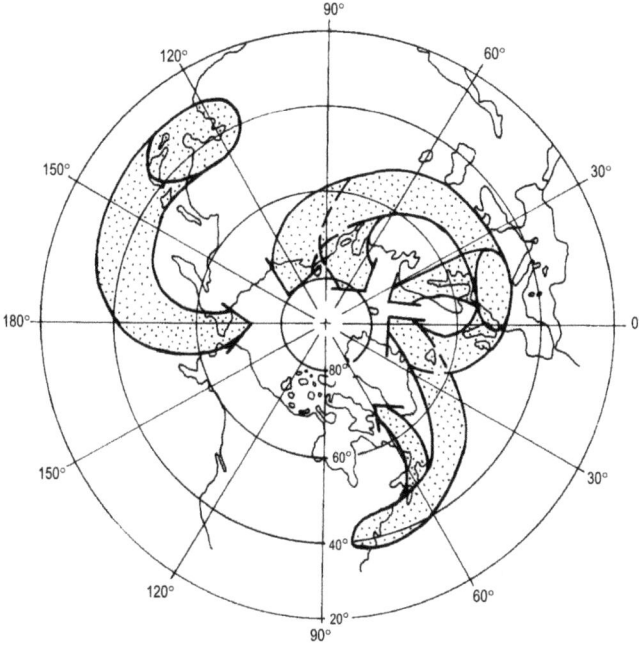

Figure 8.4. Major sources and pathways for transport of pollutants between midlatitudes and the Arctic (after Rahn and Shaw 1982).

Moreover, the strong surface-based temperature inversions frequently occurring in this time (see Chapter 4) cause the atmosphere to stabilise, which, in turn, inhibits the turbulent transfer between the atmospheric layers. As a result of this, and also of the occurrence of only light precipitation, the removal of gases and aerosols from the Arctic atmosphere is greatly weakened. In summer, the polar front is situated to the north of the most important industrial centres and the transport of polluted air is significantly smaller. In addition, the pollution which reaches the Arctic during this time will be washed out rapidly because the region has much more rain and snow in summer than in winter. For details of the climate and meteorology of the arctic air pollution see Raatz (1991).

The question still remains as to whether the haze comes mainly from one region or from a combination of regions. In the early 1980s, atmospheric scientists proposed the following scenario, which is cited here after Barrie (1992): "Eastern North American and South-East sources share similar features that make them less likely to contribute much haze pollution to the northern region. They are at lower latitudes (25°–50°N) than Eurasian sources (40°–65°N) and on the eastern side of continents immediately upwind of stormy oceans (Atlantic and Pacific). Their industrial waste blows mostly to the east, where it is washed out by rain over the oceans. In contrast, Eurasian pollution blows to the north-east over land which in winter is a cold, snow-covered polar desert. It encounters little rain or snow to wash it from the atmosphere. In summer, movement of Eurasian pollution to the Arctic is weaker, because winds blow less frequently to the north and because there is more rain to remove it on its pathway northwards." Barrie *et al.* (1989), working on the chemical transport model, found that most of the pollution in a year (96%) entered the Arctic from Eurasia; the reminder (4%) came from North America (Figure 8.5). The haze pollution coming from Eurasia was approximately evenly split between western Europe, eastern Europe, and the USSR.

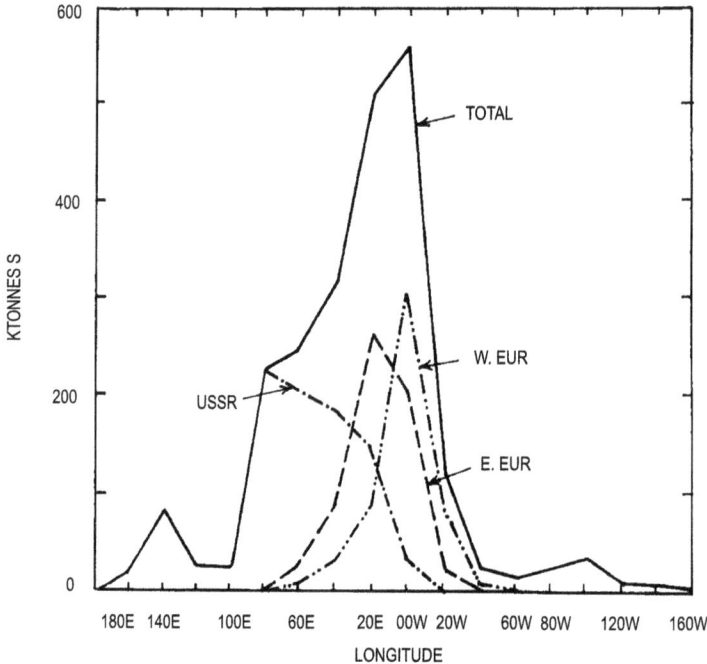

Figure 8.5. The amount of sulphur entering the Arctic circle on the wind during one year (July 1979 – June 1980) according to latitude. Results are given for all pollution and then for each of the major source regions that, combined, contribute 96 per cent of the total sulphur entering the Arctic (after Barrie 1992).

One should mention here that not all scientists share these views on the above scenario (see Jaworowski 1989; Khalil and Rasmussen 1993). Jaworowski (1989) writes that "...the current studies do not provide any evidence which unequivocally identifies the respective contributions from anthropogenic and natural emission sources, to the long-range transport of impurities into the arctic atmosphere". One of the most important arguments for the anthropogenic origin of the 'Arctic haze' is the enrichment of vanadium. Rahn and Shaw (1982) stated that the enrichment factor for vanadium (over 1.5 times) provides an extremely sensitive test of anthropogenic pollution against the natural character of aerosol. Jaworowski (1989) criticises this assumption because it is not in agreement with the results of studies indicating that many heavy metals, including vanadium, are enriched by up to several orders of magnitude in the airborne dust, both over industrialised areas and in remote ones such as the South Pole, central Greenland, and mid-oceanic localities. This means rather that this enrichment is due to natural processes (Duce *et al.* 1975; Jaworowski *et al.* 1981). Jaworowski (1989) further writes that "...also the ratios of concentrations of particular elements in aerosol samples and their distribution in variously sized fractions of particles, have only a hypothetical character and are of uncertain value for detailed identification of emission sources of arctic pollutants". Seasonal variations of air pollution in the Arctic also cannot be used as an indicator of the anthropogenic influence because such variations occur in the lower latitudes too. In addition, Jaworowski (1989) adds that the precipitation patterns in the Arctic may also contribute to the spring maximum of pollutants in the air. The surface-based temperature inversions and anticyclonic situations, both of which occur very frequently in spring, also have a similar influence on pollution. The general conclusion of the Jaworowski (1989) critique is that a quantitative estimate of the relative contributions of human and natural sources to arctic pollution with heavy metals, mineral, acids, etc. will not be possible until the long-term observations of the temporal trends of contaminants in the arctic air become available.

From the list of unresolved issues concerning the 'Arctic haze' (e.g., very incomplete knowledge about horizontal and vertical distribution and even about the components of the 'Arctic haze') by Heintzenberg (1989), one can understand the point of the Jaworowski's (1989) critique. Thus Jaworowski's view, to a certain degree, has been confirmed by the measurements and analyses of 30 gases in 'Arctic haze' and in clean Arctic air made during the Arctic Gas and Aerosol Sampling Program (AGASP) during the spring of 1983, 1986, and 1989 (Khalil and Rasmussen 1993). To look for the possible origins of the haze Khalil and Rasmussen (1993) used cluster analysis to derive regional signatures of trace gases at ground-based sites in the middle and high northern latitudes. Based on this analysis, they have argued that trace gases in the 'Arctic haze' do not come from North America and China and are unlikely to come from western Europe. Further they concluded that Arctic pollution must originate from eastern European and Russian industrial regions. For the present discussion, the most important point to note is their finding that the haze originates not at distant locations in Russia, but from within the Arctic Circle. Khalil and Rasmussen (1993) argued that industrial activity in the Kola Peninsula (with the large city of Murmansk), involving power plants, mining operations, and military-industrial products, can emit large amounts of haze-producing pollutants into the Arctic atmosphere and may thus cause the 'Arctic haze'. More recently Harris and Kahl (1994) also found that the Ni–Cu smelting complex at Norilsk may be one of the major contributors to the haze. Based on the above factors, the existing opinion, expressed recently by Raatz (1991), that local sources within the inner Arctic are usually point sources which are only of local importance and contribute little to the arctic-wide phenomenon called the 'Arctic haze', must be revised.

The climatological importance of the 'Arctic haze' results from its influence on the radiation balance of the atmosphere, in both its short-wave and long-wave components. In the solar range (0.3 – 3.0μm), haze particles both reflect part of the incoming radiation back to space and absorb it, mainly by the black element carbon (soot). In the terrestrial infrared spectrum (> 3μm) haze particles can increase the radiative cooling efficiency of the atmosphere at high relative humidities (Blanchet and List 1987, Blanchet 1991). In moist air the 'Arctic haze' forms small droplets or ice crystals that have a volume up to two orders of magnitude larger than dry aerosol particles. As a result, a substantial increase of aerosol optical depth occurs that is particularly effective in the 8–12μm window region. Because of the surface temperature inversion, the top of the haze layer can be 10–20°C warmer than the surface temperature. Since haze layers act as grey bodies to thermal radiation, they can increase the outgoing long-wave radiation by 1–2 W/m^2 (Blanchet and List 1987). Most authors (e.g., Blanchet 1991; Shaw 1995) estimate that the net effect of the 'Arctic haze' on the radiation balance of the atmosphere is positive. However, Karlqvist and Heintzenberg (1992) have concluded that "...at present it is not possible to predict with any certainty whether the effects of 'Arctic haze' are positive or negative; in other words whether this form of pollution adds to the greenhouse effect or contributes to a cooling of the atmosphere." For more details about relations between 'Arctic haze' and climate see, e.g., Blanchet and List (1987), Valero *et al.* (1988), or Blanchet (1991).

Pollutants may also get into the Arctic via river runoff and oceanic circulation. According to Gobeil *et al.* (2001) these pathways, especially the ocean, may, over a longer period of time, be more important then the atmosphere pathway. For more details see the cited paper.

Chapter 9

CLIMATIC REGIONS

From the descriptions which have been presented earlier of the different elements of the Arctic climate, one can see that their spatial changes are extremely heterogeneous. Surprisingly, the greater horizontal gradients occur in winter, when the differentiated influence of the solar energy is meagre or equal to zero (polar night). During this time the differences in observable meteorological fields are caused mainly by the atmospheric circulation and, to a significantly lesser degree, by the oceanic circulation. This is connected with the fact that in the cold half-year, the cyclonic activity is more intensive and frequent than in the warm half-year. In addition, the temperature contrast between the Arctic air and the advected air from the moderate latitudes is highest during this time. The same may also be observed in the case of oceanic circulation. In turn, the effect of the underlying surface is not large since snow and sea-ice cover almost the entire Arctic.

On the other hand, in the warm half-year, the solar radiation factor is most important and causes the greatest heterogeneity of the meteorological elements in all spatial scales: micro-, topo-, and macroclimatic. The underlying surface, which is significantly differentiated (snow, ice, tundra, and water), increases the influence of solar radiation. However, because of the attenuated influence of the atmospheric and oceanic circulation and the larger areas of the Arctic Ocean and seas not covered by sea ice (open water), the climatic differences between the regions are less in summer than in winter. On the other hand, the greatest horizontal gradients of the meteorological elements are observed in the coastal areas in the macroclimatic scale, and between glaciated and non-glaciated areas in the topoclimatic scale (Baranowski 1968; Przybylak 1992a).

There is very little literature concerning the climatic regionalisation of the Arctic. Only Prik (1960, 1971) investigated this problem for the entire Arctic. The results of her work have been published more recently in the *Atlas Arktiki* (1985). Prik (1960), using peculiarities of the atmospheric circulation and the distribution of the main meteorological elements, distinguished five main climatic regions: Atlantic, Siberian, Pacific, Canadian-Greenland, and Interior Arctic. In her next paper devoted to the climatic regionalisation of the Arctic, Prik (1971) delimited seven climatic regions. The sixth and seventh regions were distinguished by dividing the Canadian-Greenland region into three new ones: Canadian, Baffin Bay, and Greenland. The climatic regions and sub-regions (in the case of some regions) presented in the *Atlas*

Arktiki (1985) were delimited based on the analysis of the distribution of mean long-term fields of almost all climatic elements, their seasonal changes, as well as the character of variability of these elements.

For the Canadian Arctic, Maxwell (1982) distinguished a number of climatic regions. The major climatic controls which he takes into account are as follows: cyclonic and anticyclonic activity, the sea ice-water regime, broad scale physiographic features, and net radiation. In addition, the secondary sectioning was made using information about local topography, aviation weather, maritime influences, temperature, precipitation, snow cover, and wind. He delimited five climatic regions, and within each of them (except region III) at least two sub-regions. There are no similarities between Maxwell's climatic regionalisation of this part of the Arctic and that of Prik (*Atlas Arktiki* 1985). The reason for this is probably the different criteria used for the delimitation of the climatic regions and the more detailed and subjective character of Maxwell's regionalisation.

A more detailed description of the climate in the Arctic for the seven above-mentioned regions (see also Figure 1.2) is presented in the following sections.

9.1 The Atlantic Region

In the cold half-year, the most striking feature of this region is its extreme high temperatures (relative to other parts of the Arctic) related to strong and vigorous cyclonic activity and the warm ocean currents which are branches of the Gulf Stream (see Figure 4.4). For example, the mean monthly air temperatures in Spitsbergen are about 20°C higher than in the Canadian Arctic at the same latitude. The anomalies get smaller to the north, northeast, and east because the influence of cyclonic activity and warm oceanic currents is reduced. On the other hand, the cold East Greenland and East Spitsbergen currents significantly cool the areas where they occur. The intense cyclonic activity also brings exceptionally great cloudiness and heavy precipitation to the Atlantic region. The wind speeds occurring here are the highest in the entire Arctic. Also the variability of some meteorological elements (in particular the air temperature) is greatest here (see Figure 4.7). The Atlantic region is characterised by the lowest degree of climate continentality (see Figure 4.2). The ocean between Jan Mayen and Björnöya has almost a 'pure oceanic' type of climate. For example, the mean monthly temperatures (1951–1980) at Jan Mayen range between –6.1°C (February) and 4.9°C (August) with absolute maximum and minimum temperatures reaching only 18.1°C and –28.4°C (1922–1980), respectively (Steffensen 1982). Prik (*Atlas Arktiki* 1985), taking into account the spatial differentiation of the climatic conditions in this region, delimited four sub-regions: southern, western, northern, and eastern.

The southern sub-region is the largest one (see Figure 1.2) and can be characterised in brief as the warmest, cloudiest, and rainiest area in the entire Arctic. The average air temperatures of the winter months in the southern parts of this sub-region oscillate between −1°C and 0°C and in the northern part between −8°C and −10°C. Storms and heavy precipitation, often occurring in the form of wet snow and sometimes as rain, are frequent here. The highest variability of temperature in the entire Arctic (see Figure 4.7) is in the border areas of the southern and northern sub-regions, where there is the greatest changeability of thermally contrasted air masses inflowing from northern and southern sectors.

The western sub-region is significantly smaller than the southern one, but has an area similar to the two others (the northern and eastern regions). Prik (1971) has distinguished this sub-region for three reasons: 1) maximum horizontal temperature gradients, 2) significant variability in the average monthly and diurnal temperatures, and 3) clear predominance of severe northern winds. In comparison with the previous sub-region, lower temperature, cloudiness and precipitation characterize the western sub-region. On the other hand, the greatest variability of precipitation is observed here (see Figures 7.8 and 7.9).

The northern sub-region includes the eastern half of Spitsbergen, other Svalbard islands, Zemlya Frantsa Josifa, the northern parts of the Barents and Kara seas, and the part of the Arctic Ocean adjoining them. The climatic conditions are more severe here than in the southern sub-region because the influence of both atmospheric and oceanic circulation is significantly weaker and, in the case of the latter factor, it is even absent. On the other hand, cyclones reaching this sub-region have an occluded stage. This, according to Prik (1971), creates considerable fluctuations in temperature, an increase in cloudiness, and an intensification of winds. Air temperatures are lower than in the other two sub-regions and rapidly decrease from −10°C and −12°C in the south-western part to −25°C in the north-eastern part. The winds blow mainly from south-easterly and easterly directions and their speed is lower than in other sub-regions. Cloudiness is quite extensive (60–65%). The interesting feature of this sub-region is the fact that in winter it is characterised by the greatest temperature variability, while in summer the temperature variability is the lowest in the entire Arctic.

The eastern (or Kara) sub-region includes the eastern half of Novaya Zemlya, the Kara Sea up to the Taymyr Peninsula, and the part of continent that is washed by it. Climatic conditions are most severe in this sub-region because Novaya Zemlya acts as a climatic barrier. It significantly reduces the entering of warm water from the Barents Sea and also, to a lesser degree, the movement of travelling cyclones along the Iceland–Kara Sea trough. Cyclones, which reach this sub-region passing Novaya Zemlya island from its northern and southern sides, together with the cyclones developing in the south-east of this sub-region, cause quite large diurnal variation in temperature, severe winds (mainly south-westerly or southerly), an insignificant occurrence of cloudy days, and frequent, although not heavy, precipitation. Correlation analysis reveals that the temperature relationship of this sub-region is greater with the Siberian region than with the Atlantic region (Przybylak 1997b). Therefore, it is proposed that this sub-region be included within the Siberian region.

The above characterisation of the climate of the Atlantic region and its sub-regions is based mainly on winter conditions, which occur throughout most of the year. In addition, under these conditions the spatial differentiation of the climate is significantly the greatest. For this reason, the summer climate is presented only for the Atlantic region as a whole. In summer, because of the decrease in the meridional gradient of temperature between moderate and high latitudes, the cyclonic activity is less intense. On the other hand, the frequency of cyclones is only slightly lower than in winter due to the drop in air pressure in the Arctic. The influence of warm currents is also limited in summer. This is true both of their magnitude and their spatial occurrence. The most important factor differentiating the climate is the incoming solar radiation (the polar day). As a consequence, a zonal distribution of meteorological elements is observed (see Figure 4.4). For example, the temperatures decrease to the north from 8–10°C (southern part) to about –1°C and 0°C (northern part). The cold East Greenland and East Spitsbergen currents considerably reduce the air temperature. In summer, the winds blow from the opposite directions than in winter, i.e., from the north and east. Cloudiness over the ocean and seas is very high (75–85%), while on land (both in coastal and inland areas) it rapidly decreases to about 60–65%.

9.2 The Siberian Region

This region is located far from the Atlantic and Pacific oceans. Therefore, this area is characterised by one of the most extreme continental climates on earth. The winter climate is dominated by the Siberian high, which undoubtedly mostly determines the climatic regime of the region. Cyclones are a rather rare phenomenon and, if they occur, they travel mainly along the Lena and Kolyma rivers. The influence of the Siberian high is seen in the directions of winds, which are here mostly from the southern sector and have moderate speeds (about 5 m/s in the maritime areas and below 3 m/s in the continental areas). In comparison with the Atlantic region, the annual cycle of wind speed in this region is the opposite: lower winds occur in winter and higher ones in summer. Air temperatures are among the lowest in the Arctic and drop below –30°C. On the continent, they rapidly decline in direction to the centre of the Siberian high, where near Oimekon and Verkhoyansk the lowest temperatures in the Northern Hemisphere occur. However, this area lies outside the Arctic. The variability of air temperature and pressure is lower here than in neighbouring regions. As a result of the domination of anticyclonic circulation, low cloudiness (35–45%) and precipitation (< 10 mm/month) is observable.

In summer, anticyclones still prevail but cyclones also occur very often (see Figure 2.3b, d), mainly in the southern part of the Siberian region. The highest occurrence of anticyclones is noted over the Laptev and East Siberian seas; they rarely enter the continental part. Such a synoptic situation determines the prevalence of winds from the northern and eastern sectors. Thus, on the continental part, a monsoon-like change of wind directions between winter and summer is observed. The synoptic situation, which is less stable than in winter, also results in wind speeds being slightly higher in these areas, ranging from 5 to 6 m/s. Air temperatures in the maritime parts are rather low (0 – 2°C), but rapidly increase on land, reaching 10–12°C near the southern boundary of the Arctic. There is considerably more precipitation than in winter: from 20–25 mm/month over the seas to about 30–40 mm/month over the continent. Fogs are very frequent over the water areas (up to 30% in the North), while over the continent they are a rather rare phenomenon (5%).

9.3 The Pacific Region

Cyclonic activity in winter in the Pacific region is significantly lower than in the Atlantic and Baffin Bay regions and, according to the calculations of Serreze *et al.* (1993), their frequency is lower than 3%. The low frequency of cyclones is caused by the occurrence of orographic barriers (zonally stretching mountains: the Koryak, Chukchi, and Brooks Ranges). As a result, cyclones can enter to the Pacific region only through the narrow Bering Strait. The narrowness of this strait also limits the warmth carried by the ocean current. Despite the above-mentioned obstacles, their influence on the climate of this region is discernible. This means that the cyclones entering the Pacific region must transport large amounts of warmth and moist air. As a result, a considerable increase in air temperature and precipitation during such events is observed. The close proximity of contrasted and well-developed air pressure systems (Aleutian low and Siberian high) result in large pressure gradients, conditioning a dominance of severe winds, mainly from the northern sector. This transitional location of the Pacific region, in the case of shifts in those pressure systems mentioned, must cause a rapid change of wind speed and alters the physical characteristics of the air. Weak winds bringing cold and dry air occur when the Siberian high steers the weather, while strong winds bringing warm and humid air occur when the Aleutian low assumes control. As a result, the variability of air pressure, temperature, and wind speed here is high. Prik (1971) relates this to the frequent occurrence of cyclones, which – as was shown by Serreze *et al*. (1993) – is not true. In contrast with neighbouring regions (Siberian and Canadian), the Pacific region is characterised by significantly higher temperatures, wind speeds, cloudiness, and precipitation. The horizontal gradients of temperature over the marine areas are not large. Average temperatures may vary from –16°C to –18°C north of the Bering Sea and decrease to –24°C to –26°C north of Chukchi Sea (Prik 1971). These values of temperature are, however, markedly lower than in the Atlantic region. On the continental parts, the temperature rapidly decreases, particularly in Asia. Cloudiness is moderate and ranges from 55–60% in the southern region to about 40–45% in the northern parts. Precipitation amounts are equal to 25–35 mm per month. Winds blow mostly from the northern sector and their average speed is high, oscillating between 6 and 8 m/s. Storms here are frequent, on average 6 to 8 days per month.

According to Prik (1971), the cyclone frequency in summer is lower than in winter. On the other hand, results quoted by Serreze *et al*. (1993) give us the opposite picture (compare their Figures 2 and 6). Southerly winds prevail in most of the Pacific region, except in the northern part, where easterly winds are more common. Air temperatures over the water area of the seas change slightly from 1°C in the north to 6–8°C in the Bering Strait. Significantly higher temperatures are noted only on the coasts of the warmer continents. Cloudiness is very high, oscillating from 80–85% (over seas) to 70–75% (over continents). Fogs are common, with their average frequency amounting to 25–35%.

9.4 The Canadian Region

This region is one of the largest areas in the Arctic. Therefore, of course, the differentiation of climate is particularly evident here. The estimates of magnitude of these differences are, however, not identical. For example, Barry and Hare (1974) write, "The Canadian Arctic Archipelago extends over 15° of latitude but the climatic characteristics are relatively homogeneous". In turn, Maxwell (1982) noted that "...despite the northern latitudes of the Canadian Arctic Islands, the climate there is extremely diverse." Such divergences in opinion, according to Maxwell (1982), result from the fact that the majority of climatic classifications are based on data from individual stations, which are all situated in the coastal areas. This fact, as well as the use of mainly temperature and precipitation data in the process of climatic classification, introduces a bias in the results. This divergence may also result from the fact that Maxwell's (1982) work is of a regional character, so the greater detail is understandable, while other works describing climatic characteristics (Barry and Hare 1974) or presenting climatic classifications (Prik 1960, 1971; *Atlas Arktiki* 1985) for the entire Arctic must contain some element of generalisation.

In winter, as is demonstrated by the most recent data (Serreze *et al.* 1993), the frequency of both cyclones and anticyclones is similar, but rather low. The former dominate in the northern part of the Canadian Arctic, and the latter are prevalent in the western part. It is also important to add that the frequency of both these air pressure systems is significantly greater in the summer months (see Figure 2.3). However, both cyclones and anticyclones in summer and winter have been classified by Serreze *et al.* (1993) as relatively weak. This means that in the Canadian Arctic, the variability of climatic elements should be relatively low. Prik has distinguished two sub-regions in the Canadian Arctic: northern and southern (Prik 1971; *Atlas Arktiki* 1985).

In winter months very low temperatures are observed in the northern sub-region. The coldest temperatures in the entire Arctic (with the exception of Greenland) occur in the north-eastern part. The mean monthly temperatures here drop below $-34°C$ and locally even below $-38°C$. Winds blow mainly from the northern sector and their speed is moderate or weak (in the northern part). Calms are very common and are noted for about 30% of all observations. Cloudiness is low, with the frequency of occurrence of clear and cloudy days amounting to 40–50% and 30–40%, respectively. Amounts of precipitation are also very low (< 10mm per month). The annual totals here are the lowest in the Arctic (see Figure 7.3).

The southern sub-region has higher temperatures ($-20°C$ to $-30°C$) than those occurring in the northern sub-region due to the influence of both the latitude and a greater frequency of cyclones. The greater synoptic activity also causes an increase in the variability of air temperature and other climatic elements. Wind speed, cloudiness, and precipitation, similar to air temperature, are also higher here. The mean, very low temperatures in the Canadian Arctic, as noted Barry and Hare (1974) are connected with persistent rather than extreme cold.

During the summer, the significant influence of solar radiation (polar day), together with the lower temperature contrast in the time between high and moderate latitudes, markedly reduces the differences between the northern and southern sub-regions. Mean summer temperatures oscillate from 2 –4°C (in the North) to 6–8°C (in the South). A significant increase of local differentiation of air temperature may be observed, particularly between sea areas (which are full of drifting ice and therefore cold) and coastal non-glaciated regions (which absorb great amounts of solar radiation and thus are relatively warm). The highest temperatures are, however, noted in the southernmost parts of the Canadian Arctic (>10–12°C). Winds are moderate and blow mainly from the northern sector. Cloudiness is higher than in winter. The frequency of cloudy days over the archipelago is 60–70%, decreasing over the continent to about 50%. This increase of cloudiness in summer is connected both with a greater occurrence of cyclones and with moisture provided by the open water areas and melting snow. Barry and Hare (1974) noted that local fogs and stratus clouds could occur even in the absence of cyclonic convergence since little uplift is necessary to saturate the air. Precipitation falls mainly as rain, except in the mountain areas. Most precipitation is related to cyclone passages, but orographic effects are also very important. The year-to-year variability of monthly and seasonal totals is very great and it is dependent on the frequency of passing depressions.

9.5 The Baffin Bay Region

The Baffin Bay region climatically is very similar to the Atlantic region, particularly to its southern and western sub-regions. The weather in winter in both these regions is shaped mainly by the cyclones developing over the North Atlantic. In the case of the Baffin Bay region, the cyclones move from the source areas through the Davis Strait and Baffin Bay. Cyclones bring large amounts of warmth and moist air to the areas where they enter. Therefore, air temperatures here are markedly higher than in the adjacent regions. Particularly high temperatures are noted in the eastern and the northern parts of the region, where additional heat is introduced into the atmosphere by the West Greenland Current and the stationary "North Water" polynya. In the Baffin Bay region, as a consequence of the great frequency of cyclones, high cloudiness (> 60%), precipitation (50–60 mm), severe winds, and day-to-day variability of all meteorological elements are observed. The topography of the areas surrounding Baffin Bay and the Davis Strait (the mountain relief of Baffin Bay and Greenland with glaciation zones) limits the spatial development of cyclones. As a result, the air temperature significantly decreases on land. The greatest horizontal gradients are noted in the coastal areas, particularly in Greenland. Due to air circulation in the cyclones, northerly and north-westerly winds dominate in the western part of the Baffin Bay region, while in the eastern part, south-easterly winds are prevalent. Such a pattern of wind leads to the occurrence of lower temperatures and precipitation on the coast of Baffin Island than along the coast of Greenland.

In summer, the above-mentioned temperature pattern also occurs but it is mainly connected with the oceanic circulation (a cold current in the western region and a warm one in the eastern part). While cyclones are present here, their frequency and strength is lower than in winter. Open waters increase the occurrence of low cloudiness and fogs. Winds have a moderate strength and are less stable over the sea area of the Baffin Bay region. Along the coasts of Baffin Island and Greenland, the easterly and westerly winds dominate, respectively. Fogs are very rare here.

9.6 The Greenland Region

An ice sheet and peripheral glaciers cover more than 80% of Greenland. The plateau of the ice sheet generally exceeds 1200 m a.s.l., and in the highest parts rises to over 3000 m a.s.l. Coastal mountains (see Introduction) also reach this elevation. Besides these important climatic factors, the atmospheric circulation also plays a very important role, particularly in the low elevated areas. The quasi-permanent Greenland anticyclone mainly influences the weather in the northern part, while cyclones coming from the Icelandic depression influence the southern part of Greenland. The cyclones enter Greenland from the southwest; some of the deepest and most vertically developed can cross the southern part of Greenland, sharply changing the weather conditions. However, most often they travel along the western or eastern coasts. The Greenland region, which includes almost the entire island with the exception of the coastal areas not covered by ice, is the coldest part of the Arctic. In the winter months the average temperatures oscillate around –40°C, with minimum temperatures dropping below –60°C. The ice sheet climate is dominated by a surface temperature inversion averaging 400 m in depth. During cyclones the temperature can increase by 20°C to 30°C due to both their transport of warmth and the disturbance of the temperature inversion. This advection of maritime air into the interior is reflected in large interdiurnal temperature changes. In the northern part of Greenland a very low cloudiness and precipitation connected with the dominance of anticyclonic activity and with topographic conditions (on the lee side of the ice sheet) is observed. Both cloudiness and precipitation are higher in the southern half of Greenland, particularly in areas which are elevated and well exposed to the main air streams (the western, southern, and eastern slopes of the Greenland Ice Sheet). On the slopes of the ice sheet, the katabatic winds reaching the maximum speed in the marginal parts of the slopes are a very important feature of the climate.

In summer, according to the investigation made by Serreze *et al.* (1993), the cyclones tend not to cross Greenland. The differentiation of climatic elements and their variability is therefore lower at this time than in winter. Temperatures are, however, very low and in the northern part the monthly means mostly oscillate between −10°C and −12°C. Extreme temperatures can drop to almost −30°C. The katabatic winds are stable, as in winter, but they are weaker and do not reach the coastal areas.

9.7 The Interior Arctic Region

The horizontal gradients of meteorological elements here are the lowest in the Arctic. However, some differences in pattern distribution of these elements exist, which allow a distinction to be drawn between two separate sub-regions: the sub-Atlantic and the sub-Pacific (Prik 1971; *Atlas Arktiki* 1985).

The sub-Atlantic area quite often falls under the influence of the North Atlantic cyclones and therefore the temperatures here are higher than in the sub-Pacific sub-region, where anticyclones dominate. The winter temperature varies here from −24°C to −26°C in the southern part to about −32°C near the Pole. Absolute minimum temperatures can drop below −50°C. According to the map presenting the mean annual fields of sea level pressure (Serreze *et al*. 1993), southerly winds dominate in this sub-region. Wind speeds are significantly lower here than in the Atlantic region and oscillate between 5.5 m/s and 6.5 m/s but their range of variation is considerable: from extremely weak to very strong. Cloudiness in winter is significantly lower than in summer and ranges from 60% in the southern part to about 50% near the Pole. Precipitation is very frequent but its intensity is light. Monthly totals do not exceed 10–15 mm, and yearly 200–250 mm.

The sub-Pacific sub-region, due to the dominance of anticyclone circulation, has a more severe climate in winter than the previous sub-region. Temperatures are, on average, 6–10°C lower and their spatial changes are very small. Minimum temperatures generally do not drop below −53°C while maximum temperatures do not exceed −3°C to −5°C. The dominance of anticyclones results in the occurrence of unstable and weak or moderate winds. Their speeds are lower here than in the sub-Atlantic sub-region and oscillate between 4.5 m/s and 5.5 m/s. Also there is less cloudiness and precipitation here.

In summer, the meteorological regime is similar in both the sub-regions. The mean temperature approaches 0°C because it is limited by the melting process. The range of temperature oscillates between −6°C and 4–6°C. The small daily contrast of incoming solar radiation (polar day), very high cloudiness (80–90%), and the melting of snow and sea ice result in daily variations of temperature being exceptionally low (on average 0.4°C to 0.5°C). The wind directions are very changeable and speeds are not particularly high (4.5–5.5 m/s). Storm winds (> 15 m/s) are very rare in the summer months and, according to

Prik (1971), only two to five cases every 10 years are observed. Cloudiness is exceptionally high (about 90% of cloudy days and only 4–8% of clear days) and occurs mainly in the form of low clouds. Precipitation, although quite frequent, gives small totals because its intensity is very small. A characteristic phenomenon of the summer in the interior Arctic region is the frequent occurrence of fog (25–40%).

Chapter 10

CLIMATIC CHANGE AND VARIABILITY IN THE HOLOCENE

Polar regions play a very important role in shaping the global climate. Both empirical and modelling studies show that these are the most sensitive regions to climatic changes. As a consequence, warming and cooling epochs should be significantly more distinct here that in the lower latitudes. Climatic models indicate that they should also occur earlier. However, this is not always the case. It depends on the factor(s) causing the climate change (Przybylak 1996a, 2000a).

The Arctic climate system differs from the other climate systems situated in the lower latitudes firstly because it contains the cryosphere, which is present almost over the whole Arctic. The role of the cryosphere in determining climate is still not fully understood. Thus in March 2000 a new research project, within the World Climate Research Programme, called the Climate and Cryosphere (CLIC) was established (see http:/clic.npolar.no).

Future climatic change which may occur in the Arctic as a result of human activities is difficult to predict (see the next Chapter). This especially concerns its rate and magnitude. However, it is certain that man-made changes in climate will be superimposed on a background of natural climatic variations. Hence, in order to understand future climatic changes, it is necessary to have a knowledge of how and why climates have varied in the past (Bradley and Jones 1993). Therefore of particular relevance are climatic variations of the last 10–11 thousand years (the Holocene period). The climatic changes which occurred in the past will be presented for three time scales (10–11 – 1 ka BP; 1.0–0.1 ka BP; and 0.1 BP – present). For the first period and for almost the entire second period there are no instrumental meteorological observations, and thus our knowledge is mostly based on the so-called "proxy data". Until recently, the majority of paleoclimatological information has come from geological, geomorphological, and botanical studies. Since the late 1960s a new powerful source of information has been available: the ice-core analyses. For details see, e.g. Bradley (1985, 1999).

10.1 Period 10–11 ka – 1 ka BP

The start of the Holocene is estimated most often between 10 and 11 ka before present (years BP, the present being defined as 1950 A.D.). However, glaciological proxy data show that this date should be shifted to about 11.6 ka (see e.g. Johnsen *et al.* 1992 or O'Brien *et al.* 1995). As can be seen from Figure 10.1, there is a dramatic change of climate at about this time. At the Summit and Dye 3 $\delta^{18}O$ profiles, the change is equal to 3–4‰ (it gives about a 5–7°C rise in temperature). Most of the transition between the Younger Dryas and the Holocene occurred in a few decades, according to GISP2 (Greenland Ice Sheet Project) oxygen isotope ($\delta^{18}O$) ice record (Taylor *et al.* 1997). Then the warming was significantly less and the typical Holocene values reached about 10.2 ka BP.

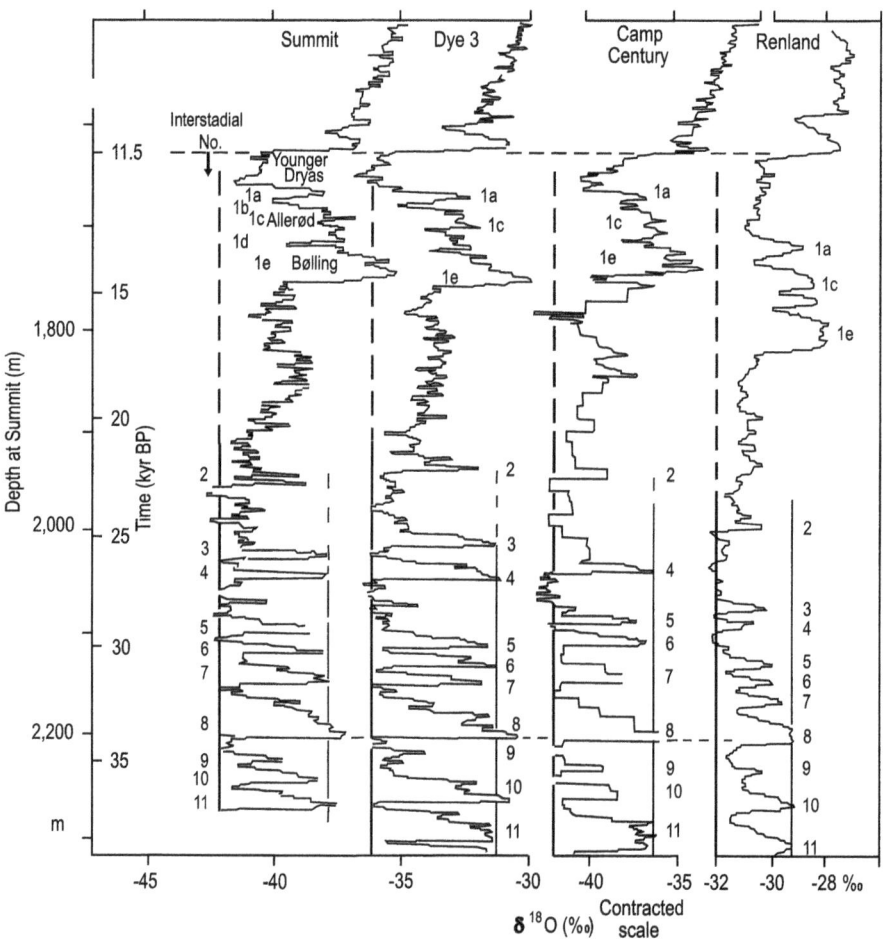

Figure 10.1. Continuous $\delta^{18}O$ profiles along sections of four Greenland ice cores from Summit (Central), Dye 3 (Southeast), Camp Century (Northwest) and Renland (East Greenland), spanning nearly the same time interval. The four records are all plotted on linear depth scales, of which only the Summit depth scale is shown to the left along with a Summit timescale. The heavy and thin vertical lines indicate estimated δ levels characteristic of late-glacial cold and mild stages respectively. The figures close to these lines define a suggested numbering of significant mid- and late-glacial interstadials (after Johnsen *et al.* 1992).

In this period a substantial increase in the annual layer accumulation was also noted (Figure 10.2) in the GISP2 deep core. Distinct changes in chemical flux values (marine and terrestrial Na, Cl, Mg, K, and Ca), which represent changes in the atmospheric composition over the Greenland Summit, are also evident (see Figure 10.3 and O'Brien et al. 1995; Taylor et al. 1997). As can be seen in Figure 10.1 it is also evident that the Holocene is a period of a relatively stable climate with mean $\delta^{18}O$ values of –34.7 ‰ and –34.9 ‰ for GISP2 and GRIP (Greenland Ice-core Project), respectively (Grootes et al. 1993). However, the small Holocene $\delta^{18}O$ fluctuations of 1–2 ‰ occur very often and are sufficient to detect changes in temperature conditions. The Holocene climate of the Arctic will be characterised here mainly according to the data from Greenland, Canadian high Arctic, and the Eurasian Arctic islands.

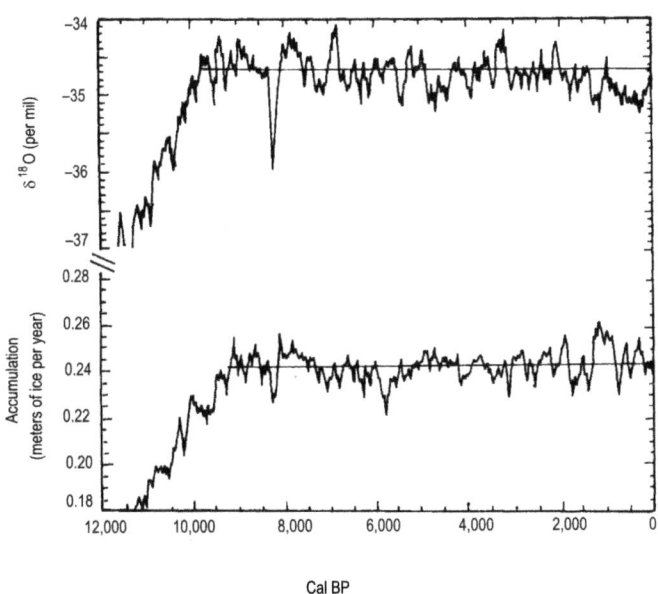

Figure 10.2. Record of 100-year smoothed accumulation and oxygen isotope profiles from the GISP2 core from 12,000 years BP to the present. Reprinted with the permission from Meese D. A., Gow A. J., Grootes P., Mayewski P. A., Ram M., Stuiver M., Taylor K. C., Waddington E. D. and Zielinski G. A., 'The accumulation record from the GISP2 core as an indicator of climate change throughout the Holocene', *Science*, 266, 1680–1682. Copyright 1994 American Association for the Advancement of Science.

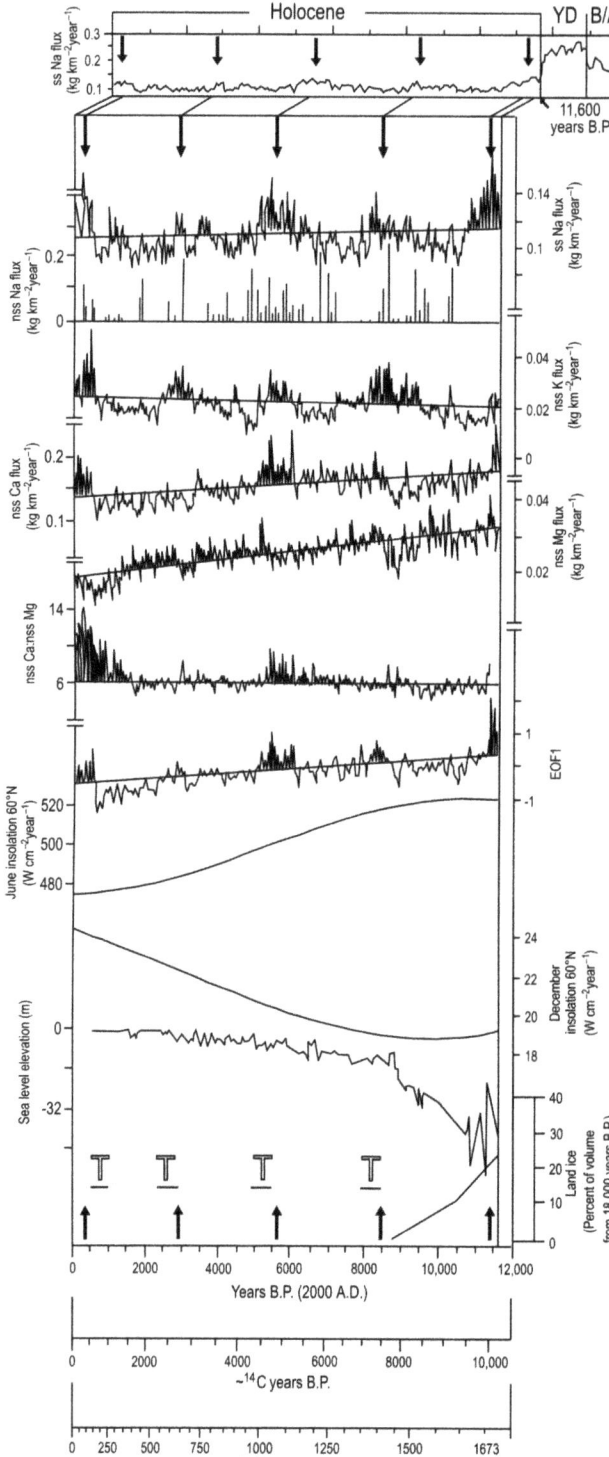

Figure 10.3. Profiles of the GISP estimated ss and nss species for the Holocene and potential climate-forcing factors. All profiles are smoothed with a robust spline (equivalent to a 100-year smooth) to be consistent with previously published GISP2 data (Mayewski et al. 1994). The ssNa profile represents the behaviour of all ss species and is illustrated through the YD and part of the Bölling–Alleröd (B/A) events to reference the relatively low Holocene glaciochemical concentrations. Increases in concentration are marked by arrows. Episodes of triple oscillations defined as $\delta^{14}C$ intervals which have Maunder- and Spörer-type patterns occurring in sets of three are also indicated (T). The most recent triple event corresponds to the Maunder, Spörer, and Wolf solar activity minima. Reprinted with permission from O'Brien S. R., Mayewski P. A., Meeker L. D., Meese D. A., Twickler M. S. and Whitlow S. I., 'Complexity of Holocene climate as reconstructed from a Greenland ice core', Science, 270, 1962–1964. Copyright 1995 American Association for the Advancement of Science.

10.1.1 Greenland

The best paleoclimatic information exists for Greenland, where during the last 30–40 years quite a large number of ice-cores have been drilled, beginning with the oldest one (Camp Century) and ending with the series of 13 drilled during the summers of 1993–1995 along the North-Greenland-Traverse. The routine analyses of ice cores include measurements of isotopic composition ($\delta^{18}O$, δD), content of greenhouse gases (CO_2 and CH_4), dust content, chemical composition, electricity conductivity, annual ice accumulation, etc. The advantage of this kind of proxy data is mainly the high time resolution, which allows researchers to investigate even seasonal changes and the wide spectrum of information available about climate and environmental changes. However, in recent years some scientists have expressed scepticism as to whether analyses of ice cores may be considered reliable because of the artificial contamination and disturbances of ice cores during the drilling (for details see Jaworowski *et al.* 1990, 1992). More recently a new source of information about past temperatures in Greenland has become available. This new source is based on temperature profiles measured down through an ice sheet in deep boreholes. This information is then used to reconstruct past surface temperatures. This is possible because temperatures down through the ice depend on the geothermal heat flow density, the ice-flow pattern, and the past surface temperatures and accumulation rates (Dahl-Jensen *et al.* 1998). Since the beginning of 1970s, this method has very often been used for the non-glaciated areas based on temperature measurements in wells (see e.g. Čermak 1971; Lachenbruch and Marshall 1986; Pollack and Chapman 1993; Majorowicz *et al.* 1999, and references therein). The main weakness of this method is its low time resolution, which decreases the further back in time one investigates. This means that the high-frequency changes are not registered. In the distant past even such prominent climatic events like Bölling/Alleröd and cold Younger Dryas periods are not resolved (see Figure 10.4).

Holocene climate history for the central part of Greenland is presented in Table 10.1 and Figures 10.2–10.4. Generally speaking, there is good correspondence in the timing of the occurrence of warm and cold periods on a millennial time scale, distinguished from different sources. Some differences are connected with different time resolutions of the sources mentioned, different sensitivity to environmental and climate-forming factors, and probably errors in the dating of some of them. The first warm period (not noted by the borehole temperatures) occurs in the early Holocene about 10.0 to 8.56 ka BP. The greatest and longest warming is clearly evident in reconstructed temperatures from GRIP borehole, lasting from 8 to 5–4 ka BP and this can be referred to as the Climatic Optimum (Dahl-Jensen *et al.* 1998). However, more high-time resolution data show that during this time at least one to three cold periods also occurred (see Table 10.1 and Figures 10.2 and 10.3). Dahl-Jensen *et al.* (1998) calculated that the surface temperature during this period was about 2.5°C warmer than the present temperature (Figure 10.4b). A period of maximum postglacial warmth between approximately 8 and 4 ka BP has been proposed from the Camp Century ice core (Dansgaard *et al.* 1971).

The next warm period had a rather short duration and began about 2.5 ka and ended between 2–1.5 ka BP. For this time there is the greatest difference in data received from the borehole temperature measurements (see Table 10.1 and Figures 10.2–10.4), which show clear cooling (0.5°C below present temperature) about 2 ka BP. Analysis of reconstructed temperature for the Dye 3 core (Dahl-Jensen *et al.* 1998) also provides similar results. All paleoclimatic proxy data reveal, on the other hand, the existence of a Medieval Warm Period, which in Greenland, as may be seen from Table 10.1 and Figures 10.2–10.4, started about 1.4–1.5 ka BP, thus significantly earlier than in Europe. According to reconstructed temperatures by Dahl–Jensen *et al.* (1998), the maximum warming occurred about 900 A.D. (Figure 10.4c) and was 1°C warmer than at present in Greenland.

Table 10.1. Warm and cold periods (ka BP) in the Holocene based on measurements of oxygen isotope $\delta^{18}O$, an ice accumulation rate and chemical fluxes at GISP2 core as well as based on borehole temperatures at GRIP core

Periods	Measured element $(\delta^{18}O)$*	Ice accumulation rate*	Chemical fluxes**	Borehole temperature at GRIP core***
Warm	9.5–8.5	9.2–8.5	10.6–9.3	
	8.0–7.6	8.1–7.3	7.9–6.3	8.0–4.2
	7.0–6.6			
	5.3–4.7	5.0–4.2		
	3.6–3.1			
	2.5–2.0	2.5–1.9	2.7–1.5	
	1.0–0.8	1.3–0.8	0.96–0.61	1.5–0.8
				0.05–0.0
Cold			> 11.3	11.6–9.5
	8.5–8.0	8.5–8.0	8.8–7.8	
	7.5–7.0			
	6.5–6.0	6.0–5.2	6.1–5.0	
	4.7–4.3			
	3.0–2.5		3.1–2.4	3.0–1.5
	1.9–1.1	1.9–1.3		
	0.8–0.0	0.8–0.0	0.61–0.0	0.7–0.1

Author interpretation based on figures presented at the following papers:
* – Meese *et al.* (1994); ** – O'Brien *et al.* (1995); *** – Dahl-Jensen *et al.* (1998)

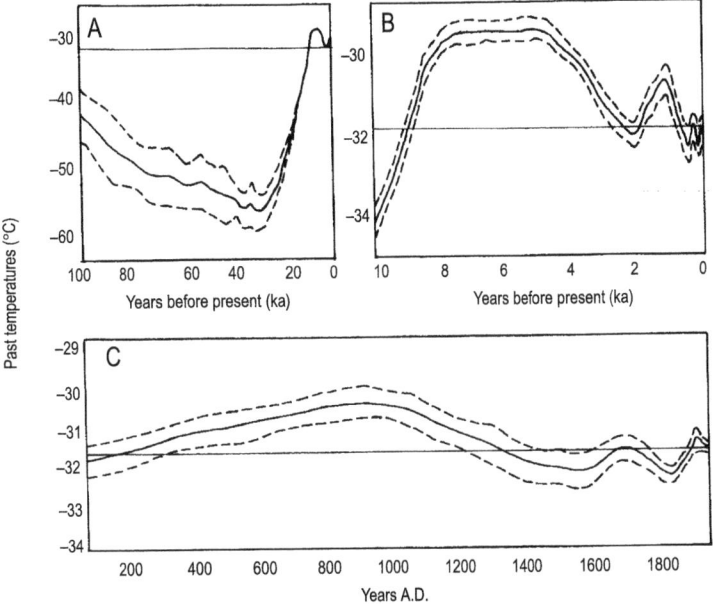

Figure 10.4. The contour plots of all the GRIP temperature histograms as a function of time describes the reconstructed temperature history (solid lines) and its uncertainty. The temperature history is the history at the present elevation (3240 m) of the summit of the Greenland Ice Sheet. The dashed curves are the standard deviations of the reconstruction. The present temperature is shown as a horizontal line. (A) – the last 100ky BP, (B) – the last 10ky BP and (C) – the last 2000 years. Reprinted with the permission from Dahl-Jensen D., Mosegaard K., Gundestrup N., Clow C. D., Johnsen S. J., Hansen A. W. and Balling N., 'Past temperatures directly from the Greenland Ice Sheet', *Science*, 282, 268–271. Copyright 1998 American Association for the Advancement of Science.

From the start of the Holocene to 1 ka BP, between 4 and 6 cold periods can be distinguished. The first one occurred in the transition period from Younger Dryas to the Holocene, when the temperatures were generally colder than average conditions in the Holocene. The second deterioration of climate occurred from 8.5 ka to 8.0 ka BP, i.e. just before the start of the Climatic Optimum. Between one and three short cold events occurred during the time of Optimum. Both the ice accumulation rate and changes in chemical fluxes show that this cooling occurred from 6 to 5 ka BP (Table 10.1 and Figures 10.2 and 10.3). On the other hand, the oxygen isotopes reveal three colder periods observed in three other periods (7.5–7.0 ka, 6.5–6.0 ka, and 4.7–4.3 ka BP). Oxygen isotope and borehole temperature data show the clear coolness of climate from 3 ka to about 1.5–1.1 ka BP with a very small warming spell between 2.5–2.0 ka BP. The remaining sources reveal not so long-lasting cooling (see Table 10.1 and Figures 10.2–10.4). The maximum cooling during this period occurred around 2 ka BP and was about 0.5°C lower than the present climate (see Figure 10.4b).

O'Brien et al. (1995) found that the cold events identified in their glaciochemical series correspond in timing to records of the world-wide Holocene glacier advances (Denton and Karlén 1973) and to cold events in paleoclimate records from Europe, North America, and the Southern Hemisphere (Harvey 1980), as determined by combining glacier advance, oxygen isotope ($\delta^{18}O$), pollen count, tree ring width, and ice core data (Figure 10.5). They also reveal quite a good correspondence between the timing of cold periods and periods of low solar output, as identified in residual tree ring radiocarbon ($\delta^{14}C$) age measurements (Stuiver and Braziunas 1989) (Figure 10.3). Moreover, they also found almost the same quasi-cycles of $\delta^{14}C$ climate (2500 years) and cold periods identified in the GISP2 record (~2600 years).

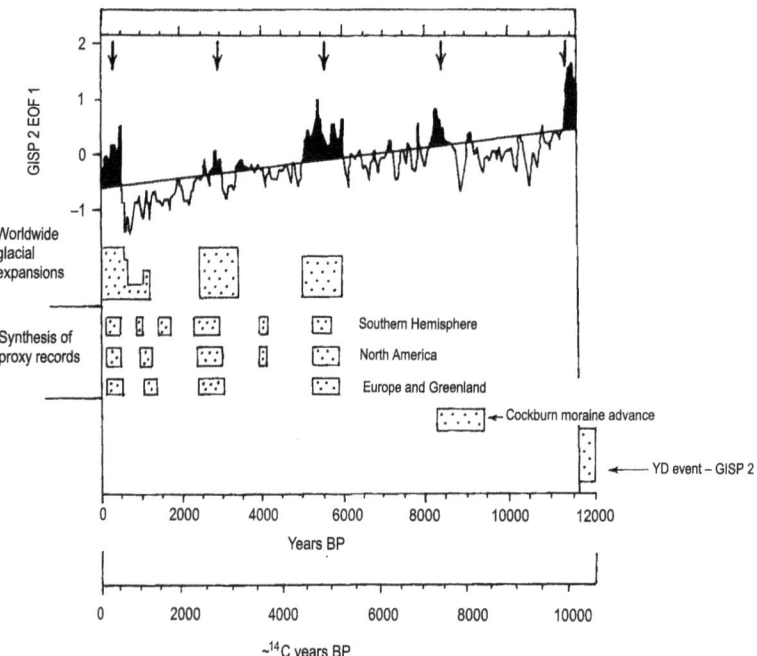

Figure 10.5. Paleoclimate cold events: GISP2 Holocene EOF1; world-wide glacial expansions and their relative magnitude (Denton and Karlén 1973); synthesis of various climate proxy records from Europe, Greenland, North America, and the Southern Hemisphere showing cold periods (Harvey 1980); the Cockburn Stade (Andrews and Ives 1972); and the YD event (Mayewski et al. 1993). Reprinted with permission from O'Brien S. R., Mayewski P. A., Meeker L. D., Meese D. A., Twickler M. S. and Whitlow S. I., 'Complexity of Holocene climate as reconstructed from a Greenland ice core', Science, 270, 1962–1964. Copyright 1995 American Association for the Advancement of Science.

From the proxy data presented here, the precipitation changes during the Holocene are best represented by the ice accumulation rate. Meese et al.

(1994) found, however, that the accumulation and oxygen isotopes correlate significantly at GISP2. From Figure 10.2 one can see that this correlation is positive. Mostly the precipitation is greater in warmer periods and lower in colder periods. There are, however, some exceptions, such as about 6.8 ka or 1.2–1.0 ka BP.

Dahl-Jensen *et al.* (1998) comparing the results presented here with those from the Dye 3 borehole (865 km further south from GRIP), found that the Dye 3 temperature is similar to the GRIP history, but has an amplitude 1.5 times greater, indicating higher climatic variability there. They concluded that the difference in amplitudes observed between the two sites is a result of their different geographic location in relation to the variability of atmospheric circulation, even on the time scale of a millennium. The importance of regional influences on environmental changes, especially in the second half of the Holocene, is also revealed by O'Brien *et al.* (1995). They concluded that this complexity in Holocene climate makes distinguishing a natural from an anthropogenically-altered climate a formidable task.

10.1.2 Canadian High Arctic

Proxy data concerning Holocene climatic change in the Canadian high Arctic largely comprises ice core, glacial, and sea-ice/ice-shelf components and less geomorphological and chronological evidence (Evans and England 1992). Ice-core analyses from the Agassiz, Meighen, and Devon Ice Caps (e.g., Koerner and Paterson 1974; Koerner 1977a, b, 1979, 1992; Paterson *et al.* 1977; Fisher and Koerner 1980, 1983; Koerner and Fisher 1985; Koerner *et al.* 1990) provide an important record of high latitude climatic change. Fisher and Koerner (1980) found for Devon Island a period of increasing postglacial warmth from 10 to 8.3 ka BP. Then the temperature showed small oscillations until 4.3 ka BP, when the maximum postglacial temperatures occurred. Since 4.3 ka BP there has been a progressive cooling. The end of the postglacial optimum occurred between 4.5 and 3 ka BP. This scheme of the climatic changes on Devon Island is in good correspondence with the reconstructed temperature history from the GRIP borehole (Figure 10.4b). Variations in the abundance of stranded driftwood, which are also used as indicator of Holocene climate change in the high Arctic, generally correlate very well with both the above-mentioned series of data (Figure 10.6). The interpretation of this figure is as follows: the greater the abundance, the warmer the summer temperatures and the lighter the sea-ice conditions, which allow for drifting of wood. Bradley (1990) summarising the proxy data for the Holocene paleoclimate of the Queen Elizabeth Islands, identified two basic climatic periods: 1) the early-mid Holocene when summer temperatures were comparable or higher than at present, and 2) the last 3500 ± 500 years over which summer temperature dropped significantly. Evans and England (1992), analysing proxy data from northern Ellesmere Island, generally give a similar reconstruction.

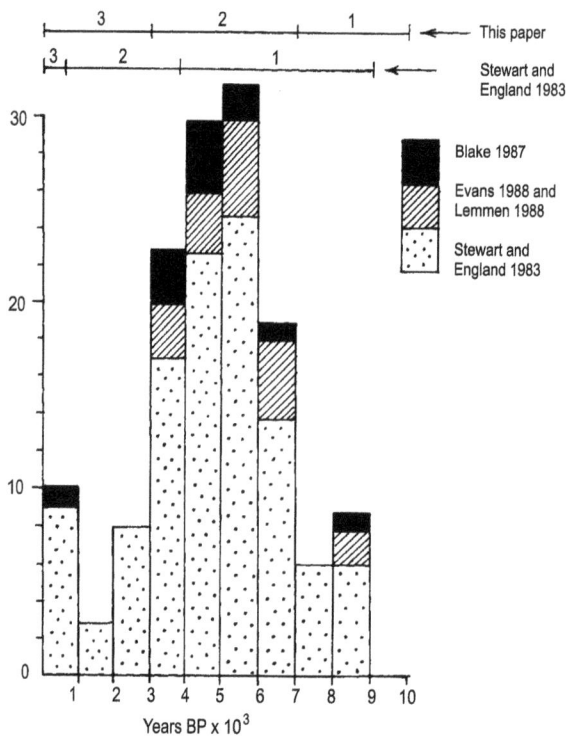

Figure 10.6. Histogram of driftwood radiocarbon dates from the Canadian and Greenland high Arctic based on data from Stewart and England (1983), Evans (1988), Lemmen (1988) and Blake (1987) (after Evans and England 1992). 1, 2 and 3 in the upper part of the figure denote the periods of significant differences in the abundance of stranded driftwood.

10.1.3 Eurasian Arctic Islands

The Greenland Ice Sheet and the Canadian high Arctic represent the typical continental climate, while on the other hand, the Eurasian Arctic islands (from Svalbard to Severnaya Zemlya), and particularly Svalbard, characterise the part of the Arctic with the most maritime climate. For this part of the Arctic there exist some ice-core analyses from Svalbard, Zemlya Frantsa Josifa, and Severnaya Zemlya, but most of them do not cover the last thousand years (Tarussov 1988, 1992; Vaikmäe 1990). Only the ice core from the Vavilov ice dome (Severnaya Zemlya) supplies proxy data for almost the entire Holocene period (Figure 10.7). Unfortunately, there are some uncertainties in the dating of the ice core; the ice age is evidently overestimated, even in the upper part of the profile (Tarussov 1992). Tarussov further notes that the interpretation of this core is problematic due to the strange absence of a correlation between the chloride and $\delta^{18}O$ curves of the same core. Based on the review of literature concerning the history of glacier advances and retreats during the Holocene in Svalbard, Novaya Zemlya, Zemlya Frantsa Josifa and Severnaya Zemlya, estimated using geomorphological and glacier investigations (e.g., Bazhev and Bazheva 1968; Szupryczyński 1968; Grossvald 1973; Baranowski 1977b; Werner 1993; Lubinski et al. 1999), and for the last thousand years also using ice-core analyses (Vaikmäe and Punning 1982; Tarussov 1988, 1992; Vaikmäe 1990), one can conclude that there exists a good correspondence between climatic changes in this region of the Arctic. Therefore, the history of the Holocene climate will be presented here using mostly data from Svalbard, for which this history is best known.

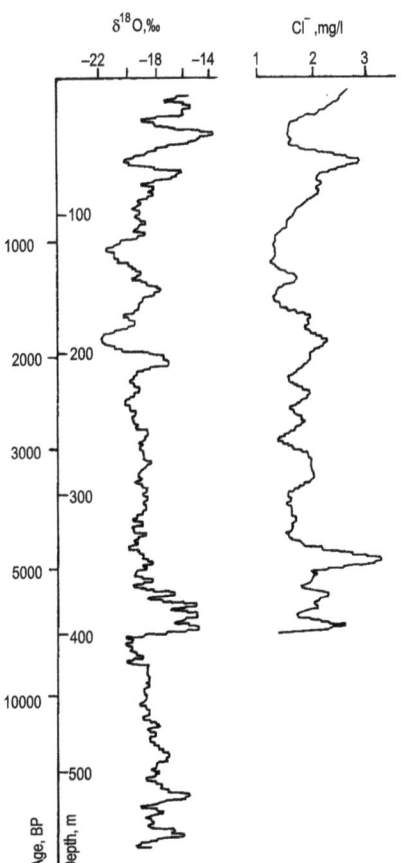

Figure 10.7. Variations in $\delta^{18}O$ and Cl⁻ concentrations for the Severnaya Zemlya ice core (after Vaikmäe 1990).

Recently, Werner (1993) has provided a review of our current knowledge concerning the climatic changes in Svalbard in the Holocene and has also supplied a new Holocene moraine chronology for central and northern Spitsbergen. He presents evidence for multiple Neoglacial advances in this area. The fragmentary moraine record indicates two Little Ice Age advances and two older Neoglacial advances. The oldest moraines had stabilised by ca. 1.5 ka BP, and a second group of moraines by ca. 1.0 ka BP. The first group of moraines, according to Werner (1993) may correspond to the advance of glaciers between 3.5 and 2.0 ka BP, reported by many authors (e.g. Szupryczyński 1968; Baranowski 1975, 1977a, b; Baranowski and Karlén 1976; Punning et al. 1976; Lindner et al. 1982; Niewiarowski 1982; Marks 1983). The advance of glaciers indicated by moraines dated to ca. 1.0 ka BP is not recognised in the stratigraphy of southern Spitsbergen. The moraine chronology proposed by Werner (1993) compares well with other proxy climate records on Spitsbergen, summarised in his Figure 10. From this, it can be seen that Climatic Optimum occurred between 7 and 4 ka BP, the same as in Greenland and in the Canadian Arctic. During this period the reduced sea ice (Haggblom 1982), increased the production of local pollen (Hyvarinen 1972) and the occurrence of thermophilus marine molluscs (Feyling–Hanssen and Olsson 1960) were observed. There are also no traces indicating an advance of the glaciers. The proxy climatic records further show evidence for late-Holocene (4–2 ka BP) climatic deterioration. In the next 1000 years, there is evidence mainly in sea ice and glacial records of some warming of the climate, similar to the reconstructed temperature histories for the GRIP borehole (Figure 10.4b). More recently, Svendsen and Mangerud (1997) have obtained a generally similar history of the Holocene climate based on investigations of sediment cores from the proglacial lake Linnévatnet, west Spitsbergen.

Summarising all the proxy data from the Arctic presented in this section we can say that:
1. The Holocene climate until 1 ka BP was warmer than today, except during the early part and the period about 2 ka BP,
2. A Climatic Optimum occurred between 8 and 5–4 ka BP with temperatures being 2–2.5°C higher than present,
3. A drop in temperature was noted between 4 and 2 ka BP (minimum) and then an amelioration of climate was observed with a temperature maximum of about 900–1000 A.D.,
4. A significant similarity of climatic changes was noted in the entire Arctic analysed as well as in the areas bordering the Norwegian and Greenland seas (Iceland, Jan Mayen, and Scandinavia) (Werner 1993). Therefore we can probably state that the remaining part of the Arctic (not presented here) also had a similar climate history during the Holocene.

10.2 Period 1 ka – 0.1 ka BP

Generally speaking in the history of the climate of the last 1 ka years, three periods have most often been distinguished: the Medieval Warm Period (MWP), the Little Ice Age (LIA) and the Contemporary Global Warming (CGW). The latter period will be described in the next section. Thus, what do the proxy data tell us about the climate in the Arctic during the first two periods? The most detailed answers are given by ice-core analyses.

10.2.1 Greenland

The best ice-core analyses from the whole Arctic are available for the Greenland Ice Sheet. In the first half of the 1990s, the two longest ice cores (GRIP and GISP2) were drilled in the Summit (central part of Greenland) as well as 13 shallow ones (covering the last 500–1000 years) along the North-Greenland-Traverse. In addition, as was mentioned in the previous section, measurements of the borehole temperatures allow the reconstruction of the surface temperature histories for GRIP and Dye 3 areas. Let us start with the analysis from the proxy data giving the most averaged history. The borehole temperatures confirm the MWP and the LIA as having existed in Greenland. As was mentioned in the previous section, the MWP occurred here earlier than in other parts of the world (this fact allows Vikings to have built settlements in the southern part of Greenland). Maximum warmth is centred between 900 and 1000 A.D., but the beginning of this period can be dated between 500 and 600 A.D. and the end about 1200 A.D. (Figure 10.4c). From this Figure it can be seen that temperatures at this time were about 1°C greater than they are at present in Greenland. This period with temperatures higher than normal in Greenland lasted from about 200 A.D. to 1300 A.D. The MWP is also clearly seen in the ice accumulation rate data (Figure 10.2). The duration is the same as is shown by borehole temperatures but the greatest maximum of ice accumulation occurred around 800 A.D. On the other hand, the secondary maximum of accumulation corresponds very well with the maximum temperature from about 900–1000 A.D. The average accumulation from A.D. 620 to 1150 was 0.26 m of ice per year, 8% higher than the average Holocene accumulation rate and the highest rate recorded in the last 12 ka years (Figure 10.8 and Meese *et al.* 1994). In coastal Greenland, the MWP began as early as 800 A.D. (Lamb 1977). Proxy data available to Lamb were, of course, not as precise as those presented here. For this reason, and because all of Greenland reacts equally to factors determining climatic changes at present (see Przybylak 1996a, 2000a), it seems that the start of the MWP in coastal parts should be shifted to about 600 A.D. The historical records, mainly from northwestern Europe, describe an MWP occurring anywhere between A.D. 800 and A.D. 1300 (Lamb 1977; Houghton *et al.* 1990, 1996) with dates varying by as much as 200 years. This means that in Europe the MWP started about 200 years later than is indicated by the GISP2 record.

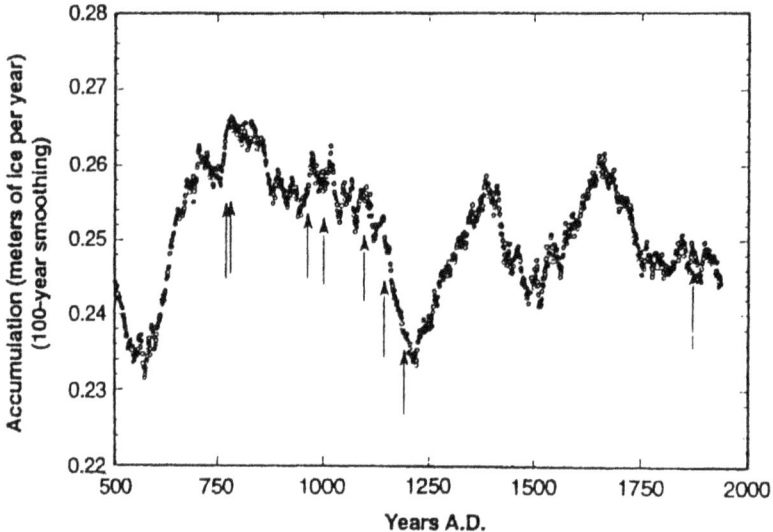

Figure 10.8. The 100-year smoothed accumulation record from the GISP2 core for the period A.D. 500 to the present. The arrows show locations of visually identified melt layers in the ice core. Reprinted with the permission from Meese D. A., Gow A. J., Grootes P., Mayewski P. A., Ram M., Stuiver M., Taylor K. C., Waddington E. D. and Zielinski G. A., 'The accumulation record from the GISP2 core as an indicator of climate change throughout the Holocene', *Science*, 266, 1680–1682. Copyright 1994 American Association for the Advancement of Science.

The LIA is not as well defined as the MWP in the literature. The first views, based on rather low-resolution proxy data, assumed that the LIA was one long, sustained cold period with dates ranging from A.D. 1200 to 1800 or A.D. 1350 to 1900 (for details see e.g. Lamb 1977, 1984; Starkel 1984; Grove 1988). The new, high-resolution, data reveal that this opinion was wrong, and that the climate during this period underwent significant fluctuations from cold to warm and warm to cold conditions. However, through most of the period a cold climate occurred. The climatic changes in the LIA period in Greenland is shown in Figures 10.4 and 10.8–10.10. The reconstructed temperature for GRIP and the mean isotope record from northern Greenland (Figure 10.4 and Figure 10.9) clearly show the existence of an LIA in Greenland. According to borehole temperatures, the LIA lasted from about A.D. 1400 to 1900. Throughout this period, except for a few decades around 1700 A.D., the temperature was colder than at present in Greenland. Dahl–Jensen *et al.* (1998) distinguished two cold periods centred at 1550 and 1850 A.D. with temperatures 0.5 and 0.7°C below the present, respectively. The mean isotope record from northern Greenland (Figure 10.9) shows that in this part of Greenland the LIA ended earlier, at about 1850 A.D. Based on an oxygen isotope record from Camp Century (Figure 1 in Johnsen *et al.* 1970), which lies in the same part of Greenland but has a longer record, we can assume that in the northern Greenland, similar to central Greenland, the start of the LIA occurred about 1400 A.D. In this record the two minima are also quite evident, but the times of their occurrence are different. The first minimum was centred around 1680 A.D. and the second around 1820–1830 A.D. The temperature during these periods was estimated to be 1°C lower than present (Figure 10.9).

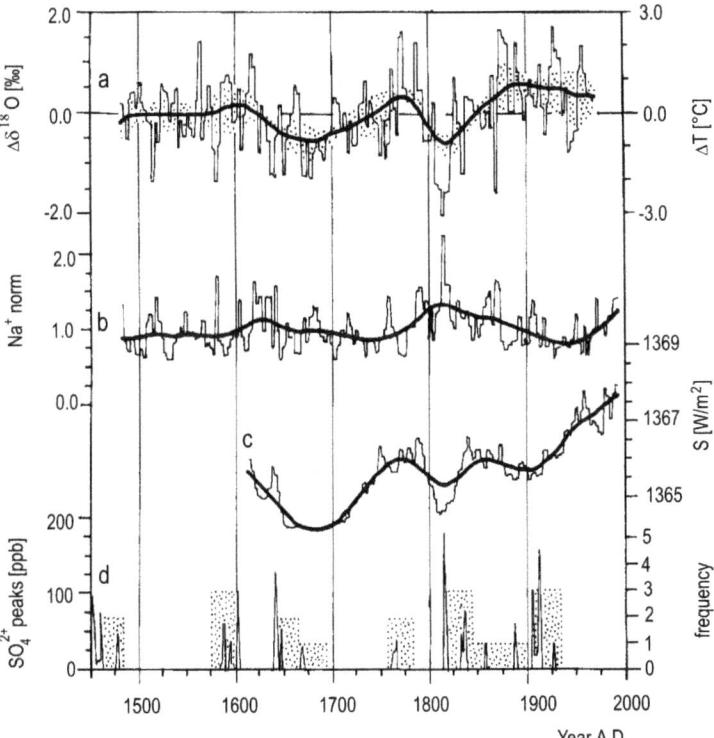

Figure 10.9. a) stacked isotope record of core B18, B21 and B29 for time span 1480–1969: Thin line represents the average of the triannual data sets, thick line and shading using dots the mean and standard deviation of the spline approximations after subtracting the core averages, b) stacked record of Na+ concentrations in core B16, B18 and B21. To allow for different absolute sea salt level in each core, which are largely caused by the different altitude of the drill sites, Na+ concentrations were normalised to the individual core average, c) three years intervals of reconstructed solar irradiance for the time span 1612 to 1913 (Lean *et al.* 1995), d) SO_4^{2-} concentration above background (thin line) and frequency in a 30 year interval (dotted bars) of stratospherically derived volcano horizons in the annual record of core B21 (after Fischer *et al.* 1998).

The break in the prolonged cooling, observed particularly from about A.D. 1600 to 1850, occurred in the second half of the 18th century. The greatest peculiarity of this record, not observed in other records (see Figures 10.4, 10.8 and 10.10), is the fact that the highest temperatures during the whole period recorded occurred at the end of the 19th century. The ice accumulation record from GISP2 (Figures 10.8 and 10.10) does not show so clearly the existence of the LIA in Greenland. However, this kind of data has a lower reliability than the two previous kinds. This is probably due to the fact that in the short-term scale (LIA) the positive correlation found by Meese et al. (1994) between precipitation (accumulation) and temperature ($\delta^{18}O$) is significantly less than in the long-term scale (the Holocene). Przybylak (1996a), working on the basis of the instrumental observations, found that in the warmer and colder periods both above and below normal precipitation can occur in the Arctic. From this account, the mean accumulation over the last 800 years was slightly higher than normal (about 3%), but there was significant variability on decadal and century time scales (Figure 10.10). Surprisingly, however, the accumulation variations agree quite well with surface temperatures reconstructed for the GRIP borehole. The greatest discrepancies are the facts that: 1) the first accumulation minimum occurred about 100 years earlier than the minimum temperature and 2) the accumulation after 1850 A.D. does not show any rise.

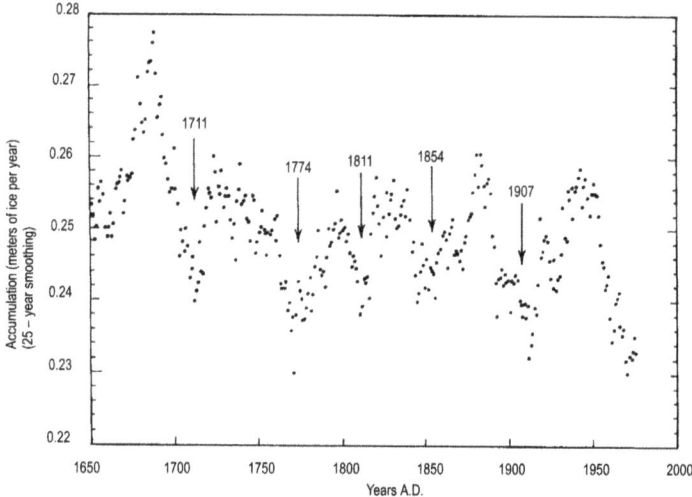

Figure 10.10. The 25-year smoothed accumulation record from the GISP2 core from A.D. 1650 to the present. The dates above the arrows correspond to years of decreased accumulation that correlate with dated glacial advances or cold periods in Greenland and elsewhere (Grove 1988). Reprinted with the permission from Meese D. A., Gow A. J., Grootes P., Mayewski P. A., Ram M., Stuiver M., Taylor K. C., Waddington E. D. and Zielinski G. A., 'The accumulation record from the GISP2 core as an indicator of climate change throughout the Holocene', *Science*, 266, 1680–1682. Copyright 1994 American Association for the Advancement of Science.

10.2.2 Canadian High Arctic

A review of the ice-core literature and data (e.g. Koerner and Fisher 1981; Bradley 1990; Koerner 1992) presenting the history of the climate in the Canadian Arctic in the Holocene period and particularly in last two to three millennia shows that in this part of the world the MWP is rarely distinguished. Generally most researchers indicate a steady decrease in temperature from 2000–3000 BP until the LIA period. Similar results have also been obtained based on glacial geology records (Blake 1981, 1989) as well as on peat, driftwood, whalebone, and mollusc studies (Blake 1975, Dyke and Morris 1990; Dyke *et al.* 1996, 1997). Recently, however, Koerner (1999) has suggested that the MWP could have occurred from A.D. 1200 to A.D. 1400. This statement is only partly documented by the $\delta^{18}O$ record (see Figure 10.11), which shows that this was really only the case during the periods A.D. 1200–1250 and A.D. 1350–1400.

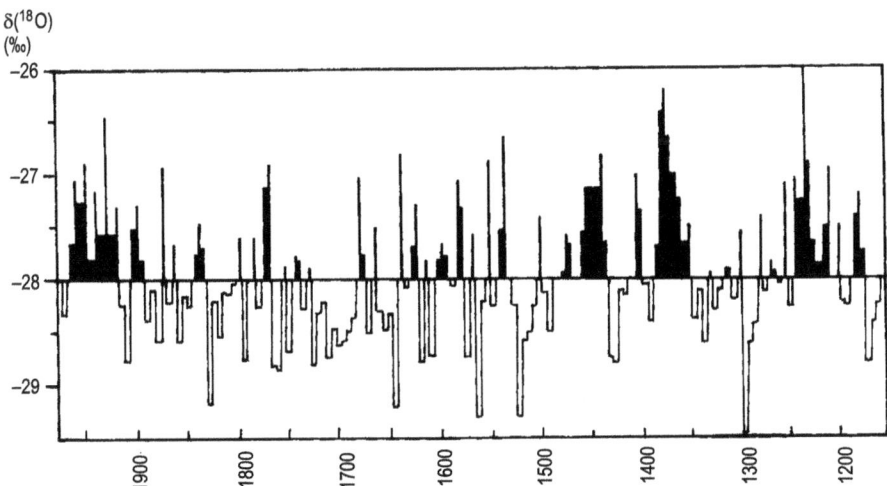

Figure 10.11. Five-year averages of oxygen isotope (δ) for the last 800 years from Devon Island ice cap (Alt 1985) (after Alt *et al.* 1992, modified).

On the other hand, the LIA period is distinctly visible in the light of the available proxy data. For example, both ice-core records (Figures. 10.11 and 10.12), from the high Canadian Arctic (Devon Island and Ellesmere Island) reveal the existence of an LIA ranging from A.D. 1400–1550 to 1900. However, given the results of Bradley's review (Bradley 1990), most researchers assume that start of the LIA probably occurred in the mid-16th century. The LIA is particularly clearly distinguished in the records showing the core area affected by melting (Figure 10.12). Only melting on the Devon Island Ice Cap had above normal values in the mid-16th century and in the short period centred around 1780 A.D. On the other hand, the isotope record shows the existence of fluctuations in climate around the long-term mean. However, the dominance of cold spells is unquestionable (Figure 10.11). The warmest periods during the LIA occurred in the mid-15th century (also seen in Agassiz Ice Cap, Figure 10.12(2)).

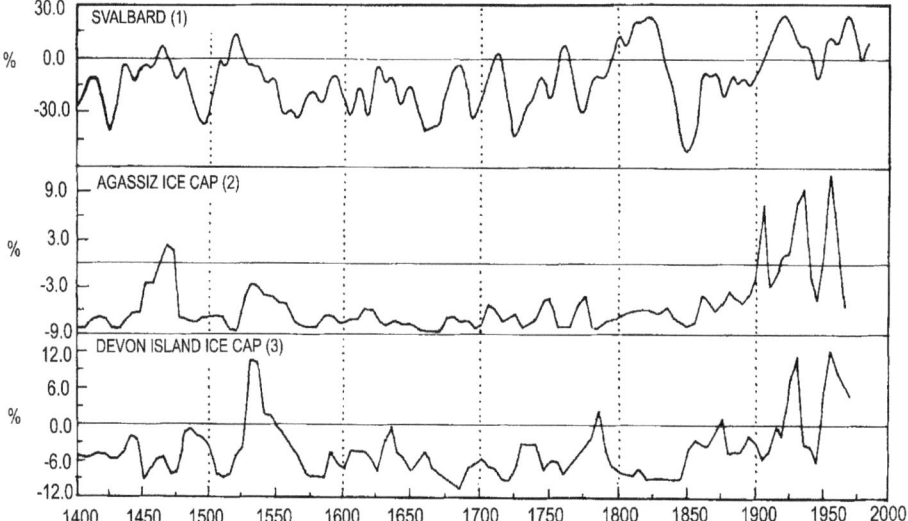

Figure 10.12. Reconstructed summer temperature anomaly for Svalbard (1), Agassiz ice cap (2) and Devon Island ice cap (3) (referenced to the mean of each series from 1860–1959). The Svalbard record is of summer melt from the Lomonosov ice cap in Svalbard (Spitsbergen – Tarussov 1992). The series from the Agassiz (Northern Ellesmere Island) and Devon Island ice caps (Canada) are based on ice core studies (Koerner 1977a, Koerner and Fisher 1990). All records show the core area affected by melting, expressed as percentage departures from the mean (after Bradley and Jones 1993).

The slight maximum of temperature around A.D. 1600 was significantly lower and interrupted by short cold spells, and probably therefore it is not registered in summer melting records. Other proxy data (lake sediment and tree-ring records) collected for the last four centuries (see Figure 2 in Overpeck *et al.* 1997) clearly indicate the presence of the LIA prior to A.D. 1850–1900. The existing documentary records from the Canadian Arctic used mainly the Hudson Buy Company Archives from remote trading posts from the 18th and 19th centuries (e.g. Moodie and Catchpole 1975; Wilson 1988, 1992; Ball 1983, 1992; Catchpole 1985, 1992a, b). All these investigations clearly confirm the exceptionally cold conditions of the early 19th century. Bradley (1990), summarising proxy data from the Queen Elizabeth Islands, has identified the LIA period from 400 to 100 years BP as particularly severe. Moreover, he writes that this period "may have been the coldest period in the entire Holocene". On the other hand, Evans and England (1992) did not find any well-recorded geomorphological evidence on northern Ellesmere Island confirming the existence of the LIA. Nevertheless, they noted further that undated dual advances by some glaciers might reflect mid-Holocene and the LIA accumulations. They add also that the formation and destruction of ice wedge polygons in sandar indicate respective reduction and increase in meltwater discharge associated with the LIA and then recent warming.

10.2.3 Eurasian Arctic Islands

Again, as in the previous section, the best proxy data for this part of the Arctic are to be found in Svalbard. Therefore, the history of the climate for the period analysed will be presented mainly using information from this area. Researchers investigating the history of the climate in the Holocene period (e.g. Tarussov 1992; Werner 1993; Svendsen and Mangerud 1997) generally do not mention the existence of the MWP. Most of them indicate late-Holocene climatic deterioration (see also Section 10.1), which began around ca. 4000 BP (Werner 1993) or 3000 BP (Tarussov 1992) and persisted to the LIA (Werner 1993) or to the 9th century (Tarussov 1992). However, there are also some researchers who indicate that the warmer period occurred between A.D. 600 and A.D. 1100 (e.g. Baranowski and Karlén 1976; Baranowski 1977b; Haggblom 1982; Svendsen and Mangerud 1997). For example, Svendsen and Mangerud (1997), analysing the rate of lake sedimentation (Linnévatnet lake), found that glacial maxima occurred around 2800–2900 BP, 2400–2500 BP, 1500–1600 BP and during the LIA. Thus between these periods warmer conditions prevailed. Baranowski (1977b) even suggested that the MWP probably lasted longer and was warmer than the contemporary warm period. The climate in the transitional period between the MWP and the LIA (i.e. in the 12th century and in the first half of the 13th century) was near the norm (Gordiyenko *et al.* 1981). From this analysis, and that presented earlier for the Canadian high Arctic, it may be concluded that late-Holocene histories of climate in both study areas are roughly similar. The greatest difference, however, concerns the time of the occurrence and the magnitude of warming during the MWP. It seems that this period was clearer in the Eurasian Arctic islands than in the Canadian high Arctic and that it occurred earlier, i.e. probably at the same time as it occurred in Greenland.

Proxy data for the LIA period present a clearer climatic picture. In Svalbard, similar to Greenland and the Canadian Arctic, the ice-core results from the Lomonosov ice cap and the Grønfjord-Fridtjof ice divide distinctly show the existence of the LIA ranging from A.D. 1300–1400 to 1900. This is very well seen both in the isotope record (Figure 10.13) and the summer melt (Figure 10.12(1)). Most of the geomorphological proxy data give the same results. Baranowski (1977b) concluded that the LIA period in Spitsbergen occurred between around 750 and 110 years BP. Similar results for this island are also presented by Punning and Troitskii (1977). The maximum advance of glaciers here occurred about 1600 A.D. and between 1750–1850 A.D. (Ahlmann 1948, 1953). Also the botanical proxy data confirm the existence of the LIA during this time with the culmination between the 17th and 19th centuries (Surova et al. 1982). Grossvald (1973) found that on Zemlya Frantsa Josifa the LIA began in the 14th century and ended about 1900 A.D. Bazhev and Bazheva (1968) gave very little information concerning the behaviour of glaciers in Novaya Zemlya during the LIA. They stated only that the start of the LIA occurred after the 16th century.

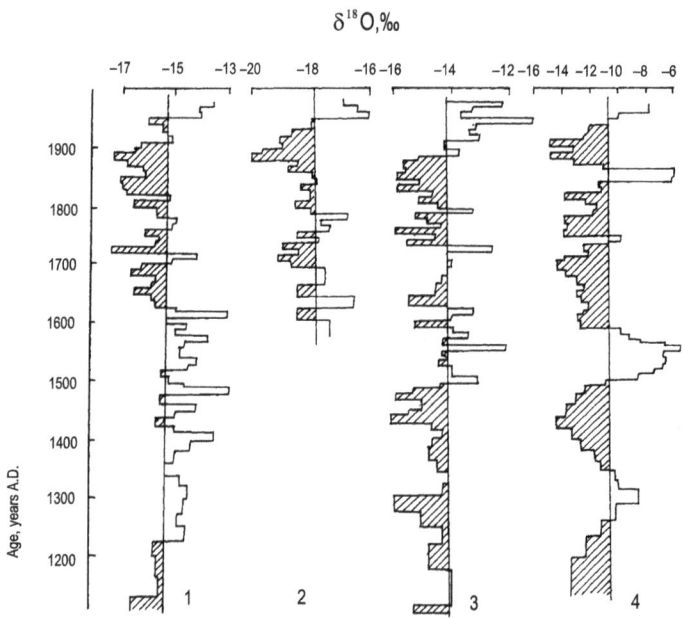

Figure 10.13. Variations in $\delta^{18}O$ for the Svalbard ice cores: (1) Westfonna, (2) Austfonna, (3) Lomonosov plateau, (4) Grönfjord–Fridtjof ice divide (after Vaikmäe 1990).

The LIA was interrupted in Spitsbergen in the 16th century. However, the pronounced warming occurred mainly in the lower located glaciers (Figure 10.13). On the other hand, on the Lomonosov plateau (1000 m a.s.l.) both warm and cold spells occurred during this period. Summer melting here was above normal only in the two first decades of the 20th century (Figure 10.12(1)). On the other hand, the LIA on the Nordaustlandet Island began with a significant delay, in comparison with most of the above-mentioned records, i.e. about

1600 A.D. (see Figure 10.13). The greatest temporal asynchronicity of glacioclimatic conditions can be clearly noted between Spitsbergen and Nordaustlandet in the period 1200–1500 A.D. However, there is full agreement that the culmination of the LIA occurred between the 17th and 19th centuries, although during this period some warming phases have also been observed. For example, surprisingly high values of summer melting were noted in the first two decades of the 19th century on the Lomonosov plateau, while in the Canadian Arctic (as has been mentioned) and also in Greenland very severe conditions prevailed. However, such an opposite tendency of air temperatures in the parts of the Arctic mentioned is also occurring at present (see Przybylak 1997b). The isotopic record also shows slightly lower than normal temperatures, but they are significantly higher than in the mid-19th century. An excellent agreement between summer melt and isotopic data (see Figures 10.12 and 10.13) exists for the mid-19th century. The area affected by melting was almost two times lower than normal. These exceptionally cold conditions in the Svalbard caused the significant advances of glaciers. As a result, the LIA moraines are typically the most extensive and best preserved (e.g. Szupryczyński 1968; Liestøl 1969; Niewiarowski 1982; Pękala and Repelewska-Pękalowa 1990; Werner 1990, 1993; Elverhøi et al. 1995; Svendsen and Mangerud 1997). At the Austfonna (Figure 10.13(2), Nordaustlandet) this mid-19th century cooling was shifted to the turn of the 19th and 20th century.

Werner (1993) also found a second period when moraines were deposited (ca. 650 years BP). It was probably connected with the evident climate deterioration in the northern part of Spitsbergen between A.D. 1250 and 1350 (see Figure 10.13(3), Lomonosov plateau). On the other hand, the southern part of Spitsbergen and also Nordaustlandet had a slightly warmer than normal climate during this period.

Summarising the results presented, one can conclude that most proxy data from the Arctic indicates that the LIA period occurred between 1300–1400 and 1900 A.D. In accordance with recent findings, this cold period was interrupted by shorter or longer warm periods, which were observed in different periods, but mainly before 1800 A.D. The majority of the proxy data presented show that the greatest cooling in the Arctic occurred in the first half (the Canadian Arctic) and around the mid- or in the second half (Svalbard and probably other Eurasian Arctic islands) of the 19th century.

10.3 Period 0.1 ka – Present

It is clear from Section 4.1. that a reliable estimate of areally averaged Arctic temperature can only be offered from circa 1950. Nonetheless, in the literature many works can be found which also provide an areally averaged

'Arctic' temperature for earlier years (e.g. Kelly and Jones 1981 a–d, 1982; Kelly et al. 1982; Jones 1985; Alekseev and Svyashchennikov 1991; Dmitriev 1994). 'Arctic' is used here in inverted commas because, in reality, these series represent the temperature changes in selected northern latitude bands which, as follows from Figure 1.1, significantly differ from the Arctic as defined in this book. Until 1911 the only Arctic stations for which it has been possible to compute these series were located in Greenland. Other data used in these analyses were taken from stations located in the Subarctic and even in the mid-latitudes. Moreover, the station coverage of these regions was very low, especially in the 19th century and was biased towards the lower latitudes. For example, Jones (1985) states that the 'Arctic' temperature was computed from grid points (5°x10°) covering only 6%, 10%, and 20% of the latitude band 65°N–85°N in the years 1851, 1874, and at the end of the 19th century, respectively. This author opposes the definition of such series as 'Arctic' because such definitions lead inevitably to the identification of misleading estimates of Arctic air temperature tendencies. This is very well illustrated in Figure 10.14, from which it can be seen that a warming in the 1930s was most pronounced in the real Arctic (see the top curve which represents the real Arctic in the greatest degree). This warming is reduced when more areas from the Subarctic and from the mid-latitudes are included in the Arctic. The second phase of contemporary warming (after 1975) in the real Arctic series is not seen, while in the other series it is distinct. For the whole Northern Hemisphere (bottom curve) the warming in last decades is even greater than in the 1930s.

10.3.1 Temperature Variations Prior to 1950

Przybylak (2000a) chose six stations to illustrate the variation of Arctic air temperature prior to 1950 (Figure 10.15). All of them represent the analysed climatic regions and offer long series. Figure 10.15 shows slightly rising temperatures in Greenland prior to 1920. After this time, the rate of warming significantly increases. This trend was noticed very early in Greenland and in the Atlantic Arctic region and has been described by various authors (e.g. Knipovich 1921; Scherhag 1931, 1937, 1939; Hesselberg and Birkeland 1940; Vize 1940; Weickmann 1942; Lysgaard 1949). The maximum temperature occurred in the 1930s and was higher by about 2–5°C than those occurring prior to the 1920s. The most pronounced rise in temperature occurred in the Atlantic region and throughout the Arctic in winter. During this season, the mean temperature rose locally by up to 9°C (Przybylak 1996a, 2002a). Since the 1930s a statistically significant decrease in temperature has been noted.

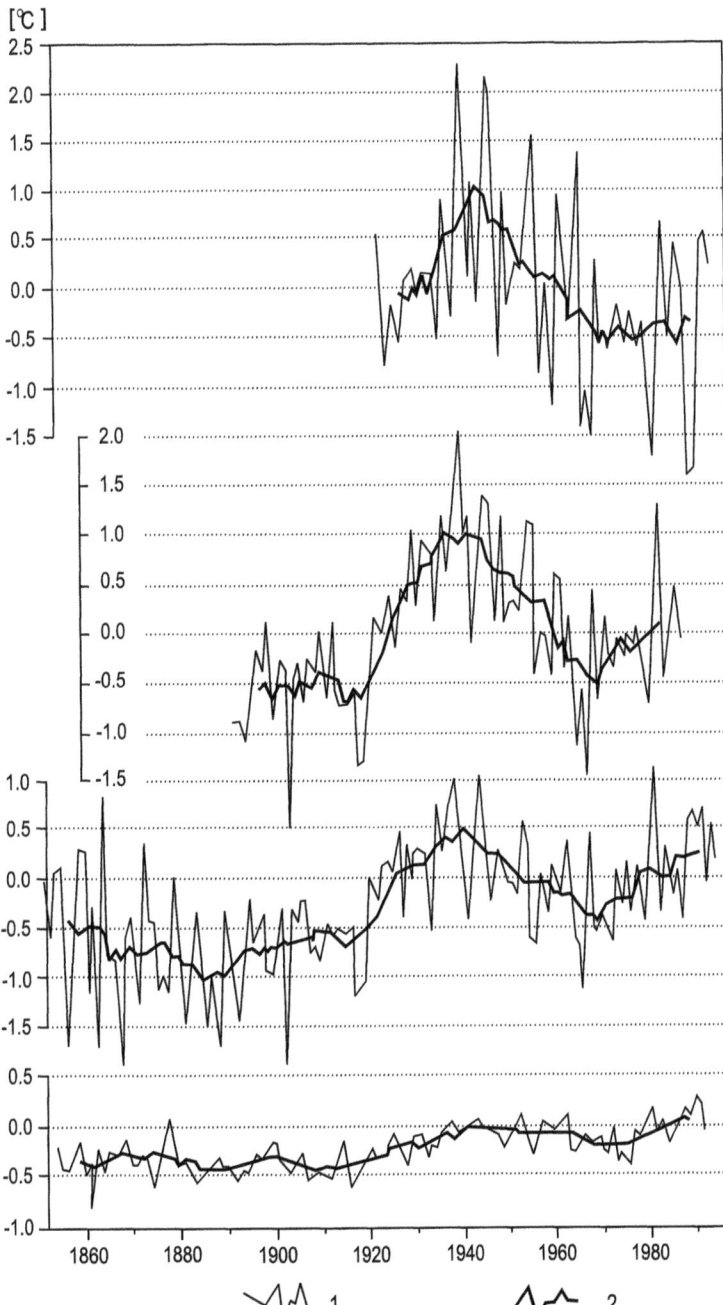

Figure 10.14. Year-to-year course of annual (1, solid line) and 5-year running (2, heavy solid line) mean anomalies of air temperature for the zones (after Przybylak 1996a): (a) 70°N–85°N (after Dmitriev 1994); (b) 65°N–85°N (after Alekseev and Svyashchennikov 1991); (c) 60°N–90°N (after Jones 1995, personal communication); (d) 0°N–90°N (after Jones 1994).

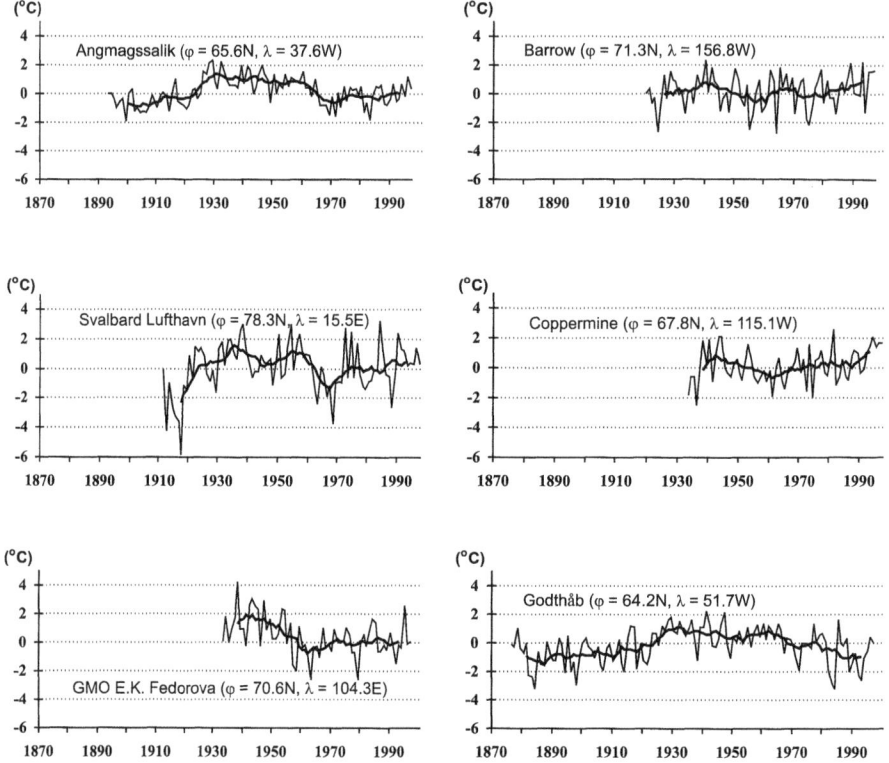

Figure 10.15. Year-to-year course of the annual (solid line) and 5-year running (heavy solid line) mean anomalies of air temperature in the Arctic stations having the longest observational series (after Przybylak 2000a).

All stations (except Barrow and Coppermine, representing only a small part of the Arctic) show the greatest warming in the 1930s. The reason most often given for this warming wave is a change in atmospheric circulation (see e.g. Scherhag 1931; Weickmann 1942; Petterssen 1949; Lamb and Johnsson 1959; Girs 1971; Lamb 1977; Lamb and Morth 1978; Kononova 1982) which is now thought to have been at least partly related to the North Atlantic Oscillation (NAO). In this decade, as has recently been reported by Slonosky and Yiou (2001) the values of the NAO index were comparable to those occurring in the late 20th century. However, patterns of temperature changes in the two periods differ, particularly in the area of western Greenland, where warming also occured in the 1930s (see Figure 10.15). The NAO and its influence on the Arctic climate is described in sub-chapter 10.3.3. A secondary air temperature maximum can be seen in the central part of the Atlantic region (Svalbard Lufthavn) in the 1950s and in Greenland in the 1960s. This is not present in the Siberian region. Spatial coherency in Arctic temperature changes was significantly greater before the 1950s than it was afterwards (Figure 10.15).

10.3.2 Temperature Variations After 1950

Detailed research into air temperature tendencies in the Arctic using data from 33 to 35 stations in the periods from 1951 to 1990 (Przybylak 1996a, 1997a, 2002a) and from 1951 to 1995 (Przybylak 2000a) revealed the predominance of negative trends, even though most of them were not statistically significant. Similar results have also been obtained by Chapman and Walsh (1993); Kahl et al. (1993a, b); Walsh (1995); Born (1996); Førland et al. 1997 and others.

The areally averaged seasonal and annual Arctic temperatures computed using data from 30 grid-boxes (after Jones 1994, updated) located in the study area are in good agreement with the above results obtained from the stations (see Przybylak 2000a). This data set, however, represents the Arctic without almost the entire region of Siberia. In addition, one should add that the quality of this type of data in its present state is significantly lower than the stations' data used by Przybylak (2000a). In contrast to grid-box data, the temperature series from stations have no gaps. Taking these factors into account, identification of characteristics of long-term Arctic temperature variations should still be based on stations' data.

A comparison of temperatures calculated from stations and grid-boxes (Przybylak 2000a) indicates that the general patterns of seasonal and annual Arctic temperature variation are roughly similar. The greatest differences occur in winter and autumn. Correlation coefficients computed between these series entirely confirm these conclusions. The highest correlation was found for summer ($r = 0.90$) and spring ($r = 0.82$), and the lowest for winter ($r = 0.55$) and autumn ($r = 0.66$). For the annual values, the correlation coefficient is equal to 0.74. All these correlations are statistically significant at the level of 0.001.

Slight increases in air temperature in the Arctic have been prevailing in the recently observed "second phase of contemporary warming" (after 1975). However, they are up to four times smaller for the areally averaged Arctic air temperature than for the analogous series for the Northern Hemisphere (land + ocean). Such a situation occurred, for example, in the period 1976–1995 (Przybylak 2000a).

These results raise the following question: What are the causes of the lack of warming in the Arctic in the above period? According to Przybylak (1996a, 2000a, 2002a), this situation may result from:

(i) A delay in the reaction of the Arctic climatic system, which has considerable inertia because of large water masses, along with sea and land ice. One may liken the Arctic to a large refrigerator. To warm such a refrigerator, a significantly greater amount of energy must be supplied than would be necessary to warm a lower latitude region to the same degree. This means that the warming in the Polar regions connected with the increasing radiation forcing will occur later (not earlier as is commonly assumed) than in lower latitudes. This conclusion is consistent with results presented by Aleksandrov and Lubarski (1988). Analysing observational evidence, they found that in the phase of global warming, the increase of air temperature in the Arctic was occurring later than in the lower latitudes. On the other hand, in the phase of global cooling, the opposite relation exists. It may be said that this conflicts with the warming in 1920–1940, which occurred earlier in the Arctic than in other parts of the world. This is correct, but the main reason for the latter warming was a change in atmospheric circulation. As such, the reaction of climate to a change of forcing is immediate. The considerable inertia of an Arctic climate system should also significantly delay the start of positive feedback mechanisms (such as sea-ice – albedo – temperature feedback) which are responsible for a significant portion of Arctic greenhouse warming.

(ii) The influence of natural factors (mainly a change in atmospheric circulation) which, while leading to a cooling of the Arctic, considerably reduces or completely removes the warming caused by the greenhouse effect. Przybylak (1996a, 2002a) shows that since the mid-1970s there have occurred significant increases in the frequency of the occurrence of the zonal macrotype of circulation (W) and decreases in the occurrence of the eastern macrotype of circulation (E), according to the typology of Vangengeim–Girs (see e.g. Girs 1948, 1971, 1981; Vangengeim 1952; Barry and Perry 1973). The first macrotype gives negative temperature anomalies in the Arctic and the second gives positive ones. This means that the described circulation changes lead to the cooling of the Arctic. Other natural factors should also cause Arctic cooling, e.g. the statistically significant decrease of solar irradiance in the Arctic reported by Stanhill (1995) and the downward trend of solar activity observed since 1957 when the secular maximum occurred. Voskresenskiy *et al.* (1991) found decreasing Arctic temperatures in the periods of lower solar activity.

(iii) The influence of a rising concentration of anthropogenic sulphate aerosols. Santer *et al.* (1995) found that the anti-greenhouse effect made by sulphate aerosols since pre-industrial times is greater in most of the Arctic than the greenhouse effect connected with the rise of CO_2 during the same period.

(iv) The combined effect of these factors.

The above situation rapidly changed due to the pronounced warming of the Arctic between 1996 and 2000. Przybylak (2002a), using data from 46 stations (37 from the Arctic and 9 from the Subarctic), reported that the greatest warming occurred in the Canadian Arctic and in Alaska, where 5-year anomalies fluctuated most often from 1–2°C above the 1951–1990 mean. Significant warming also occurred in the Norwegian Arctic. The warming was clearly weakest in the Russian Arctic and on the western coast of Greenland. For the majority of the analysed stations, the pentad 1996–2000 has been the warmest since 1951. This is true of all stations in the Canadian Arctic and most of the stations in Pacific region (PACR). In the remaining area of the Arctic, the warmest pentad was usually that from the 1950s.

Table 10.2. Anomalies of mean seasonal and annual air temperatures from the decade 1991–2000 (in °C) in the Arctic referred to the mean 1951–1990

Area	DJF	MAM	JJA	SON	ANNUAL
Atlantic region	0.6	1.4	**−0.1**	0.5	0.6
Siberian region	**−0.1**	0.3	0.5	0.2	0.2
Pacific region	0.2	1.7	0.9	0.9	1.0
Canadian region	0.2	1.3	0.8	1.4	1.0
Baffin Bay region	**−1.2**	**−0.4**	0.1	0.3	**−0.2**
Arctic 1	0.2	1.0	0.4	0.7	0.6
Arctic 2	0.7	1.0	0.4	0.6	0.7
NH (land+ocean)	0.5	0.4	0.3	0.3	0.4

Bold numbers denote the negative 10-year anomalies of air temperature, Arctic 1 – areally averaged temperature based on data from 37 Arctic stations, Arctic 2 – areally averaged temperature for 60–90°N latitude band (after Jones et al. 1999, updated), NH (land + ocean) – areally averaged temperature for Northern Hemisphere (after Jones et al. 1999, updated)

Air temperature in the 1990s was higher than normal in a significant area of the Arctic (Table 10.2, Figures 10.16 and 10.17). Anomalies calculated for the annual air temperature for this decade reveal that the greatest warming (> 1.0°C) occurred in the northwestern and the northeastern parts of the Canadian Arctic and on the northern coast of Alaska (Figure 10.16). It was also significant in the Norwegian Arctic where air temperature anomalies reached 1.0°C. In this decade, cooling only occurred in the southern part of the Baffin Bay region (BAFR) and, most probably, in the southwestern part of the Greenland region (GRER). Areally averaged annual air temperature for the Arctic in this decade exceeded the norm by 0.6°C (Table 10.2). In the period 1951–2000, it was the warmest decade in the Arctic. Mean air temperatures for the climatic regions analysed in the present work revealed that they were warmest in the Canadian region (CANR), PACR (anomalies of

188   *The Climate of the Arctic*

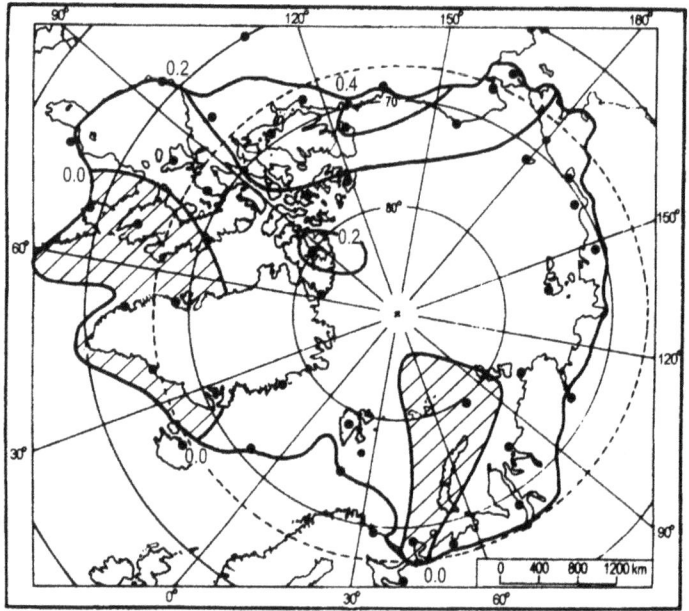

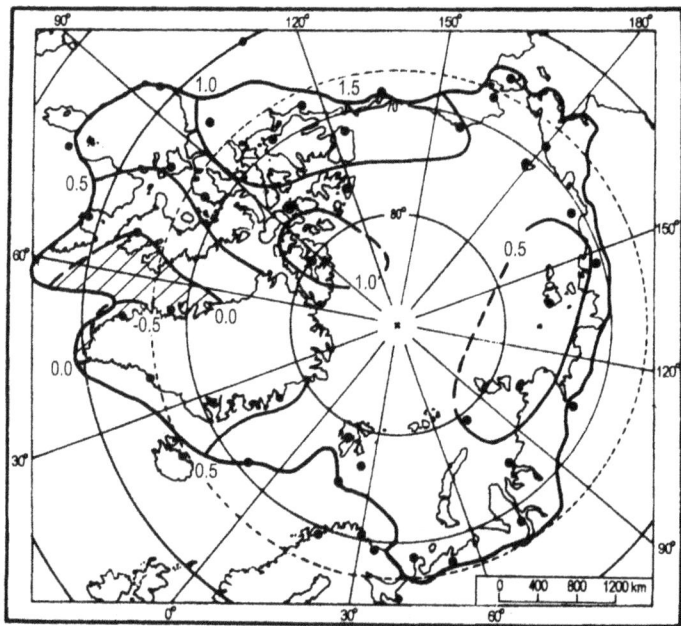

Figure 10.16. The spatial distribution of the mean annual trends in air temperature (°C/10 years, upper map) over the period 1951–2000 and the anomalies of mean annual 10-year (1991–2000) air temperature, with the 1951–1990 mean (°C, lower map) in the Arctic.
Key: negative trends (anomalies) are hatched; dashed contours over the Arctic Ocean indicate that the data are extrapolated from the coastal stations.

1.0°C), and in Atlantic region (ATLR, 0.6°C). An air temperature which was slightly lower (−0.2°C) than the norm was characteristic of BAFR. In all the analysed seasons during the 1990s, air temperature in the Arctic was higher than in the previous forty years (Table 10.2, Figure 10.17). During this decade, spring and autumn air temperature increased most (by 1.0°C and 0.7°C, respectively), while, as has been mentioned earlier, a significantly weaker warming occurred in winter (only by 0.2°C) (Table 10.2). Such a pattern of changes was observed in ATLR, PACR, and CANR; however, a slightly greater warming occurred in CANR in autumn. In comparison to the mean air temperature for the period 1951–1990, the greatest warming in the Siberian region (SIBR) and in BAFR occurred in summer (by 0.5°C) and in autumn (by 0.3°C), respectively.

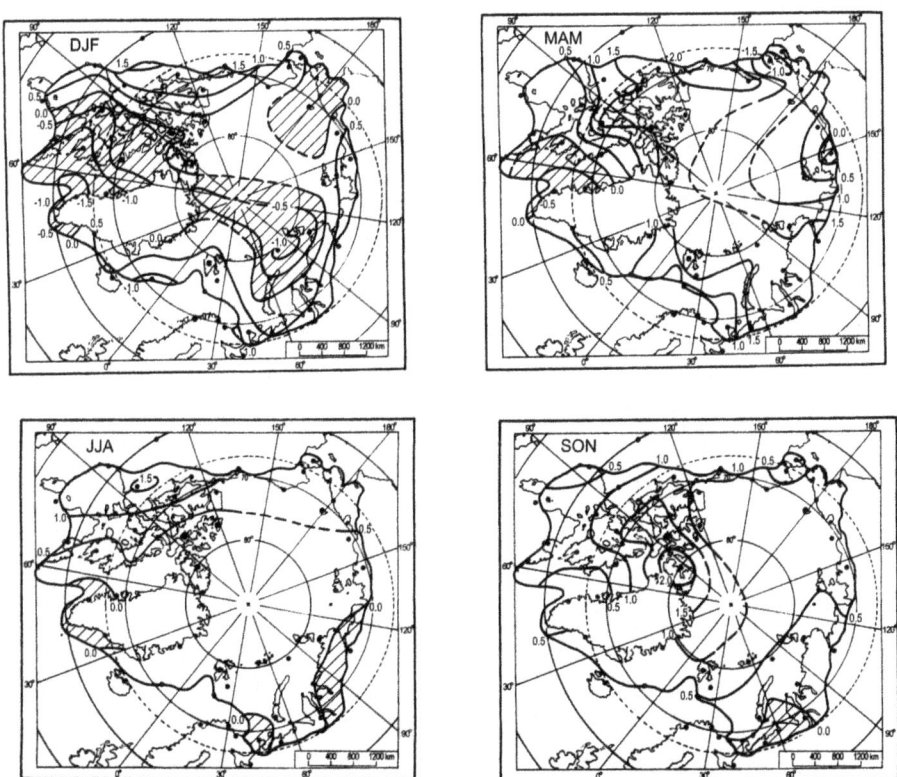

Figure 10.17. The spatial distribution of the anomalies of mean seasonal 10-year (1991–2000) air temperature, with the 1951–1990 mean (°C) in the Arctic. Key as in Figure 10.16.

An analysis of the spatial distribution of seasonal anomalies of air temperature in the decade 1991–2000 (Figure 10.17) fully confirms the conclusions obtained on the basis of areally averaged air temperature. The picture shows that the warming was most common in spring and in autumn. In comparison to the anomalies calculated for the decade 1981–1990 (see Figure 11 in Przybylak 1996a or Figure 5.5 in Przybylak 2002a), the most significant changes in the 1990s occurred in autumn. These changes were particularly significant in the northwestern part of the Canadian Arctic and in the Norwegian Arctic. In the 1980s (negative) and the 1990s (positive) anomalies of air temperature occurred in the greater part of the Arctic in all seasons. What is surprising is that, in the context of the greatest changes in winter air temperature in the Arctic that were predicted by climatic models, the area covered by negative anomalies in this season showed no signs of becoming any smaller. Similar to the 1980s, these negative anomalies are present in BAFR and in the eastern part of CANR, while in the Norwegian Arctic the area of negative anomalies of winter air temperature in the decade 1981–1990 moved further east (see Figure 5.5 in Przybylak 2002a). A new area with negative anomalies appeared in the northeastern part of the Russian Arctic (Figure 10.17). Such a spatial distribution of the anomalies of winter air temperature is, to a large degree, consistent with the distribution of air temperature anomalies that are caused by the influence of changes in atmospheric circulation. These changes may be determined by the NAO index (see Figure 12 in Przybylak 2000a). One should also notice the significance of the occurrence of major warming in summer, especially in the southwestern Canadian Arctic and in Alaska. By contrast, this warming was weak in the central Arctic (usually ≤ 0.3°C). In the 1990s, the summer cooled slightly in the western part of the continental Russian Arctic and around southern Greenland (Figure 10.17). This result differs significantly from those obtained by Chapman and Walsh (1993), and by Rigor *et al.* (2000) for the periods 1961–1990 and 1979–1997, respectively. They concluded that summer warming did not occur in the Arctic.

For all seasons except winter, and for particular years during the decade 1991–2000, changes in areally averaged air temperature in the Arctic (Arctic 1) correlate well with the changes in Northen Hemisphere air temperature (land + ocean) and with the changes in air temperature (only land stations) in the zone stretching between 60–90°N (Arctic 2) (Table 10.2). The greatest consistency of anomalies occurs in summer. As a result, during this season there was a much greater warming in Subarctic regions than in the real Arctic.

Since about the mid-1990s the rate of warming in the real Arctic became greater than the increasing rate of Northen Hemisphere air temperature (Figure 10.18). Earlier, such a situation had occurred in the 1950s, the period ending the warming phase of the Arctic which had begun in the 1920s. In the years to come, the temperature in the Arctic may reach the level of the warming that occurred in the 1930s and 1940s – the greatest warming of the 20th century.

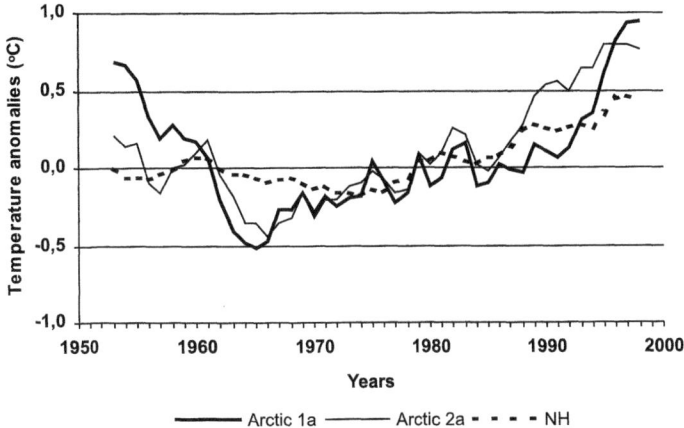

Figure 10.18. Running 5-year mean annual anomalies of air temperature in the Arctic (Arctic 1 and Arctic 2) and the Northern Hemisphere (NH) over the period 1951–2000.
Key: Arctic 1 – areally averaged air temperature based on data from 37 Arctic stations (see Table 9.1 or Figure 9.1 in Przybylak 2002a), Arctic 2 – areally averaged air temperature for 60–90°N latitude band (after Jones *et al.* 1999, updated), NH – combined land+ocean areally averaged air temperature for Northern Hemisphere (after Jones *et al.* 1999, updated).

In comparison both with the period 1951–1990 (Table 5.11, Figures 5.20–5.21 in Przybylak 2002a) and with the period 1951–1995 (Table I, and Figures 5–8 in Przybylak 2000a), the inclusion of the data from the whole of the 1990s exerted a significant influence on the values of the trends of air temperature (Figures 10.16, 10.19, 10.20, and 10.21). From 1951 to 1990, air temperature in the Arctic revealed negative trends for all seasons of the year and for annual means. These trends were statistically significant only in autumn. According to annual means, the greatest cooling of the Arctic occurred in BAFR (where trends were statistically significant), CANR, and ATLR. PACR was the only region that revealed a positive trend during this period. In the subsequent five years almost all the Arctic (except BAFR and the southwestern part of CANR) warmed slightly; however, taking this period into account did not lead to any major changes. Even though negative trends of annual air temperature still dominated in BAFR, CANR, and ATLR, their values decreased (with the exception of BAFR). The values of trends increased significantly in PACR, and thus became statistically significant (Table I in Przybylak 2000a). Areally averaged Arctic air temperature continued to reveal a negative trend (–0.04°C/10 years).

The inclusion of the 1990s in the calculations changed the trends of areally averaged air temperature for all the Arctic and for particular regions (Figures 10.19 and 10.20), along with the spatial distribution of air temperature in this area (Figures 10.16 and 10.21). In the period 1951–2000, the trend of areally averaged annual air temperature in the Arctic (Arctic 1) is already positive (0.08°C/10 years) (Figure 10.19). Positive trends also occurred in all seasons (Figure 10.20). The highest increase in air temperature was observed in spring (0.15°C/10 years), while the lowest occurred in winter and in summer (0.04°C/10 years). However, it should be emphasised that neither seasonal nor annual trends were statistically significant. These trends were significantly (usually 2–3 times) lower than in the area referred to as Arctic 2. Except for spring and autumn, these trends are also lower than those that occurred in the last 50 years in the Northern Hemisphere, which were statistically significant in all individual seasons and for the year as a whole, usually at the level of 0.001 (Table 10.2).

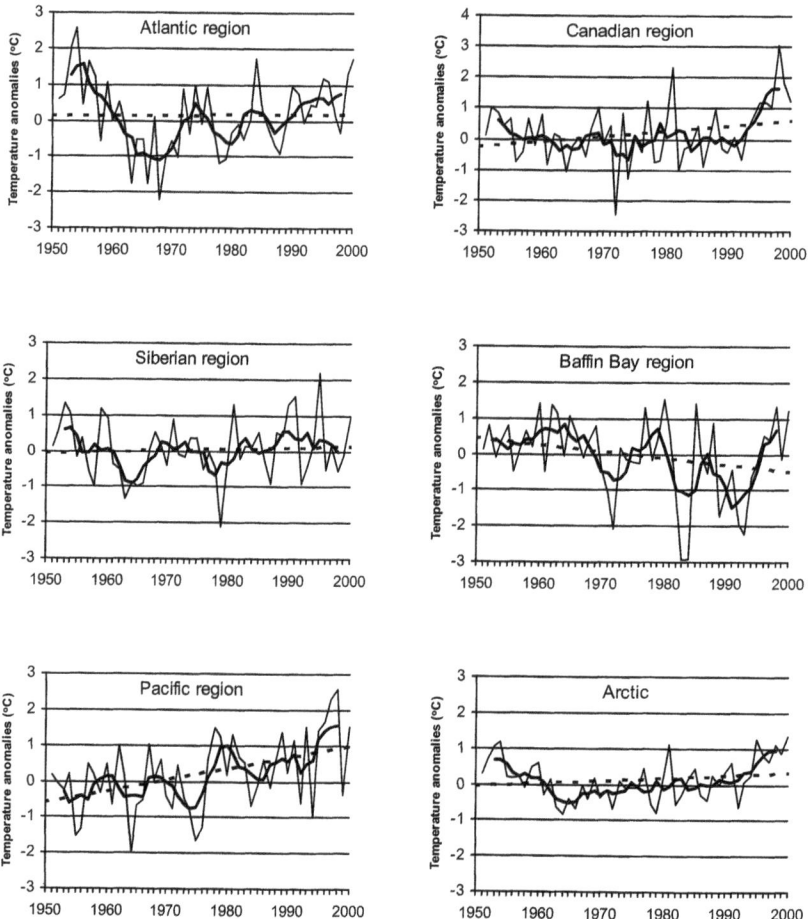

Figure 10.19. Year-to-year courses of mean annual anomalies of air temperature and their trends in the climatic regions of the Arctic and for the Arctic as a whole over the period 1951–2000 (based on data from 37 stations).
Key: solid lines – year-to-year courses, heavy solid lines – running 5-year mean, dashed lines – linear trends.

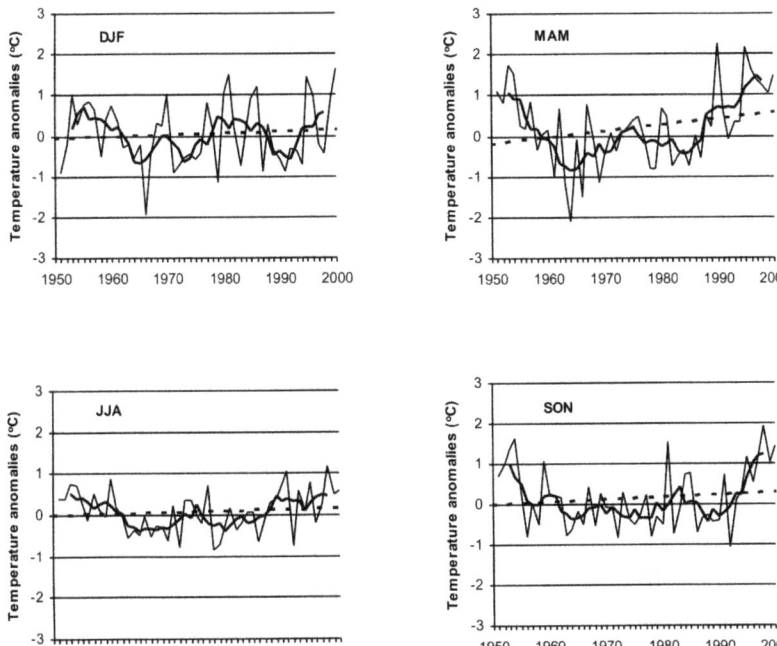

Figure 10.20. Year-to-year courses of mean seasonal anomalies of air temperature and their trends in the Arctic over the period 1951–2000 (based on data from 37 stations). Key as in Figure 10.19.

In comparison to the period 1951–1995, the greatest changes in trend values were observed for areally averaged temperatures in ATLR, CANR, and BAFR. However, the period 1996–2000 did not significantly influence the trends of air temperature in SIBR and PACR. In the period 1951–2000, the highest increase in annual air temperature occurred in PACR (0.33°C/10 years) and was statistically significant. Positive trends were also observed in CANR and SIBR, though these were not statistically significant. ATLR did not reveal changes in air temperature, and there was a cooling in BAFR. With the exception of these two regions and SIBR in autumn, mean seasonal trends of air temperature in the remaining areas are positive. However, it was only in PACR that statistically significant trends occurred (excluding autumn).

In the period 1951–2000, trends in annual air temperature in the Arctic were positive throughout the research area, except for the southeastern part of CANR, the southern part of BAFR, and the southwestern and eastern parts of ATLR. The greatest increases in air temperature occurred in the southwestern part of the Canadian Arctic and in Alaska, where a particularly high number of stations revealed statistically significant trends (see Table 9.3 in Przybylak 2002a). Apart from Eureka station, trends greater than 0.2°C/10 years did not occur outside this region.

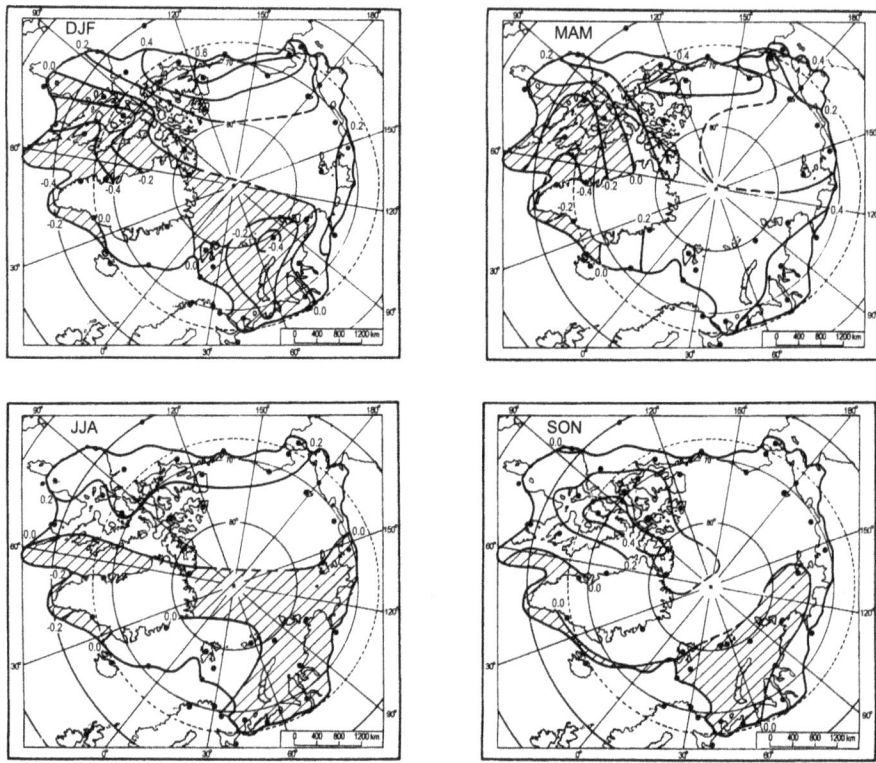

Figure 10.21. The spatial distribution of the mean seasonal trends in air temperature (°C/10 years) in the Arctic over the period 1951–2000. Key as in Figure 10.16.

An analysis of the spatial distribution of the trends of air temperature for particular seasons in the Arctic (Figure 10.21) confirmed the earlier assumption that the greatest warming occurred in spring and in autumn. It also follows from Figure 10.21 that warming was most common in the Arctic in these particular seasons. In spring, negative trends were noticed only in the southeastern part of the Canadian Arctic, in the area around Greenland, and, most probably, in the southern part of GRER. In autumn, negative trends also occurred in this region, but they were quite limited and encompassed only the areas around southern Greenland. In this season, negative trends also occurred in the southern and eastern parts of ATLR and in the western part of SIBR. In spring, the highest increase in air temperature (> 0.4°C/10 years) was observed in the southwestern part of the Canadian Arctic, in Alaska, the Chukchi Peninsula, and, primarily, in the western part of the Russian Arctic (Figure 10.21). In autumn, the greatest trends, which also exceeded 0.4°C/10 years, occurred only in the central part of CANR. Regions where positive and negative trends of air temperature occurred in summer and in winter are similar to one another, except for small areas in the Norwegian, Canadian, and Russian Arctic (Figure 10.21). Interestingly enough, the range of values of the trends of air temperature differs considerably throughout the area of the Arctic. Both negative and positive trends are greater in winter than in summer. In these two seasons, negative trends occur in southeastern part of the Canadian Arctic (they cover a larger area in winter), in BAFR, and in the western and eastern parts of ATLR. In summer, negative trends were also observed in the western part of SIBR. In both seasons, the greatest warming (> 0.2°C/10 years) occurred in the southwestern part of the Canadian Arctic and in PACR. In the latter region, the trends were statistically significant (Table 10.2).

Mean trends of seasonal and annual air temperature, calculated for 34 Arctic stations over the period 1976–2000, are usually greater than analogous trends calculated for the period 1951–2000 (see Table 9.3 in Przybylak 2002a). They are also often statistically significant. Positive trends of air temperature dominate in all seasons and in annual means. Negative trends in mean air temperature for spring and summer were observed at only two stations. As regards mean autumn and annual air temperature, such a situation occurred at four stations, while in winter a cooling was observed over a considerable area of the Arctic. The cooling occurred in BAFR, CANR (except its southwestern part), PACR (except its northeastern part), and in isolated areas in the western part of the Russian Arctic. Trends of areally averaged air temperature in this season were negative in almost all regions except ATLR. A significant decrease was observed in BAFR (–0.85°C/10 years) and in PACR (–0.38°C/10 years). In the remaining seasons, the negative trend occurred only in BAFR in spring. The majority of statistically significant trends were noticed in this season (in three regions). According to mean annual air temperature for the examined period, the greatest warming occurred in CANR (0.68°C/10 years) and in ATLR (0.55°C/10 years), while the lowest was in BAFR (0.04°C/10 years). Mean Arctic air temperature (Arctic 1) increased most in spring (0.80°C/10 years) and in autumn (0.60°C/10 years), while the lowest increase was in winter (0.11°C/10 years). Mean air temperature for all seasons (except winter) and annual air temperature are statistically significant at the level of at least 0.01. It is worth emphasising that the trends of air temperature are greater here than in whole Northern Hemisphere (NH) and in its northern part (Arctic 2). This spatial distribution of the trends of air temperature in the Northern Hemisphere has now become generally consistent with the expected changes in air temperature connected with the increasing concentration of CO_2 and other trace gases. The greatest disparity concerns winter air temperature that, according to the prognoses based on climatic models, should have warmed most. As has been mentioned earlier, winter thermal conditions in the Arctic are probably still shaped mostly by the atmospheric circulation that has been revealing a strong increase in zonal circulation (high values of the NAO and the Arctic Oscillation (AO) indices have been observed since the end of the 1980s).

10.3.3 The Influence of Atmospheric Circulation on Temperature

It is not possible to investigate the reasons for recent air temperature variations without discussing atmospheric circulation changes. It is widely known that the importance of circulation in the formation of climate is much greater here than at lower latitudes (see Alekseev *et al.* 1991, their Table 1). Alekseev *et al.* (1991) also found that the advection of warmth from lower latitudes by atmospheric and oceanic circulation provides more than half the energy annually available in the Arctic climate system. This makes such advection more important than solar irradiance flux. The share of advection is especially large in the cold season, when there is only a negligible inflow of solar irradiation. During the polar night it is equal to 100%. As mentioned in Chapter 1, atmospheric circulation provides as much as 95% of warmth advection to the Arctic, while oceanic circulation provides only 5%. Vangengeim (1952, 1961) found that changes of synoptic processes in the Arctic are about 1.5 times faster than in moderate latitudes. As a consequence, it is possible to conclude that the Arctic is significantly more sensitive and vulnerable to atmospheric circulation changes (such as those, for example, driven by the AO and the NAO phenomena) than any other area.

Przybylak (1996a, 2002a) determined the relations between atmospheric circulation and air temperature in the Arctic using daily data. He found that changes observed after 1975 in atmospheric circulation led to the Arctic cooling in the period 1976–1990 (as was mentioned earlier). For example, the intensification of zonal circulation in mid-latitudes, which is noted since 1975 (see Kożuchowski 1993; Jönsson and Bärring 1994), accounts for Arctic cooling in all seasons except spring (Table 10.3). Cooling is significant mainly in the cold half-year, but only in autumn is it present in all climatic regions. The greatest influence of atmospheric changes represented by the zonal index on temperature is noted in the Baffin Bay and Canadian regions, where statistically significant negative correlations in winter and autumn were found. On the other hand, the observed intensification of zonal circulation leads to the Northern Hemisphere warming in all seasons except summer. Statistically significant correlations were computed for winter and annual values (Table 10.3).

In the previous sub-chapters, some relations of the Arctic climate with the NAO were suggested. More recently a new term has been introduced into the literature: the Arctic Oscillation (AO) (Thompson and Wallace 1998). The NAO and the AO are dominant patterns of atmospheric circulation variability over the North Atlantic and over the Northern Hemisphere poleward of 30°N, respectively (Hurrell 1995; Hurrell and van Loon 1997; Thompson and Wallace 1998, 2000; Houghton et al. 2001; Mysak 2001). The NAO/AO patterns can be obtained as the leading empirical orthogonal functions of the sea-level pressure (SLP) fields over the domains mentioned above. It is widely known that the largest north-south air mass exchanges associated with the AO occur over the Atlantic Ocean. That is why the NAO is often regarded as the regional representative of the AO (Delworth and Dikson 2000; Mysak 2001). Moreover, Deser (2000) found that the AO time series is nearly indistinguishable from the leading structure of variability in the Atlantic sector (the NAO). The correlation coefficient of monthly SLP anomalies during November-April 1947-97 is 0.95. According to Wang and Ikeda (2000), the main difference between the AO and the NAO is that the AO operates seasonwide and correlates to the surface air temperature, also seasonwide, while the NAO (index) correlates to surface air temperature anomalies in winter (strongest), spring, and autumn, but not in summer. The AO is strongly coupled with atmospheric fluctuations at the 50-hPa level on the intraseasonal, interannual, and interdecadal time scales and therefore, as Thompson and Wallace (1998) write, "can be interpreted as the surface signature of modulations in the strength of the polar vortex aloft".

Table 10.3. Correlation coefficients between areally averaged seasonal (DJF, MAM, JJA and SON) and annual air temperature for regions, the Arctic, the Northern Hemisphere and zonal index over the period 1951-1990 (after Przybylak 1996a)

AREA	DJF	MAM	JJA	SON	ANNUAL
Atlantic region	0.11	0.13	−0.08	−0.01	0.10
Siberian region	−0.07	0.25	−0.32	−0.13	0.15
Pacific region	−0.07	0.18	−0.16	−0.06	0.21
Canadian region	−0.40*	−0.11	−0.08	−0.40*	−0.28
Baffin Bay region	−0.48**	−0.18	−0.25	−0.55***	−0.46**
Arctic	−0.31	0.09	−0.14	−0.33*	−0.05
NH (land+ocean)	−0.43**	0.31	−0.06	−0.20	−0.39*

*,**,*** – Correlation coefficients statistically significant at the levels of 0.05; 0.01 and 0.001, respectively; Arctic – areally averaged temperature based on data from 33–35 Arctic stations; NH (land+ocean) – areally averaged temperature for Northern Hemisphere (source: Jones 1994, updated)

The above patterns of atmospheric circulation significantly influence the Arctic climate system. Researchers have investigated the relationships between the NAO and the AO indices, on the one hand, and, on the other hand, factors such as surface air temperatures (e.g. Hurrell 1995; Hurrell and van Loon 1997; Thompson and Wallace 1998, 2000; Przybylak 2000a; Rigor et al. 2000; Wang

198   *The Climate of the Arctic*

and Ikeda 2000; Broccoli *et al.* 2001; Slonosky and Yiou 2001), atmospheric precipitation (Hurrell and van Loon 1997; Przybylak 2002a, b), and sea-ice area (including its extent) and sea-ice motion (Kwok and Rothrock 1999; Yi *et al.* 1999; Dickson *et al.* 2000; Hilmer and Jung 2000; Kwok 2000; Wang and Ikeda 2000; Vinje 2001). Here, the influence of these indices only on surface air temperature in the Arctic is briefly presented mainly using the results of investigations obtained by Przybylak (2000a) and Rigor *et al.* (2000).

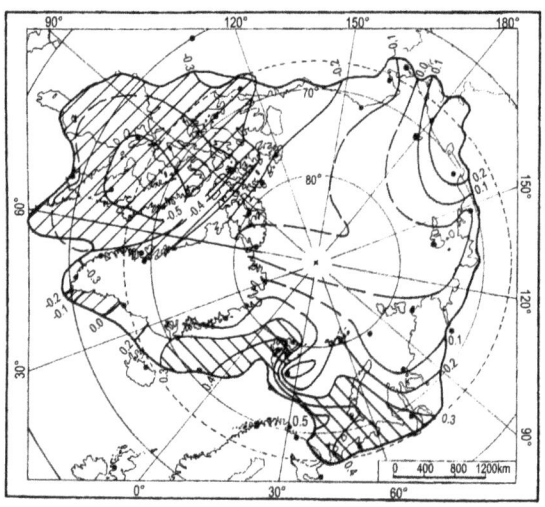

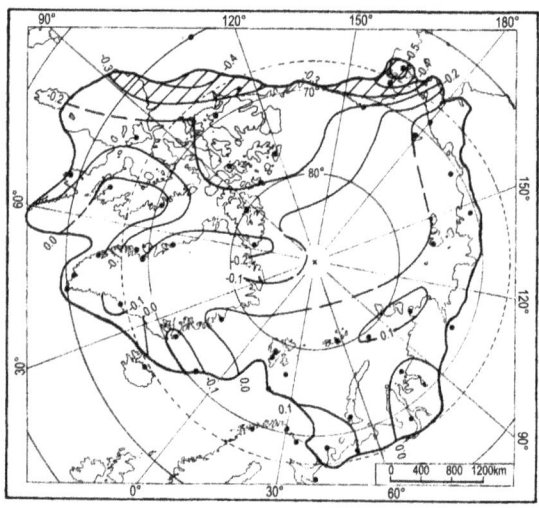

Figure 10.22. Spatial distribution of the coefficients of correlation between mean annual air temperatures in the Arctic and the NAO (upper map) and NP (lower map) indices over the period 1951–1995 (after Przybylak 2000a). Statistically significant correlations are hatched. Other key as in Figure 10.16.

Rigor *et al.* (2000) estimated the contribution of the AO to trends in winter (Dec–Feb) surface air temperature over the Arctic. They found that the AO explains more than 50% of the temperature trends in a large portion of the Arctic (60–70%) – see Figure 14d in Rigor *et al.* (2000). The relationship between these two elements is especially strong in the eastern part of the Arctic, where the AO accounts for as much as 74% of the warming during winter. Thompson *et al.* (2000) have associated ca. 30% of the recent wintertime warming of the extratropical Northern Hemisphere with the multidecadal trend in the AO. Thus, this means that the influence of the AO is greater on the climate in the Arctic than in the moderate latitudes.

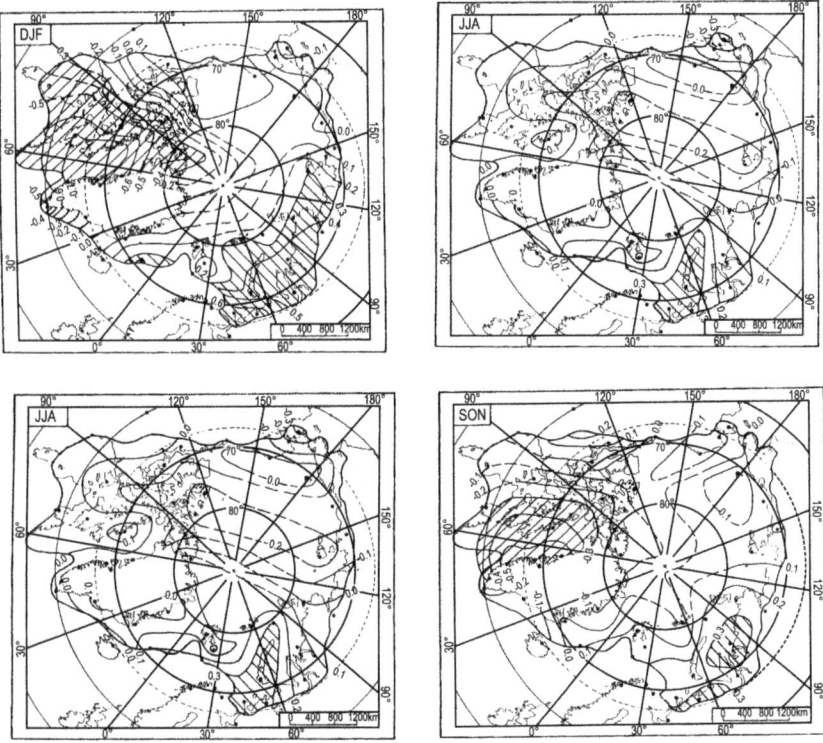

Figure 10.23. Spatial distribution of the coefficients of correlation between mean seasonal temperatures in the Arctic and the NAO index over the period 1951–1995 (after Przybylak 2000a). Statistically significant correlations are hatched. Other key as in Figure 10.16.

Przybylak (2000a) found that the relations between the NAO indices (after Hurrell 1995 and after Jones *et al.* 1997) computed as the normalised SLP differences between series taken from Lisbon/Gibraltar (Iberian Peninsula) and Stykkisholmur/Akureyri (Iceland), respectively, and the Arctic air temperature (Figures 10.22–10.24) are roughly similar to those presented earlier for the zonal index. The strongest statistically significant relations exist with mean annual air temperatures in the Baffin Bay and the Canadian Arctic (negative correlations) and in the southern part of Atlantic regions (positive correlation) (Figure 10.22). Changes in the NAO index here explain about 10–25% of the air temperature variance. As was mentioned above, the relationships between changes in atmospheric circulation in the North Atlantic and temperature, not only in the Arctic, are strongest in the winter months (Hurrell 1995, 1996; Jones *et al.* 1997). Correlation coefficients computed for particular seasons confirm this finding (Figure 10.23). In winter, as for annual values, negative correlations occur in the Baffin Bay and Canadian regions. However, they are significantly greater and, for example, in the central part of the Baffin Bay region, explain as much as 40–50% of winter air temperature variance. Statistically significant positive correlations are present mainly in the eastern parts of the Atlantic region and in the western part of the Siberian region (Figure 10.23).

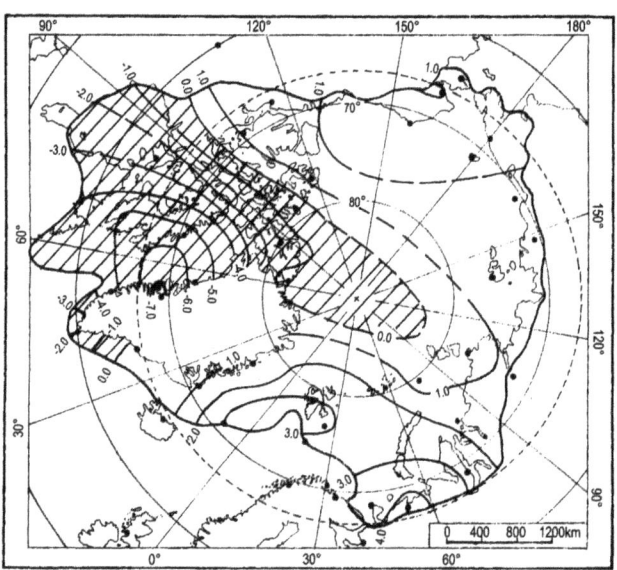

Figure 10.24. Differences of air temperature (in °C) between the most extreme 7-year run of NAO+ winters (December–February 1989–1995) and NAO– winters (December–February 1963–1969) (after Przybylak 2000a). The NAO+ and NAO– winters were taken after Dickson *et al.* (1997). Negative differences are hatched. Other key as in Figure 10.16.

Changes in winter air temperatures between 7 years with positive modes and 7 years with negative modes of the NAO index are in very high agreement with those presented above (compare Figure 10.24 with Figure 10.23). Strong cooling connected with the highest positive values of the NAO index occurred in the Baffin Bay region (4–7°C) and in the eastern part of the Canadian Arctic (1–5°C). On the other hand, warming was observed in the southern and eastern parts of the Atlantic region reaching 1–4°C. These results are in agreement with those presented by Hurrell (1995, Figure 3) and Hurrell (1996, Figure 3). In light of the results presented by Serreze et al. (1997, their Figure 6) concerning the positive minus negative NAO index difference field of cyclone events in the cold season, the causes of warming in the Barents and Kara seas regions are difficult to explain. Serreze et al. (1997) found a decrease in cyclone events. Recent results published by Dickson et al. (1997, 2000) may help to resolve this issue. Dickson et al. (1997) found that the "increasingly anomalous southerly airflow that accompanies such a change over Nordic seas is held responsible for a progressive warming in the two streams of Atlantic water that enter the Arctic Ocean across the Barents Sea shelf and along the Arctic Slope west of Spitsbergen". The temperatures of these two Atlantic-inflow streams were between 1°C and 2°C higher than normal in the late 1980s and early 1990s. Alekseev (1997) and Zhang et al. (1998) also presents similar results.

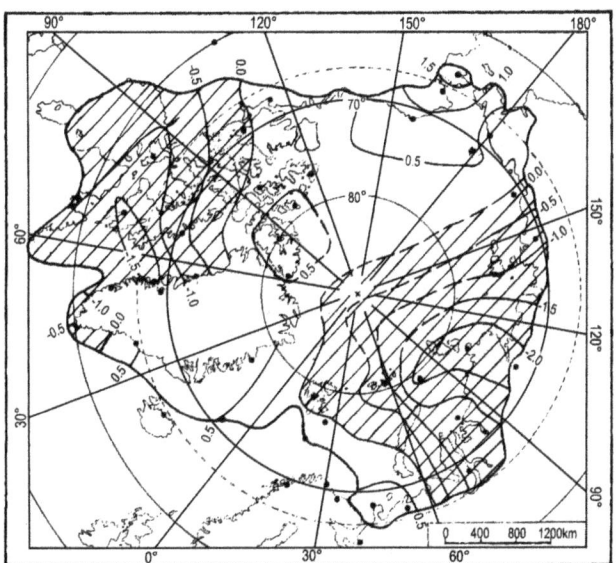

Figure 10.25. Differences of mean winter (December–February) air temperature (in °C) between 10 years with the strongest El Niño phenomena and 10 years with the strongest La Niña phenomena (after Przybylak 2000a). Negative differences are hatched. Other key as in Figure 10.16.

Przybylak (2000a) has also used the North Pacific (NP) index to check for possible Pacific influences on Arctic air temperature. This index is calculated as the area-weighted mean SLP over the region 30°N to 65°N, 160°E to 140°W (Trenberth and Hurrell 1994). The NP index signal-strength dominates that of the NAO index only in some fragments of the Pacific region and in the south-western part of the Canadian region (Figure 10.22). Only here are the correlation coefficients statistically significant. A roughly similar situation also occurs in all seasons (Przybylak 2000a). Changes in atmospheric circulation in the North Pacific have the greatest influence on Arctic air temperature in winter and the lowest in summer (as with the NAO index).

The influence of ENSO (El Niño – Southern Oscillation) on Arctic air temperature is significantly lower than that associated with circulation changes in the North Atlantic (NAO index) – compare Figures 10.24 and 10.25. During the El Niño phenomena, decreases of winter temperature may be observed in the Kara Sea region (by about 2°C), in the Baffin Bay region, and in the eastern part of the Canadian Arctic (0.5°C – 1.5°C). On the other hand, significant warming is present only in Alaska. In other seasons, the general patterns of air temperature differences are similar to that for winter but these differences are significantly less. It is clear that the influence of ENSO on Arctic air temperature is indirect and occurs mainly through changes in atmospheric circulation in the North Pacific and North Atlantic. Hurrell (1996) found a statistically significant correlation ($r = 0.51$) between the NP index and the Southern Oscillation (SO) index (the index is calculated as the normalised SLP difference between series taken from Tahiti (French Polynesia) and Darwin (Australia)) but no correlation between the NAO and SO indices based on data for the period 1935–1994. The present author has repeated Hurrell's calculations and has found significantly lower correlations between the NP and SO indices ($r = 0.19$). Similar results ($r = 0.23$ and $r = 0.11$) have been obtained for other periods: 1899–1995 and 1951–1995, respectively. In the 1899–1995 period, the correlation coefficient is statistically significant at the level of 0.05. However, after 1975 significant and more consistent changes for all these indices are observed (see Figure 2 in Hurrell 1996). Computations of correlation coefficients between the analysed indices for the periods 1956–1975 and 1976–1995 confirm this conclusion. For example, correlation coefficients between SO and NAO indices for the winter (December–March) are equal to $r = 0.00$ and $r = -0.23$, respectively. It follows that in the last two decades the ENSO could to a greater degree cause the changes in atmospheric circulation also noted in the North Atlantic.

Chapter 11

SCENARIOS OF THE ARCTIC CLIMATE IN THE 21ST CENTURY

Both observations and model studies have shown that the Arctic is a region of high climate sensitivity to increased concentration of greenhouse gases (see e.g. Houghton *et al.* 1990, 1992, 1996, 2001). Most climate model simulations suggest that a doubling of CO_2 will cause a rise in global mean surface air temperature from 1.4°C to 5.8°C (Houghton *et al.* 2001) with a two- to three-fold amplification in the Arctic. Atkinson (1994) gives the following reasons for this enhanced warming in the Arctic: 1) snowline-albedo feedback, 2) release of CO_2 and CH_4 from soil carbon and methane hydrates from sea bed sediments, 3) strong surface inversion which gives reduced vertical mixing, 4) increased absolute humidity which reduces radiative cooling, 5) low atmospheric H_2O increase which enhances the CO_2 effect and 6) Arctic haze which enhances springtime warming. The first three reasons are the most important. The final three factors play lesser, contributory roles.

How can we predict climatic change in the Arctic? Most scientists distinguish two main approaches which are used to estimate a climatic change associated with a warmer world (see e.g. Jäger and Kellogg 1983; Palutikof *et al.* 1984; Palutikof 1986; Wigley *et al.* 1986; Salinger and Pittock 1991). The first approach uses the general circulation models (GCMs) or global chemical models to construct scenarios of the future climate. The second one uses past warm periods as analogues of a future, warm, high carbon dioxide world. In the present paper the results based on these approaches are presented.

11.1 Model Simulations of the Present-day Arctic Climate

One measure of the level of confidence in results generated by GCMs is the degree to which they reproduce the current global climate. A review of the first three IPCC reports (Houghton *et al.* 1990, 1992, 1996) and other works (e.g. Gates *et al.* 1996, 1999) shows that the largest disagreement between coupled climate model simulations of present-day climate is in the Polar regions. It is worth adding, however, that in more recent models this disagreement is still observable, though it is lower (Boer *et al.* 2000; Flato *et al.* 2000; Giorgi and Francisko 2000a, b; Houghton *et al.* 2001; Lambert and Boer 2001). Lambert and Boer (2001) also found that the intermodel scatter of the simulated climate variables is the largest here and over mountains.

According to Randall et al. (1998) the large degree of disagreement among the models reflects both the weakness of our current understanding of Arctic climate dynamics and the sensitivity of the Arctic climate to different formulations of various physical processes.

Recently a number of papers have been published in which researchers have concentrated on simulating the Arctic climate with the GCMs (see e.g. Walsh and Crane 1992; Cattle and Crossley 1995; Tao et al. 1996; Walsh et al. 1998; Randall et al. 1998; Weatherly et al. 1998; Zhang and Hunke 2001) and with limited-area models (Walsh et al. 1993; Lynch et al. 1995; Dethloff et al. 1996, 2001; Rinke et al. 1997, 1999a, b, 2000; Dorn et al. 2000; Rinke and Dethloff 2000; Görgen et al. 2001).

What are the results presented in these papers? Figure 11.1 shows the annual mean fields of SLP simulated by the five most widely known atmospheric GCMs (GFDL, GISS, NCAR, OSU, and UKMO), and the "observed" field based on the NCAR sea level pressure analyses for 1952–1990.

Significant differences between modelled and observed fields may be clearly seen. The best simulation is given by the GFDL model. Pattern correlations computed by Walsh and Crane (1992) between simulated and observed fields based on data for the zone 70–90°N entirely confirm this conclusion. The highest correlations were noted for winter and spring (0.909 and 0.908, respectively) and the lowest for summer (0.568). Other models have significantly lower correlations. It is worth noting that even such marked baric centres as the Icelandic and Aleutian lows vary widely from model-to-model. An analysis conducted using a new model (the NCAR Climate System Model) confirms the existence of significant differences between the annual mean field of SLP simulated by the model and the "observed" field obtained from the ECMWF (European Centre for Medium-Range Weather Forecasts) analyses (Weatherly et al. 1998). On the other hand, as recent works show, the NAO and the AO, which significantly influence the Arctic climate in the wintertime, are simulated quite well by the majority of the coupled climate models (for details see e.g. Delworth 1996; Broccoli et al. 1998; Laurent et al. 1998; Saravanan 1998; Fyfe et al. 1999; Osborn et al. 1999; Shindell et al. 1999; Houghton et al. 2001).

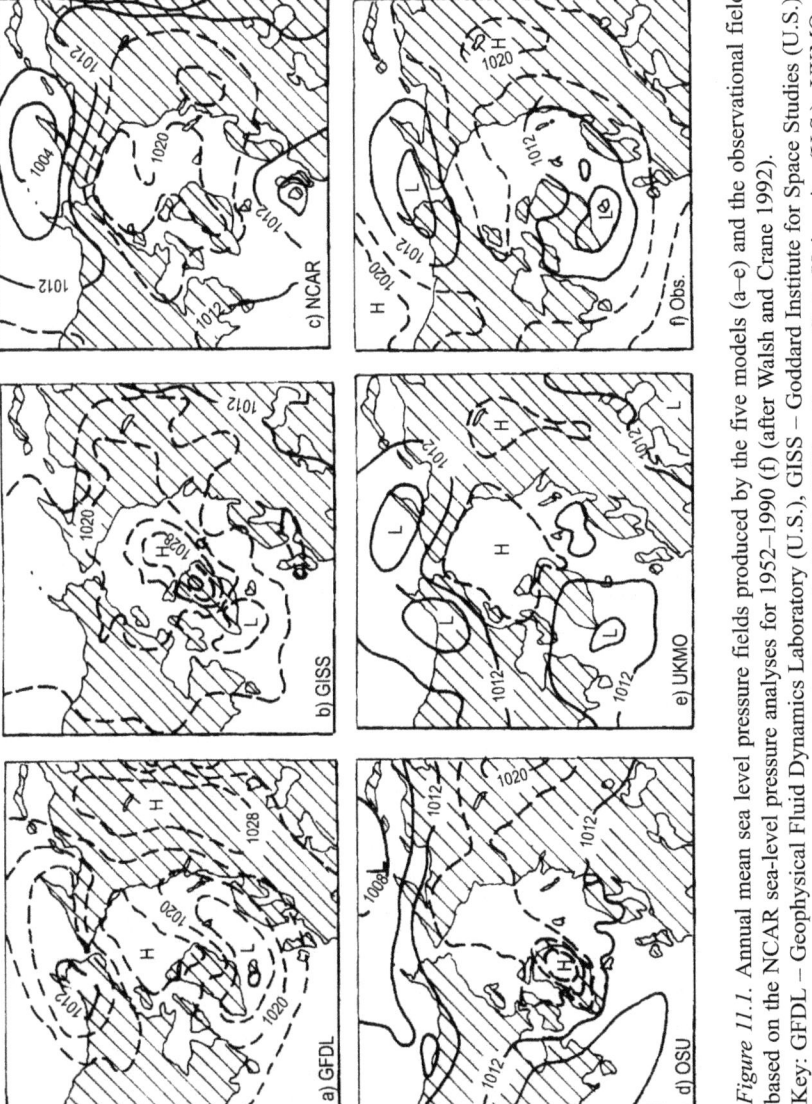

Figure 11.1. Annual mean sea level pressure fields produced by the five models (a–e) and the observational field based on the NCAR sea-level pressure analyses for 1952–1990 (f) (after Walsh and Crane 1992). Key: GFDL – Geophysical Fluid Dynamics Laboratory (U.S.), GISS – Goddard Institute for Space Studies (U.S.), NCAR – National Center for Atmospheric Research (U.S.), OSU – Oregon State University (U.S.), UKMO – United Kingdom Meteorological Office (U.K.).

Walsh and Crane (1992) also present the winter fields of surface air temperature simulated by four of the models and the "observed" climatology based on the data of Crutcher and Meserve (1970). Although all the models have temperature minima between –35°C and –45°C, their locations vary considerably from model-to-model. It is very interesting and surprising that the GFDL model, which most successfully simulates the Arctic sea level pressure, is the least successful at simulating the Arctic air temperature. The bias in this model in mean winter and autumn air temperature is about 10°C (the model is too "cold" in comparison with observations). The reason for this is the fact that the GFDL model incorporates a poor sea-ice model (Mysak, personal communication). The best results were obtained using the UKMO model. The average air temperature differences between model and observed values did not exceed 1°C for winter, spring, and autumn. Only in summer was the model "colder" by about 2°C. Tao *et al.* (1996) give the results of Arctic air temperature for the 10-year period 1979–1988 simulated by 19 numerical models participating in the AMIP (the Atmospheric Model Intercomparison Project). For more details of this project see, for example, Gates (1992). In winter, summer, and autumn, the majority of models give the areal mean air temperature in the Arctic Ocean lower than observations (see Figure 11.2). On the other hand, in spring the models' areal means show the warm bias, with the exception of four of them. Model-to-model differences of simulated areal mean air temperature are very high (in winter and spring up to about 15°C, in autumn more than 10°C and in summer about 6°C).

Walsh *et al.* (1998), using 24 climate models participating in the AMIP, compared model simulations of precipitation and evaporation with observational estimates of these elements. Figure 11.3 presents the decadal (1979–1988) annual mean precipitation and annual mean values of precipitation minus evaporation from the AMIP models, as well as the corresponding observational estimates after Bryazgin (1976a), Legates and Willmott (1990), Jaeger (1983) and Vowinckel and Orvig (1970). From Figure 11.3a, it can be seen that almost all models simulate excessive precipitation. Aside from the "outlier" model of SUNYA, the wettest models are CCC, UIUC, MPI and NCAR, while the JMA and UGAMP models simulate the least precipitation. Significantly poorer simulations are evident both for shorter periods and smaller regions. For example, the monthly precipitation simulated for the Arctic Ocean differs by a factor of 2 among the various AMIP models (see Figure 3 in Walsh *et al.* 1998). Roughly similar results were obtained for the precipitation minus evaporation values (Figure 11.3b).

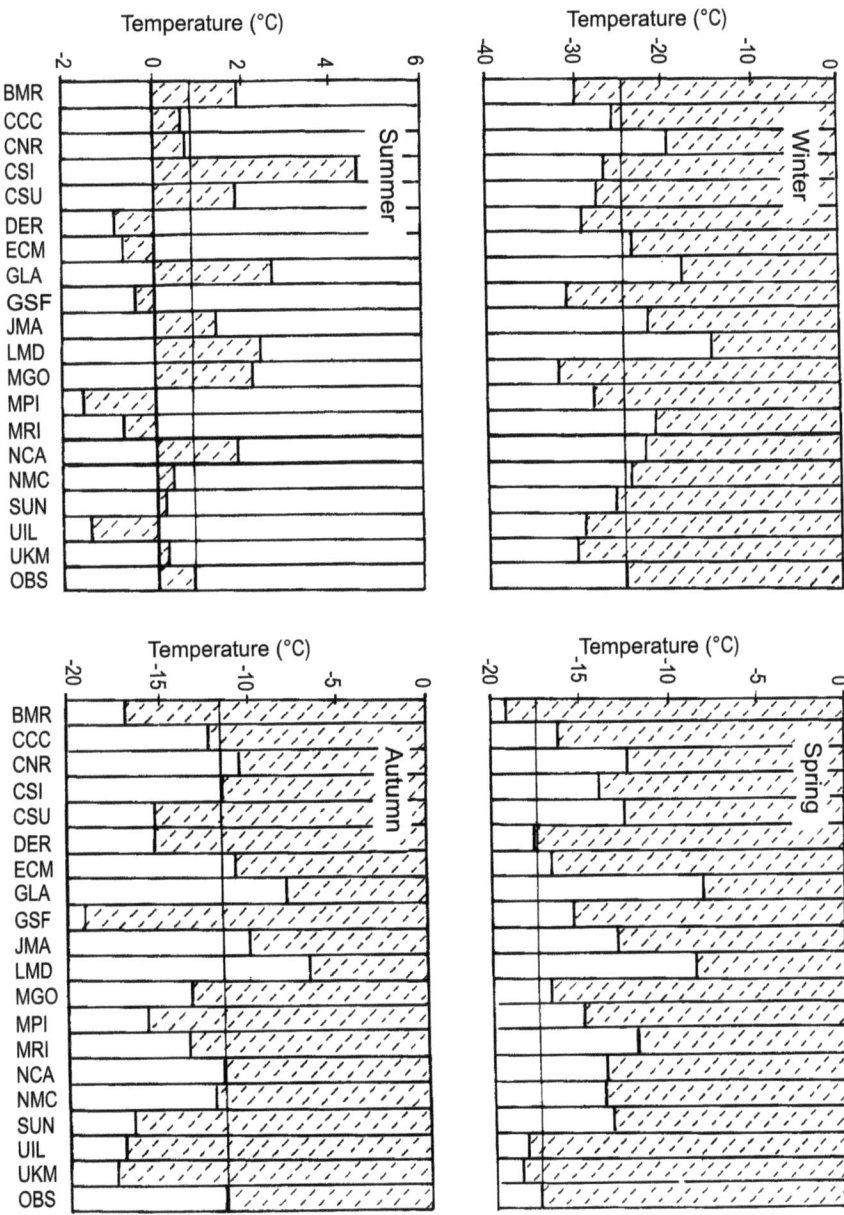

Figure 11.2. Seasonal mean air temperatures (°C) for the Arctic Ocean domain as computed from observed data (OBS, after Crutcher and Meserve 1970) and the 19 AMIP models. Temperatures are shown for winter (DJF), spring (MAM), summer (JJA), and autumn (SON) (after Tao et al. 1996).
Key: AMIP models: BMRC (Australia), CCC (Canada), CNRM (France), CSIRO (Australia), CSU (U.S.), DERF/GFDL (U.S.), ECMWF (Europe), GLA (U.S.), GSFC (U.S.), JMA (Japan), LMD (France), MGO (Russia), MPI (Germany), MRI (Japan), NCAR (U.S.), NMC (U.S.) SUNYA (U.S.) UIL (U.S.), UKMO (U.K.).

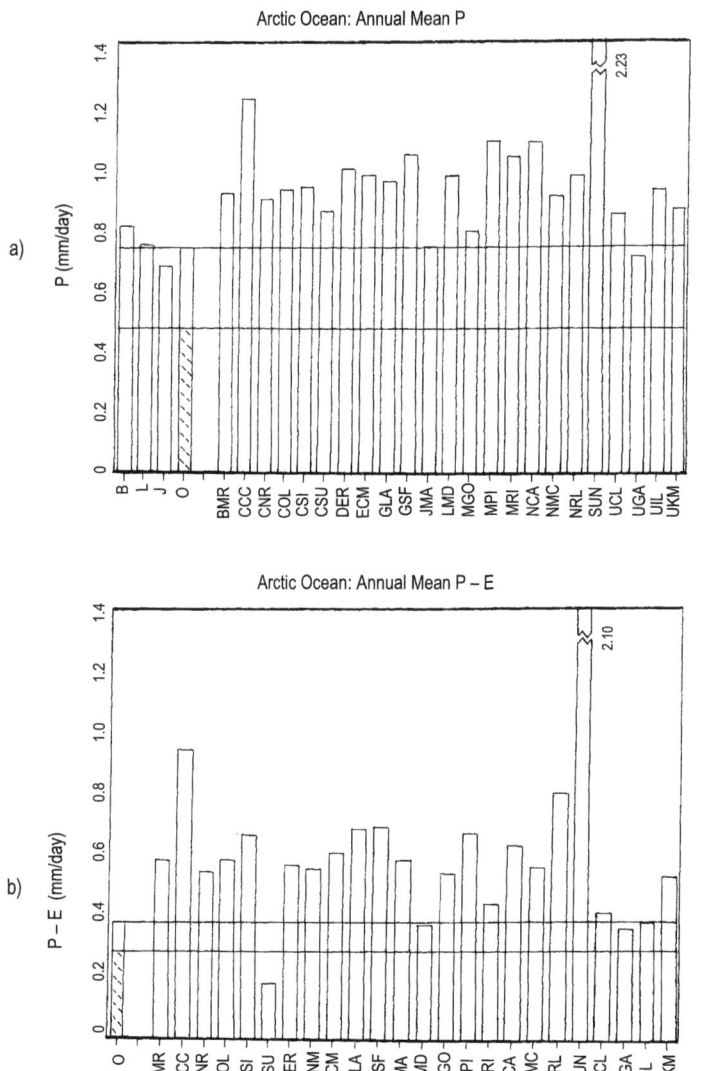

Figure 11.3. Annual mean rates of (a) precipitation, P, and (b) precipitation minus evaporation, P – E, for the Arctic Ocean as evaluated from observational estimates (bars at left) and from AMIP models (after Walsh *et al.* 1998).
Observational sources are: B = Bryazgin (1976a), L = Legates and Willmott (1990), J = Jaeger (1983), and O = Vowinckel and Orvig (1970) (hatched bar is for domain excluding Barents-Norwegian Seas). Other key as in Figure 11.2 plus COLA (U.S.), DNM (Russia), GFDL (U.S.), NRL (U.S.), UCLA (U.S.), UGAMP (U.K.), UIUC (U.S.).

The simulation of the total cloudiness varies tremendously from model-to-model (Figure 11.4). During summer, for example, the simulated cloudiness over the Arctic Ocean varies from 30% to 98%, while its value, according to observations, is equal to about 82–84%. An intriguing issue is the fact that majority of the models do not show the summertime increase of cloudiness. Moreover, some of them simulate even the lower cloudiness in this season (e.g. CSI, GLA, LMD, and SUN).

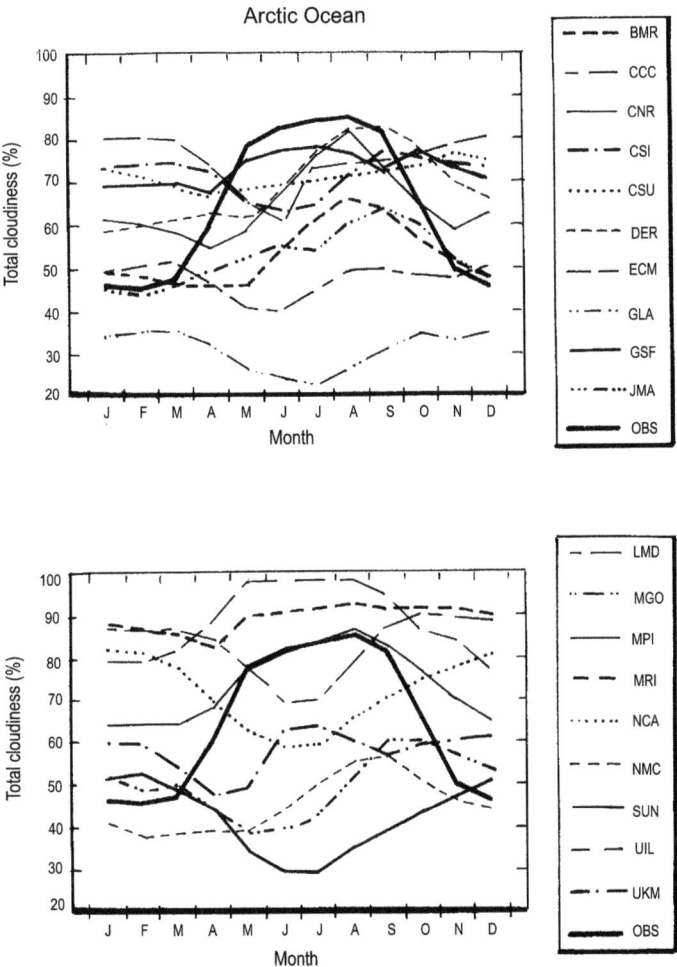

Figure 11.4. Annual cycle of monthly mean total cloudiness (%) over the Arctic Ocean as simulated by the 19 AMIP models. Heavy lines show the annual cycle for 75–85°N derived from observations made at drifting ice stations (Vowinckel and Orvig 1970, p. 150) (after Tao *et al.* 1996). Key as in Figure 11.2.

The above review reveals that the confidence of the GCMs in simulating the present-day climate on a regional scale (here for Arctic climate) is as yet rather unsatisfactory. According to Lynch *et al.* (1995) the model biases can be attributed mainly to inadequate topographical representation due to low horizontal resolution and the inadequate representation of cloud and sea ice distribution. Moreover, they state that GCMs appear to be inadequate for Arctic climate simulation and prediction. Therefore, to overcome this problem, they propose the use of limited-area models (so-called Regional Climate Models (RCMs)). The development of RCMs was initiated by Dickinson et al. (1989) who nested the NCAR regional model MM4 with a resolution of 60 km for the western United States in a GCM. Walsh et al. (1993) and Lynch et al. (1995) presented the first RCM (called the Arctic Region Climate System Model (ARCSyM)) for the Western Arctic. This model is based on the NCAR RCM and is a significant step in the development of a fully coupled regional Arctic model of the atmosphere-ice-ocean system. Furthermore, a team of German and Danish scientists have recently applied the RCM (called HIRHAM) to the whole Arctic (Dethloff et al. 1996; Rinke et al. 1997).

The above models, among others, were used to simulate monthly (January and July) fields of different meteorological elements. The results obtained are better than in the case of GCMs, but are still not satisfactory. For example, in winter the simulated air temperature in some parts of the Arctic is higher than observations by up to 10°C (Western Arctic, see Lynch et al. 1995) and 12°C (central Arctic, see Dethloff et al. 1996). In summer the differences between model and observations are much smaller than in winter, with the largest deviations being up to 4°C over the centre of the Arctic Ocean (Dethloff et al. 1996). Here one should add that Dethloff et al.'s (1996) model simulations are compared with the ECMWF analyses, which are based on the ECMWF model. The ECMWF model shows a systematic air temperature bias to be excessively low at the surface over sea ice during winter. Similar results for summer were also obtained by Lynch et al. (1995), who found a cold bias (i.e. the model is "colder") of 1–2°C in the mountainous regions and a warm bias of 3–5°C over the tundra in the Western Arctic. Significant differences in the simulated precipitation and cloud fields with the observation data were also noticed.

Rinke et al. (1997) found that model outputs greatly overestimate the incoming short-wave flux and significant differences in the net radiation over the Arctic, especially in July (up to 100 W/m^2 and ± 70 W/m^2, respectively). It is worth adding, however, that the model experiments described in Lynch et al. (1995), Dethloff et al. (1996) and Rinke et al. (1997) have shown the potential for realistic simulations of climate processes of the Arctic troposphere and lower stratosphere meteorological fields in limited-area climate models.

The results presented here come from the first version of RCMs, which still contain many shortcomings. The improvements in the physical parameterisation packages of radiation and in the description of sea-ice thickness and sea-ice fraction introduced recently into the new version of RCMs, significantly reduce the model biases (Rinke et al. 1999a, 2000; Rinke and Dethloff

2000; Dethloff *et al.* 2001). For example, in the case of surface air temperature, the differences between HIRHAM model simulations and "observed" values (gridded 2 m air temperature for the Polar Exchange at the Sea Surface (POLES), climatology of the Legates and Willmott (1990) or ECMWF analyses) generally do not exceed 5°C in winter. In June these differences are clearly lower, although locally they also reach 5°C (see Rinke *et al.* 2000, their Figure 11). The worst results have been obtained by applying the ARCSyM model. Rinke *et al.* (2000), comparing the simulated near-surface air temperature in January, found that ARCSyM temperatures are higher than the HIRHAM simulation, to the order of 5–15°C. On the other hand, both models gave quite similar results for June.

In the case of atmospheric precipitation, the situation is opposite. Both models better simulate precipitation for January than they do for June (see Rinke *et al.* 2000). Moreover, in June the differences in monthly totals between HIRHAM and ARCSyM simulations are much larger. Monthly mean surface energy balance components (sensible heat flux, latent heat flux, and net radiation) are mostly underestimated in January and June in both models (see Figures 6 and 8 in Rinke *et al.* 2000) in comparison with the NCEP reanalyses. The differences in the case of net radiation do not exceed 30 W/m² in January and 40 W/m² in June and are significantly smaller than in previous simulations (Rinke *et al.* 1997). Sensible heat flux differences are smaller than 10 W/m² in both months. A similar size of the differences is noted for latent heat fluxes in January. On the other hand, in June the differences are larger, especially in case of the HIRHAM model, reaching more than 30 W/m². Taking into account fact that the NCEP reanalyses of the energy balance components for the high latitudes are overestimated (see Gupta *et al.* 1997), the real biases are probably lower.

Thus one must conclude that the RCMs, especially their new versions, more reastically simulate the present-day Arctic climate than do the GCMs and hence they should constitute a reliable tool for climate change studies in the Arctic. Up to now there have been no such studies, but probably in the near future these kinds of models will be used to simulate the Arctic climate change with the doubling of CO_2. That is why, in the next sub-chapter, scenarios of the Arctic climate for this century are presented based only on the GCM outputs and the analogue method.

11.2 Scenarios of the Arctic Climate in the 21st Century

11.2.1 The GCM Method

Two types of GCMs can be distinguished: equilibrium and transient models. The first group simulate changes of climate for CO_2 doubling occurring rapidly, while the second group compute the same for a gradually increasing CO_2 (most often of a 1% compounded increase per year). The second approach is more realistic and resembles the contemporary changes of CO_2 concentrations.

The first equilibrium experiments conducted using atmospheric general circulation models simulated a very large increase of temperature in the Arctic (up to 10–16°C in winter), especially in the northern parts of the Atlantic and Pacific Oceans (see Washington and Meehl 1984, or Meehl and Washington 1990). In the rest of the Arctic the projected rise of temperature varies between 4°C and 6°C in winter and about 2°C in summer. This geographical pattern of temperature changes is in disagreement with recent observational changes of temperature in the Arctic (see e.g. Chapman and Walsh 1993, their Figure 1, or Przybylak 1996a).

More sophisticated high-resolution equilibrium simulations of the 2 x CO_2 climate, which use GCMs coupled with mixed layer oceans, give significantly more reliable results for some parts of the world, but probably not for the Arctic (see Figure 5.4 in Houghton et al. 1990). Results from the three models presented in this Figure (CCC, GHHI, and UKHI) show that the warming in the Arctic is highest in late autumn and winter. The projected warming in winter in the case of the first two models exceeds more than 10°C. However, the regional and local differences between winter surface air temperature responses simulated for the doubled-CO_2 climate by these models show a 10°C range. A review of current model results (see sub-chapter 11.1) shows that the largest disagreement between coupled climate model simulations of present-day climate is still in the Polar regions.

Limited space prevents me from providing a more detailed presentation of the changes of other meteorological elements simulated by the equilibrium GCMs for the 2 x CO_2 case. Generally, an increase in the cloudiness and precipitation and a decrease in the sea ice extent and its thinning in the Arctic is predicted. For more information see, for example, Washington and Meehl (1984); Schlesinger and Mitchell (1987); Houghton et al. (1990); Meehl and Washington (1990); Przybylak (1993) or Kożuchowski and Przybylak (1995).

The main weakness of these kinds of models (except for the leap rise in CO_2 concentration) is the fact that in simulations they neglect the thermal inertia of the deep ocean, and therefore give an exaggerated response, particularly in regions of deep mixing. A comparison of transient and equilibrium responses of surface air temperature with the doubling of CO_2 (Figure 11.5) using the same model (GFDL) entirely confirms the above conclusion. From this figure, it can be seen that the transient response of the surface air temperature is particularly low over the northern North Atlantic and over the circumpolar ocean of the Southern Hemisphere, where the deep vertical mixing of water is the greatest and, as a consequence, the effective oceanic thermal inertia is very large. According to Manabe et al. (1991), a relatively small surface warming in the northern North Atlantic Ocean is caused also by the reduction in the near-surface advection of warm water from South (the excess of precipitation over evaporation in high latitudes leads to a weakening of the thermohaline circulation). In the Arctic, the transient model experiment shows about 1.4–2.0 times lower warming than in the case of the equilibrium response (Figure 11.5c). According to the GFDL model results (Figure 11.5a), the annual air temperature in the Arctic should increase by 4–5°C for the period of CO_2 doubling. In winter and summer (not shown) the warming should be equal to 5–8°C and 0–2°C, respectively.

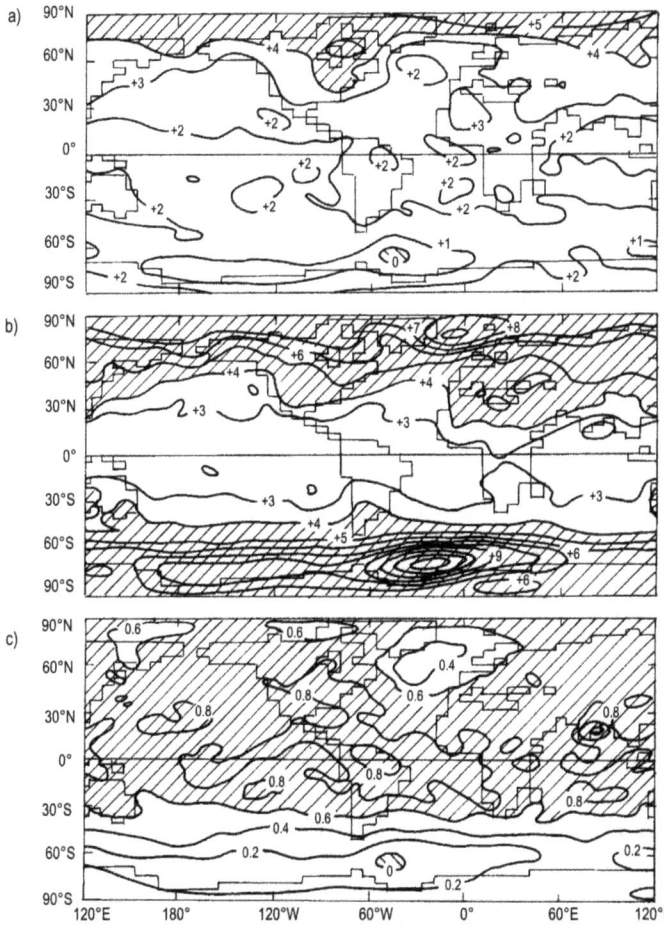

Figure 11.5. (a) The transient response of the surface air temperature of the coupled ocean-atmosphere model to the 1%/year increase of atmospheric carbon dioxide. The response (°C) is the difference between the 20-year (60th to 80th year) mean surface air temperature (1%/year increase of CO_2) and 100-year mean temperature (CO_2 constant). (b) The equilibrium response of surface air temperature to the doubling of atmospheric carbon dioxide. (c) The ratio of the transient to equilibrium responses (after Manabe *et al.* 1991).

More recently, Cattle and Crossley (1995) published results of the simulation of the Arctic climate change for the doubling of CO_2 using the UKMO model. They found that maximum changes of air temperature in winter (more than 10°C) are associated with the marginal ice zone in the Atlantic sector and the regions of the shelf seas (Figure 11.6). The warming of Greenland lies between 2–5°C. In summer, the change of air temperature over the Arctic is small and mainly oscillates between 0–2°C (see Figure 11.6). The introduction of a simple parameterisation of the effects of sulphate aerosols significantly reduces the magnitude of the warming, but changes the overall pattern very little (Figure 11.7). In winter, the predicted warming varies between 2–5°C. The lowest increase of air temperature should occur in the Atlantic sector of the Arctic. In summer, most Arctic areas show very small warming that oscillates between 0–1°C. The greatest warming should occur in Greenland and Alaska (1–4°C). Regionally, however, even the cooling may occur (see Figure 11.7).

Mean annual air temperature changes in the Arctic, according to the model developed by Mitchell *et al.* (1995), vary from 4°C to 6°C. For temperature simulation, the concentration of aerosols from 1795 to 2030–2050, based on the most probable scenario IS92a proposed by Houghton *et al.* (1992), was used. A recent IPCC report (Houghton *et al.* 2001) also presents annual patterns of air temperature change, but for the period 2071–2100 relative to the period 1961–1990 using a newly introduced set of scenarios (SRES scenarios, for details see Houghton *et al.* 2001). For two scenarios (SRES A2 and B2) the multi-model ensemble projects the warming ranges from 6°C to 10°C and from 5°C to 8°C, respectively for the greater part of the Arctic (see Figure 9.10d and e in Houghton *et al.* 2001). The greater warming, in comparison to the model of Mitchell *et al.* (1995), is caused by the fact that the future sulphur dioxide emissions for the six SRES scenarios are much lower, compared to the IS92 scenarios (see Figure 17 in Houghton *et al.* 2001).

Doubled-CO_2 climate simulation of the troposphere in the Arctic shows a warming between 1–1.5°C. On the other hand, the cooling should occur throughout almost the whole stratosphere. In the lower stratosphere the temperature will be nearly the same as today. In the middle stratosphere the predicted decrease of temperature oscillates between 2–3°C.

Most transient GCMs simulate the increase of precipitation with the doubling of CO_2 (see e.g. Manabe *et al.* 1992; Cattle and Crossley 1995; or Houghton *et al.* 1996, 2001). However, a careful examination of geographical patterns of precipitation changes shows significant differences between model predictions. Here, I present the results published by Cattle and Crossley (1995). According to their model, winter precipitation should increase slightly over the central Arctic basin with higher local increases over the surrounding landmasses (see Figure 11.8). A general decrease of precipitation is shown over

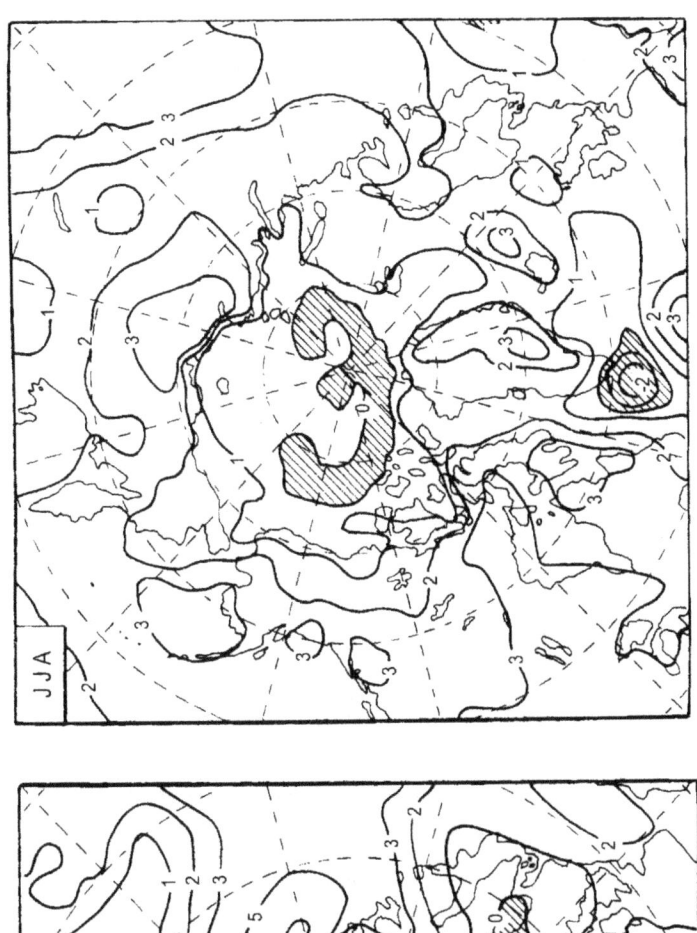
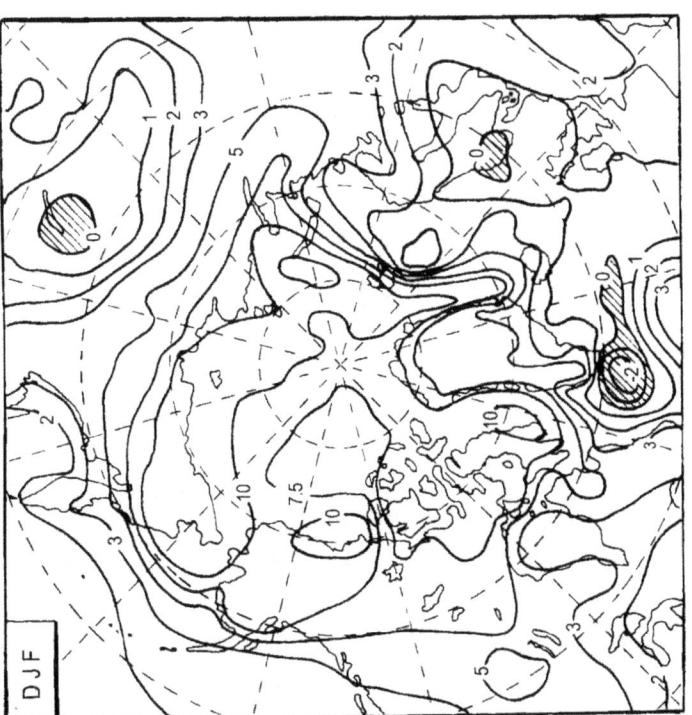

Figure 11.6. Air temperature change (in °C) over the Arctic for the decade of doubling of carbon dioxide from a run of the Hadley Centre model with transiently increasing greenhouse gases: (a) winter (DJF) and (b) summer (JJA) (after Cattle and Crossley 1995).

the region of the Greenland-Iceland-Norwegian Sea, south-eastern Greenland, Iceland, Spitsbergen and the northern parts of European and the Russian Arctic west of 50°E. In summer (Figure 11.8), a tendency towards reduced precipitation should occur in most parts of the Arctic Ocean, over some fragments of the Greenland and Norwegian seas, over almost whole Barents Sea including its surrounding islands (Novaya Zemlya, Zemlya Frantsa-Josifa and Spitsbergen), as well as over the central part of the Russian Arctic. With the inclusion of aerosol forcing there is seen an increase in precipitation in the areas of reduced precipitation in winter and more generally in summer (Cattle and Crossley 1995).

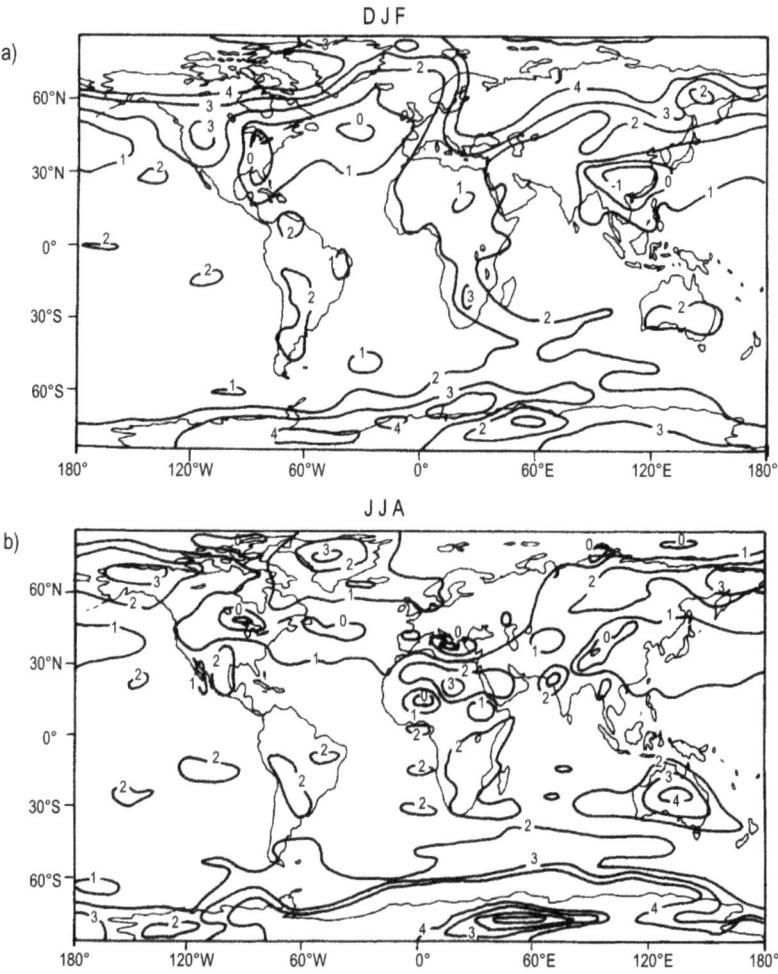

Figure 11.7. Seasonal change in surface air temperature from 1880–1889 to 2040–2049 in simulations with aerosol effects included. (a) winter (DJF) and (b) summer (JJA) (after Kattenberg *et al.* 1996).

Reviewing the literature presenting the transient experiment results using GCMs, I did not find any information about cloudiness changes in the Arctic with the doubling of CO_2. What is the current state of knowledge about other meteorological elements and components of Arctic climate system? Wild *et al.* (1997) computed changes in the zonal mean 10-m wind speed for the whole globe with the doubling of CO_2. In the Arctic, the average wind speed in summer should be higher by about 0.2–0.5 m/s. On the other hand, in winter a reduction of 10-m wind speed up to 0.5 m/s around the Pole and up to 0.4 m/s in the latitude belt 60–75°N should be observable. Winter cyclone frequencies simulated for the Arctic using the CCC GCM (equilibrium model) correspond very well with these results (Lambert 1995). Comparison of Figures 2 and 3 presented in his paper shows that the total number of cyclones decreases in the Arctic in a warmer world. However, one should add that this tendency is limited mainly to weak lows, because intense cyclones show an increased frequency.

Manabe *et al.* (1991) present surface heat balance components averaged for the "doubled" CO_2 case simulated by the transient GFDL GCM (Figure 11.9). It can be seen that the net radiation is positive in the whole Arctic (ocean and continental parts). The sensible heat is negative everywhere in the Arctic, except the southern oceanic parts, while the latent heat is negative mainly in the latitude band 60–80°N. The decreases of latent heat fluxes are significantly higher over continental parts of the Arctic (up to 2–3 W/m²) (Figure 11.9b).

A very important component of the Arctic climate system is sea ice. The UKMO GCM predicts that in a warmer world (a "doubled" CO_2 case) the sea ice thickness should be reduced by over 1m in both summer and winter, with maximum changes occurring in area covered by the thickest ice in the control run (Figure 11.10). Similar results have also been found by Manabe *et al.* (1992) and Ramsden and Fleming (1995) using the GFDL GCM and coupled ice-ocean Arctic ocean model forced using an output from the CCC model of the atmosphere, respectively. Ramsden and Fleming (1995) concluded that the Arctic ice field appears to act as a regulator of climate change, rather than as an accelerator.

Ramsden and Fleming's (1995) model predicts an increase in the ocean surface temperature with the doubling of CO_2 by about a degree, with the largest increase expected in summer, reflecting the amount of open water. Their model also foresees a slight decrease in surface salinity in the Arctic Ocean.

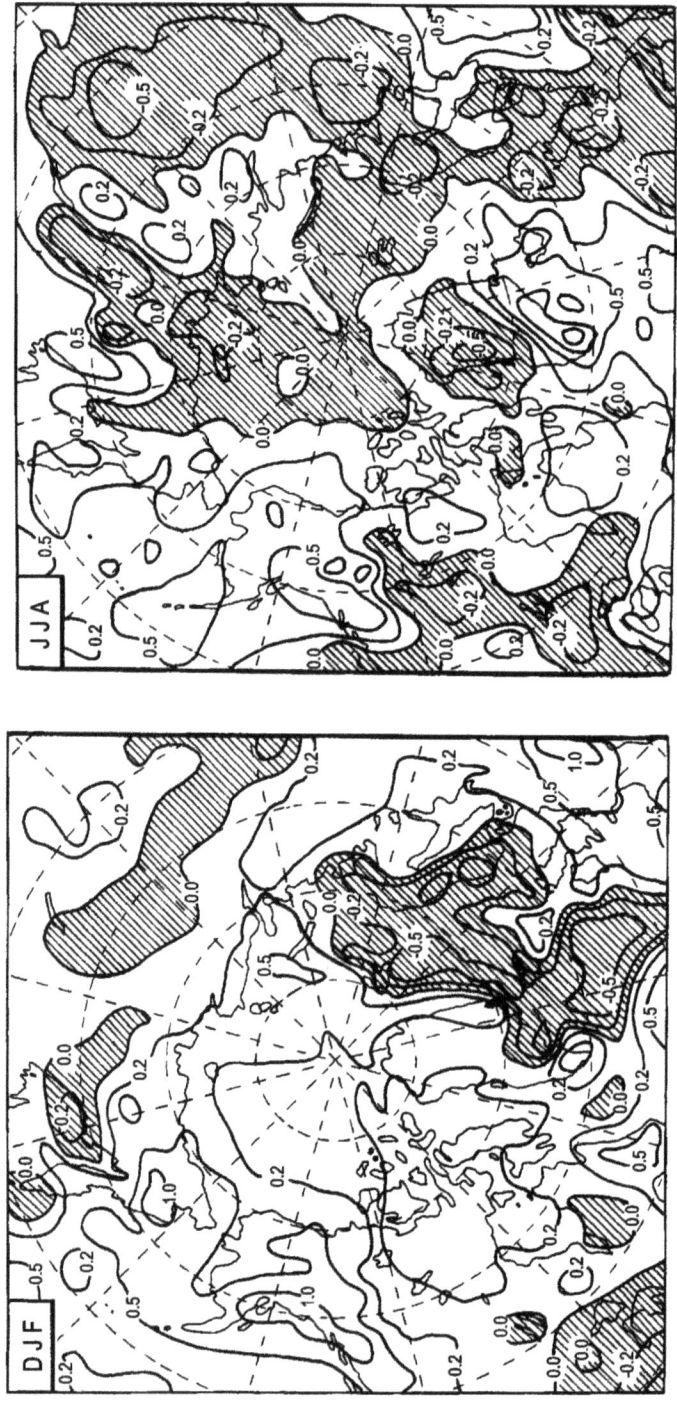

Figure 11.8. Precipitation change (mm/day) over the Arctic for the decade of doubling of carbon dioxide from a run of the Hadley Centre model with transiently increasing greenhouse gases: (a) winter (DJF) and (b) summer (JJA) (after Cattle and Crossley 1995).

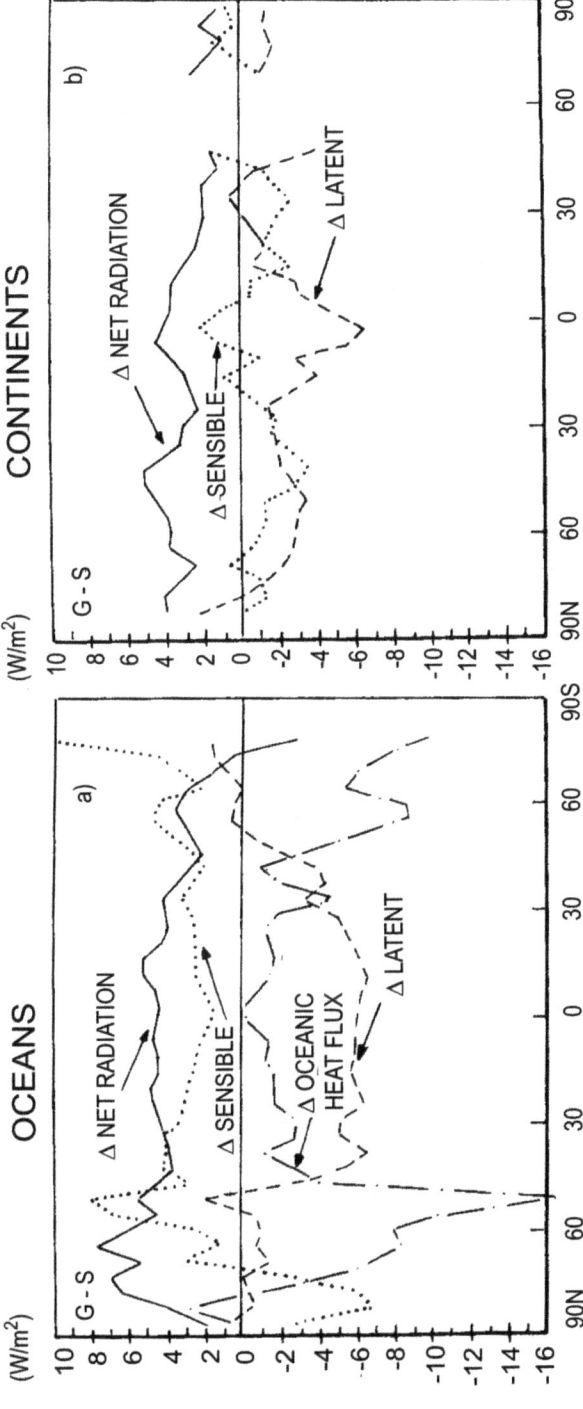

Figure 11.9. The latitudinal profile of zonal-mean surface heat balance components over oceans (a) and continents (b) between the G (1%/year increase of CO_2, averaged over the 60th to 80th year period) and S (CO_2 constant) integrations (after Manabe et al. 1991).

220 *The Climate of the Arctic*

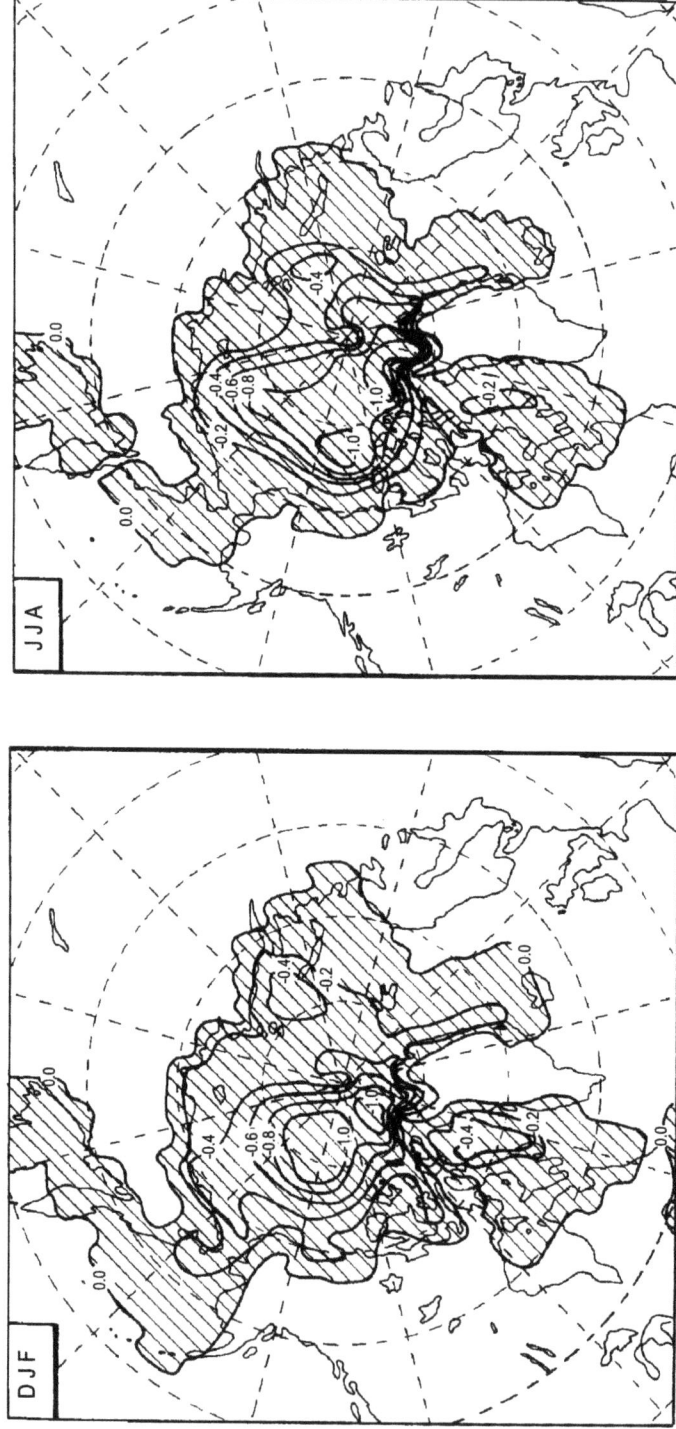

Figure 11.10. Sea ice thickness change (in m) over the Arctic for the decade of doubling of carbon dioxide from a run of the Hadley Centre model with transiently increasing greenhouse gases: (a) winter (DJF) and (b) summer (JJA) (after Cattle and Crossley 1995).

11.2.2 The Analogue Method

The analogue studies use as analogues of a high-CO_2 world warm periods taken from either paleoclimatological reconstruction of, for example, the Medieval Warm Epoch, the mid-Holocene, and the last interglacial (Eemian) or instrumental period. For the Arctic, our knowledge concerning the climates of the above periods (other than the instrumental) is not sufficient. Therefore, only the second approach is acceptable. The major advantage of scenarios based on the instrumental records, according to Palutikof (1986), is that it allows the construction of very detailed regional and seasonal scenarios constrained only by the density of the observing network. Scenarios of the Arctic climate (air temperature and precipitation) using this method have only been constructed by Przybylak (1995, 1996a, 2002a). One can also find some information about the climate change in the Arctic in a high-CO_2 world in papers analysing greater areas (Jäger and Kellogg (1983) for the southern fragment of the Arctic and Palutikof (1986) for the Canadian Arctic and for the southern part of Greenland). The major disadvantage of this method used for the Arctic and for other parts of the world is the fact that a warm-cold difference in the instrumental record for which a dense and accurate recording network is available is much smaller than even the most conservative estimates of CO_2-induced warming. Therefore, the scenarios presented below can only be treated as indicative of conditions in the early years of CO_2-induced warming, i.e. for the early decades of the 21st century.

According to scenarios presented by Przybylak (1995, 1996a, 2002a) in a warmer world the greater part of the Arctic also shows warming. The pattern of this warming is very similar in winter, spring, and for the year as a whole. The largest increases of air temperature occur in the eastern part of the Arctic, especially over the Barents and Kara seas. The autumn air temperature exhibits the most peculiar behaviour. In this season less than half of the Arctic shows the warming. The cooling should occur mostly in its western part (Greenland, Canadian Arctic and Alaska) and the Chukhotsk Peninsula, with the largest decreases over Alaska. Areally the greatest warming occurs in summer. Decreases of air temperature are found only over south and west coasts of Greenland, over Baffin Sea, and some small parts of the Russian Arctic. Winter shows the greatest extreme increases (or decreases) of air temperature in a warmer world (in both cases more than 1°C) while summer displays the smallest (mostly below 0.3°C). But the largest mean seasonal warming in the Arctic (calculated from 27 stations) is found in spring (0.31°C), with a much lower mean in winter (0.17°C), and the lowest mean in autumn (0.01°C). A comparison of the mean seasonal and annual warming of the Arctic with the hemispheric warming gives very interesting results. The intensity of the Arctic warming is more than twice as great as for the Northern Hemisphere in spring and summer, while it is only a little greater in winter, and is much smaller in autumn. The mean annual warming of the Arctic is 1.6 times greater.

The patterns of precipitation in the warmer world are more complex than those for air temperature (see Figure 1 in Przybylak 1995). This is connected with the greater spatial variability of precipitation. In all seasons except spring, the mean precipitation in the Arctic (computed from 27 stations) is lower in a warmer world. The increase occurs only in spring, which, as we remember, shows the most distinct warming. On the other hand, the largest decreases of the mean Arctic totals of precipitation are found in autumn, which is characterised by the lack of warming. However, the greater part of the Arctic shows a decrease of precipitation under warm-world conditions in winter, though these decreases are smaller than in autumn. In all seasons more than half of the Arctic shows decreases in precipitation. The winter precipitation is expected to increase only over the greater part of the Atlantic region of the Arctic and the Canadian Arctic. In spring the pattern is very similar; the main difference is the reduced area of precipitation increases. In summer, the increases and decreases of precipitation contain equal areas. The largest area of the precipitation increase includes Alaska, the Canadian Arctic, the eastern coast of Greenland, and the Greenland Sea. In autumn, in a warmer world there is a domination of the precipitation decrease, with the largest values over south-western Greenland and the Chukhotsk Peninsula (up to 40–50 mm). An increase is expected only over the Barents Sea and the adjoining islands. The annual precipitation shows a decrease over two areas: the largest one includes central and Russian Arctic and the smaller one is the southern part of Greenland with adjoining seas.

In conclusion, as pointed out by Przybylak (1995, 1996a, 2002a), a small warming and a decrease of precipitation connected with a rise in CO_2 in the first period of global warming is expected in the Arctic. He found also that there is no direct relation between the behaviour of air temperature and precipitation. Increases and decreases of precipitation in the Arctic are expected to occur in the regions which show both warming and cooling.

This review reveals that both Arctic climatologists and climate modellers still have extensive work to do. On the one hand, our knowledge concerning the climatology of the majority of the meteorological elements in the Arctic, as well as some components of the Arctic climate system, is not sufficient to reliably check the validity of the numerical models. On the other hand, the existing climate models (both RCMs and especially GCMs) describe many Arctic processes inaccurately (e.g. cloud-radiative interactions, local surface-atmosphere interactions, sea ice distribution, stratiform clouds, and Arctic haze). As a consequence, the largest disagreement between climate model simulations of the present-day climate is in Polar regions. This also means that the reliability of the predictions of the Arctic climate in 21st century using these models (at present only GCMs) is still not satisfactory. The biases are especially high on regional and local scales. On the other hand, it was shown that both kinds of models are capable of correctly simulating the large-scale climate patterns. As follows from sub-chapter 11.1, the more recent versions of the RCMs, if they are to be used in the coming years to provide climate simulations, are capable of giving significantly better and more confident simulations than are GCMs. To date, RCM simulations have been mostly aimed at evaluating models and processes rather than producing projections of future climate and, as such, they have been relatively short (10 years or less) (Houghton *et al.* 2001).

According to the latest GCMs results (transient models involving sulphate aerosol), the air temperature increase in the Arctic connected with the doubling of CO_2 varies between 2–5°C in winter and 0–1°C in summer. It is worth adding that the transient model experiment shows about 1.4–2.0 times lower warming in the Arctic than in the case of the equilibrium response. Most transient GCMs simulate an increase of precipitation in the Arctic with the doubling of CO_2. The winter precipitation increase should be higher and more general than in summer. The inclusion of aerosol forcing additionally enlarges precipitation. According to the analogue method, at the beginning of the 21st century a small warming and a decrease of precipitation in the Arctic should occur. This means that the greatest difference between model and analogue scenarios becomes apparent in the case of precipitation in the Arctic.

REFERENCES

AGASP (Arctic Gas and Aerosol Sampling Program), 1984, *Geophys. Res. Lett.*, **11** (5).
Ahlmann H.W., 1948, *Glaciological Research on the North Atlantic Coasts*, Roy. Geogr. Soc. Res. Ser. 1, London, 83 pp.
Ahlmann H.W., 1953, *Glacier Variations and Climatic Fluctuations*, Bowman memorial lectures, ser. 3, Amer. Geogr. Soc., New York, 51 pp.
Aleksandrov E.I. and Lubarski A.N., 1988, 'Stabilisation of "norms" under climate monitoring', in: Voskresenskiy A.I (Ed.), *Monitoring Klimata Arktiki*, Gidrometeoizdat, Leningrad, pp. 33–39 (in Russian).
Alekseev G.V., 1997, 'Arctic climate dynamics in the global environment', in: Proceedings Conference on Polar Processes and Global Climate, Part II of II, Rosario, Orcas Island, Washington, USA, 3–6 November 1997, pp. 11–14.
Alekseev G.V., Podgornoy I.A., Svyashchennikov P.N. and Khrol V.P., 1991, 'Features of climate formation and its variability in the polar climatic atmosphere-sea-ice-ocean system', in: Krutskikh B.A. (Ed.), *Klimaticheskii Rezhim Arktiki na Rubezhe XX i XXI vv.*, Gidrometeoizdat, St. Petersburg, pp. 4–29 (in Russian).
Alekseev G.V. and Svyashchennikov P.N., 1991, *The natural variation of climatic characteristics of the Northern Polar Region and the Northern Hemisphere*, Gidrometeoizdat, Leningrad, 159 pp. (in Russian).
Alisov B.P., Berlin I.A. and Mikhel V.M., 1954, *A Course of Climatology*, part 3, Gidrometeoizdat, Leningrad, (in Russian).
Alt B.T., 1985, 'A period of summer accumulation in the Queen Elizabeth Islands', in: Harington C.R. (Ed.), *Critical Periods in the Quaternary Climatic History of Northern North America*, Climatic Change in Canada 5, Syllogeus 55, 461–479.
Alt B.T., Fisher D.A. and Koerner R.M., 1992, 'Climatic conditions for the period surrounding the Tambora signal in ice core from the Canadian High Arctic Islands', in: Harington C.R. (Ed.), *The Year Without a Summer? World Climate in 1816*, Ottawa, Canadian Museum of Nature, pp. 309–325.
Andersson T., 1969, 'Annual and diurnal variation of fog', Meteorologiska Institutionen, Uppsala Universitet, *Meddellande*, **Nr 102**, Uppsala, 36 pp.
Andrews R.H., 1964, 'Meteorology and heat balance of the ablation area, White Glacier', *Axel Heiberg Island Research Reports, Meteorology*, **No. 1**, McGill Univ., Montreal, 107 pp.
Andrews J.T. and Ives J.D., 1972, 'Late- and Postglacial events (< 10,000 BP) in the eastern Canadian Arctic with particular reference to the Cockburn moraines and break-up of the Laurentide ice sheet', in: Vasari Y., Hyvärinen H. and Hicks S. (Eds.), Climatic Changes in Arctic Areas During the Last 10,000 years, *Acta Univ. Ouluensis*, Series A, Scientiae Rerum Naturalium 3, Geologica 1, Univ. of Oulu, Oulu, Finland, 149–174.
Arctic Climate System Study, 1994, WCRP-85, WMO/TD-No. 627, 66 pp.
Ariel N.Z., Bartkovskiy R.S., Biutner Z.K., Kucherov N.W. and Strokina L.A., 1973, 'About computation of monthly mean heat and humid fluxes over the ocean', *Meteorol. i Gidrol.*, **5**, 3–11 (in Russian).

Aristova L.N. and Gruza G.V., 1987, *Data on the Structure and Variability of Climate. Total Cloudiness on Satellite Observations. Northern and Southern Hemispheres*, ASRIHMI-MCD, Obninsk, 248 pp.

Armstrong T., Rogers G. and Rowley G., 1978, The Circumpolar North: A Political and Economic Geography of the Arctic and Sub-arctic, Methuen & Co. Ltd., London, 303 pp.

Atlas Arktiki, 1985, Glavnoye Upravlenye Geodeziy i Kartografiy, Moscow, 204 pp.

Atkinson K., 1994, 'The Canadian Arctic and global climatic change', in: Atkinson K. and McDonald A.T. (Eds.), *Environmental Issues in Canada: Canadian Studies Workshop*, February 1994, University of Leeds, Leeds, pp. 67–86.

Bader H., 1961, *The Greenland Ice Sheet*, U.S. Army, Corp. Engr. Cold. Regions Res. Eng. Lab., Res. Rept., I-B2, 17 pp.

Badgley F.I., 1961, 'Heat balance of the surface of the Arctic Ocean', in: *Proc. of the Western Snow Conference*, 11–13 April 1961, Spokane, Washington, pp. 101–104.

Baird P. D., 1964, *The Polar World*, Longmans, London, 328 pp.

Ball T.F., 1983, 'Preliminary analysis of early instrumental temperature records from York Factory and Churchill Factory', in: Harington C.R. (Ed.), *Climatic Change in Canada 3*, Syllogeus No 49, National Museum of Natural Sciences, National Museums of Canada, Ottawa, pp. 203–219.

Ball T.F., 1992, 'Historical and instrumental evidence of climate: western Hudson Bay, Canada, 1714–1850', in: Bradley R.S. and Jones P.D. (Eds.), *Climate Since A.D. 1500*, Routledge, London, pp. 40–73.

Baranowski S., 1968, 'Thermic conditions of the periglacial tundra in SW Spitsbergen', *Acta Univ. Wratisl.*, **68**, 74 pp.

Baranowski S., 1975, 'Glaciological investigations and glaciomorphological observations made in 1970 on Werenskiold Glacier and its forefield', Results of the Polish Scientific Spitsbergen Expeditions 1970–1974, *Acta Univ. Wratisl.*, **251**, 69–94.

Baranowski S., 1977a, 'Results of dating of the fossil tundra in the forefield of Werenskioldbreen', *Acta Univ. Wratisl.*, **387**, 31–36.

Baranowski S., 1977b, 'The subpolar glaciers of Spitsbergen, seen against the climate of this region', *Acta Univ. Wratisl.*, **393**, 167 pp.

Baranowski S. and Karlén W., 1976, 'Remnants of Viking age tundra in Spitsbergen and Northern Scandinavia', *Geogr. Ann.*, **58A**, 35–40.

Barrie L.A., 1986a, 'Arctic air pollution: an overview of current knowledge', *Atmos. Environ.*, **20**, 643–663.

Barrie L.A., 1986b, 'Arctic air chemistry: an overview', in: Stonehouse B. (Ed.), *Arctic Air Pollution*, Cambridge University Press, pp. 5–23.

Barrie L.A., 1992, 'Arctic air pollution', *WMO Bull.*, **41**, 154–159.

Barrie L.A., Olson M.P. and Oikawa K.K., 1989, 'The flux of anthropogenic sulphur into the Arctic from mid-latitudes in 1979/80', *Atmos. Environ.*, **23**, 2505–2512.

Barry R.G., 1964, *Weather Conditions at Tanquary Fiord, Summer 1963*, Canada Defense Research Board, Report D. Phys. R(G) Hazen 23, 36 pp.

Barry R.G., 1981, *Mountain Weather and Climate*, Methuen, London and New York, 313 pp.

Barry R.G., 1985, 'Snow and ice data', in: Hecht A.D. (Ed.), *Paleoclimate Analysis and Modeling*, John Wiley & Sons Inc., New York, pp. 259–290.

Barry R.G., 1989, 'The present climate of the Arctic Ocean and possible past and future states', in: Herman Y. (Ed.), *The Arctic Seas: Climatology, Oceanography, Geology, and Biology*, Van Nostrand Reinhold Company, New York, pp. 1–46.

Barry R.G., 1995, 'Land of the midnight Sun', in: Ives J.D. and Sugden D. (Eds.), *Polar Regions: The Illustrated Library of the Earth*, Readers's Digest Press Australia, Sydney, pp. 28–41.

Barry R.G. and Chorley R.J., 1992, *Atmosphere, Weather and Climate*, Routledge, London and New York, 392 pp.

Barry R.G., Crane R.G., Schweiger A. and Newell J., 1987, 'Arctic cloudiness in spring from satellite imagery', *J. Climatol.*, **7**, 423–451.

Barry R.G. and Hare F.K., 1974, 'Arctic climate', in: Ives J.D. and Barry R.G. (Eds.), *Arctic and Alpine Environments*, Methuen & Co. Ltd., London, pp. 17–54.

Barry R.G. and Ives J.D., 1974, 'Introduction', in: Ives J.D. and Barry R.G. (Eds.), *Arctic and Alpine Environments*, Methuen & Co. Ltd., London, pp. 1–13.

Barry R.G. and Kiladis G.N., 1982, 'Climatic characteristic of Greenland', in: *Climatic and Physical Characteristics of the Greenland Ice Sheet*, CIRES, Univ. of Colorado, Boulder, pp. 7–33.

Barry R.G. and Perry A.H., 1973, *Synoptic Climatology: Methods and Applications*, Methuen & Co Ltd, 11 New Fetter Lane London EC4, 555 pp.

Barthelet P., Terray L. and Valcke S., 1998, 'Transient CO_2 experiment using the ARPEGE/OPAICE non flux corrected coupled model', *Geophys. Res. Lett.*, **25**, 2277–2280.

Baskakov G.A., 1971, 'Sea boundary of the Arctic', in: Govorukha L.S. and Kruchinin Yu.A. (Eds.), Problems of Physiographic Zoning of Polar Lands, *Trudy AANII*, **304**, 36–58 (in Russian), Translated and published also by Amerind Publishing Co., Pot. Ltd, New Delhi, 1981, 35–60.

Baur F., 1929, 'Das Klima der bisher erforschten Teile der Arktis', *Arktis*, **2**, 77–89 and 110–120.

Bazhev A.B. and Bazheva V.Ya., 1968, 'Quaternary glaciation of Novaya Zemlya', in: *Glaciation of the Novaya Zemlya*, Izd. "Nauka", Moskva, pp. 215–231 (in Russian).

Bedél B., 1956, *Les Observations Météorologigues de la Station Francaise du Groenland*, Paris.

Belmont A.D., 1958, 'Low tropospheric inversions at ice island T-3', in: Sutcliffe R.C. (Ed.), *Polar Atmosphere Symposium*, Part I, Meteorol. Section, pp. 215–284.

Benson C.S., 1962, 'Stratigraphic studies in the snow and firn of the Greenland ice sheet', *SIPRE Res. Rep.*, **70**.

Benson C.S., 1969, 'Ice fog', *Engineering and Science*, **32**, 15–19.

Bergeron T., 1928, 'Über die dreidimensional verknupfende Wetteranalyse', *Geofysiske Publikasjoner*, vol. **V**, No. 6.

Berlyand T.G. and Strokina L.A., 1980, *Global Distribution of Total Cloud Amount*, Gidrometeoizdat, Leningrad, 72 pp. (in Russian).

Bernes C., 1996, *The Nordic Arctic Environment – Unspoilt, Exploited, Polluted?* The Nordic Council of Ministers, Copenhagen, 240 pp.

Berry M.O. and Lawford R.G., 1977, *Low-temperature Fog in the Northwest Territories*, Atmos. Environ. Serv., Tech. Memo. 850, 27 pp.

Billeo M.A., 1966, *Survey of Arctic and Subarctic Temperature Inversions*, Tech. Rep. 161, Cold Regions Res. and Eng. Lab., Hanover, N.H., 38 pp.

Blake W. Jr., 1975, 'Radiocarbon age determinations and postglacial emergence at Cape Storm, southern Ellesmere Island, Arctic Canada', *Geogr. Ann.*, **57A**, 1–71.

Blake W. Jr., 1981, 'Neoglacial fluctuations of glaciers, southeastern Ellesmere Island, Canadian Arctic Archipelago', *Geogr. Ann.*, **63A**, 201–218.

Blake W. Jr., 1987, *Geological Survey of Canada Radiocarbon Dates XXVI*, Geological Survey of Canada, Paper 86–7.

Blake W. Jr., 1989, 'Application of ^{14}C AMS dating to the chronology of Holocene glacier fluctuations in the High Arctic, with special references to Leffert Glacier, Ellesmere Island, Canada', *Radiocarbon*, **31**, 570–578.

Blanchet J.–P., 1991, 'Potential climate change from Arctic aerosol pollution', in: Sturges W.T. (Ed.), *Pollution of the Arctic Atmosphere*, Elsevier Science Publ., London and New York, pp. 289–322.

Blanchet J.–P. and List R., 1987, 'On radiative effects of anthropogenic aerosol components in Arctic haze and snow', *Tellus*, **39B**, 293–317.

Boer G.J., Flato G., Reader M.C. and Ramsden D., 2000, 'A transient climate change simulation with greenhouse gas and aerosol forcing: experimental design and comparison with the instrumental record for the twentieth century', *Clim. Dyn.*, **16**, 405–425.

Bogdanova E.G., 1966, 'Investigation of precipitation measurement losses due to the wind', *Trans. of Main Geophys. Observ.*, **195**, Leningrad, 40–62 (in Russian).

Bogdanova E.G., 1997, 'Solid precipitation section', in: Kotlyakov V.M., Kravtsova V.I. and Dreyer N.N. (Eds.), *World Atlas of Snow and Ice Resources*, Russian Academy of Sciences, Moscow, p. 57.

Boggs S.W., 1990, *The Polar Regions: Geographical and Historical Data for Consideration in a Study of Claims to Severeignty in the Arctic and Antarctic Regions*, Buffalo, NY: William S. Hein & Co., 123 pp.

Born K., 1996, 'Tropospheric warming and changes in weather variability over the Northern Hemisphere during the period 1967–1991', *Meteorol. Atmos. Phys.*, **59**, 201–215.

Bradley R.S., 1985, *Quaternary Paleoclimatology: Methods of Paleoclimatic Reconstruction*, Allen and Unvin, Boston, 472 pp.

Bradley R.S., 1990, 'Holocene paleoclimatology of the Queen Elizabeth Islands, Canadian High Arctic', *Quat. Sci. Rev.*, **9**, 365–384.

Bradley R.S., 1999, *Paleoclimatology: Reconstructing Climates of the Quaternary*, second edition, Academic Press, San Diego, 613 pp.

Bradley R.S. and England J., 1978, 'Recent climatic fluctuations of the Canadian High Arctic and their significance for glaciology', *Arctic and Alpine Res.*, **10**, 715–731.

Bradley R.S. and Jones P.D., 1985, 'Data bases for isolating the effects of the increasing carbon dioxide concentration', in: MacCracken M.C. and Luther F.M. (Eds.), *Detecting the Climatic Effects of Increasing Carbon Dioxide*, DOE/ER-0235, pp. 31–53.

Bradley R.S. and Jones P.D., 1993, "Little Ice Age' summer temperature variations: their nature and relevance to recent global warming trends', *The Holocene*, **3**, 367–376.

Bradley R.S., Keimig F.T. and Diaz H.F., 1992, 'Climatology of surface-based inversions in the North American Arctic', *J. Geophys. Res.*, **97** (D14), 15,699–15,712.

Broccoli A.J., Delworth T.L. and Ngar-Cheung L., 2001, 'The effect of changes in observational coverage on the association between surface temperature and the Arctic Oscillation', *J. Climate*, **14**, 2481–2485.

Broccoli A.J., Lau N.–C. and Nath M.J., 1998, 'The cold ocean-warm land pattern: Model simulation and relevance to climate change detection', *J. Climate*, **11**, 2743–2763.

Bromwich D.H., 1988, 'Snowfall in high southern latitudes', *Rev. Geophys.*, **26**, 149–168.

Bromwich D.H., Cullather R.I., Chen Q.-s and Csatho B.M., 1998, 'Evaluation of recent precipitation studies for Greenland Ice Sheet', *J. Geophys. Res.*, **103** (D20), 26,007–26,024.

Bromwich D.H. and Robasky F.M., 1993, 'Recent precipitation trends over the Polar Ice Sheets', *Meteorol. Atmos. Phys.*, **51**, 259–274.

Bromwich D.H., Robasky F.M., Kee R.A. and Bolzan J.F., 1993, 'Modeled variations of precipitation over the Greenland Ice Sheet', *J. Climate*, **6**, 1253–1268.

Brown R.D. and Braaten R.O., 1998, 'Spatial and temporal variability of Canadian monthly snow depth, 1946–1995', *Atmos.-Ocean*, **36**, 37–54.

Brown R.N.R., 1927, *The Polar Regions: A Physical and Economic Geography of the Arctic and Antarctic*, Methuen & Co. Ltd., London, 245 pp.

Bruce W. S., 1911, *Polar Exploration*, Williams & Norgate, London, 256 pp.

Brückner E., 1893, 'Über den Einfluss der Schneedecke auf das klima der Alpen', *Zeitschr. Deutsch. Öst. Alpenver*, Bd. **24**.

Bryazgin N.N., 1968, 'Duration of the sunshine and its anomaly during the IGY and IQSY in the Arctic', *Trudy AANII*, **274**, 50–59 (in Russian).

Bryazgin N.N., 1969, 'Account of winter precipitation in the Polar regions', *Trudy AANII*, **287**, 110–122 (in Russian).

Bryazgin N.N., 1971, 'Precipitation and snow cover', in: Dolgin I.M. (Ed.), *Meteorological Conditions of the non-Soviet Arctic*, Gidrometeoizdat, Leningrad, pp. 124–142 (in Russian).

Bryazgin N.N., 1976a, 'Mean annual precipitation in the Arctic computed taking into account errors of precipitation measurements', *Trudy AANII*, **323**, 40–74 (in Russian).

Bryazgin N.N., 1976b, 'Comparison of precipitation measurements using two types of gauges and correction of monthly precipitation totals in the Arctic', *Trudy AANII*, **328**, 44–52 (in Russian).

Budyko M.I., 1956, Heat Balance of the Earth's Surface, Gidrometeoizdat, Leningrad, 255 pp. (in Russian).

Budyko M.I. (Ed.), 1963, *Atlas of the Heat Balance of the Earth*, Akademia Nauk SSSR, GGO, Moskva, 69 pp. (in Russian).

Budyko M.I., 1971, *Climate and Life*, Gidrometeoizdat, Leningrad, 470 pp. (in Russian).

Burova L.P., 1983, *The Moisture Cycle in the Atmosphere of the Arctic*, Gidrometeoizdat, Leningrad, 128 pp. (in Russian).

Burova L.P. and Gavrilova L.A., 1974, 'General rules of humidity regime in the troposphere', in: Dolgin I.M. and Gavrilova L.A. (Eds.), *Climate of the Free Atmosphere of the non-Soviet Arctic*, Gidrometeoizdat, Leningrad, pp. 145–173 (in Russian).

Burova L.P. and Lukyanchikova N.I., 1996, 'Water vapor distribution in the Arctic atmosphere in clear and overcast sky conditions', *Russian Met. and Hydrol.*, **1**, 25–31.

Businger S., 1987, 'The synoptic climatology of polar low outbreaks over the Gulf of Alaska and the Bering Sea', *Tellus*, **39A**, 307–325.

Businger S. and Reed R.J., 1989, 'Polar lows', in: Twitchell P.F., Rasmussen E.A. and Davidson K.L. (Eds.), *Polar and Arctic Lows*, A. Deepak Publishing, Hampton, pp. 3–45.

Calanca P., 1994, 'The atmospheric water vapour budget over Greenland', *Zürcher Geographische Schriften*, **55**, Zürcher, 115 pp.

Calanca P., Gilgen H., Ekholm S. and Ohmura A., 2000, 'Gridded temperature and accumulation distributions for Greenland for use in cryospheric models', *Ann. Glaciol.*, **31**, 118–120.

Carsey F.D. (Ed.), 1992, 'Microwave remote sensing of sea ice', *Geophys. Monogr.*, **68**, Amer. Geophys. Union, 462 pp.

Catchpole A.J.W., 1985, 'Evidence from Hudson Buy region of severe cold in the summer of 1816', in: Harington C.R. (Ed.) *Climatic Change in Canada 3*, Syllogeus No 49, National Museum of Natural Sciences, National Museums of Canada, Ottawa, pp. 121–146.

Catchpole A.J.W., 1992a, 'Hudson's Bay Company ships' log-books as sources of sea ice data, 1751–1870', in: Bradley R.S. and Jones P.D. (Eds.), *Climate Since A.D. 1500*, Routledge, London, pp. 17–39.

Catchpole A.J.W., 1992b, 'River ice and sea ice in the Hudson Bay region during the second decade of the nineteenth century', in: Harington C.R. (Ed.), *The Year Without a Summer? World Climate in 1816*, Canadian Museum of Nature, Ottawa, pp. 233–244.

Cattle H. and Crossley J., 1995, 'Modelling Arctic climate change', *Phil. Trans. R. Soc. Lond. A*, **352**, 201–213.

Central Intelligence Agency, 1978, *Polar Regions Atlas*, National Foreign Assessment Center, C.I.A. Washington, DC, 66 pp.

Čermak V., 1971, 'Underground temperature and inferred climatic temperature of the past millenium', *Palaeogeogr. Palaeoclimatol. Palaeoecol.*, **10**, 1–19.

Cess R.D. and 32 co-authors, 1991, 'Interpretation of snow-climate feedback as produced by 17 General Circulation Models', *Science*, **253**, 888–892.

Chapman W.L. and Walsh J.E., 1993, 'Recent variations of sea ice and air temperature in high latitudes', *Bull. Amer. Met. Soc.*, **74**, 33–47.

Chernigovskii N.T. and Marshunova M.S., 1965, *Climate of the Soviet Arctic (Radiation Regime)*, Gidrometeoizdat, Leningrad, 199 pp. (in Russian).

Clayton H.H., 1928, 'The bearing of polar meteorology on world weather', in: Joerg W. and Louis G. (Eds.), *Problems of Polar Research*, Amer. Geogr. Paper, Amer. Geogr. Soc. Special Publ., No. 7, New York, pp. 27–37.

Coachman L.K. and Aagaard K., 1974, 'Physical oceanography of Arctic and Subarctic seas', Ch. 1, in: Hermann Y. (Ed.), *Marine Geology and Oceanography of the Arctic Seas*, Springer Verlag, Berlin–Heidelberg–New York, pp. 1–72.

Cohen J. and Rind D., 1991, 'The effect of snow cover on the climate', *J. Climate*, **4**, 689–706.
Colony R. and Thorndike A.S., 1984, 'An estimate of the mean field of Arctic sea ice motion', *J. Geophys. Res.*, **89** (6), 10,623–10,629.
Cracknell A.P. (Ed.), 1981, *Remote Sensing in Meteorology, Oceanography and Hydrology*, Ellis Horwood Limited, Chichester, 542 pp.
Crane R.G. and Barry R.G., 1984, 'The influence of clouds on climate with a focus on high latitude interactions', *J. Climatol.*, **4**, 71–93.
Crutcher H.J. and Meserve J.M., 1970, *Selected Level Height, Temperatures, and Dew Points for the Northern Hemisphere*, NAVAIR 50-1C-52 (revised), Chief of Naval Operations, Naval Weather Service Command, Washington, D.C., 420 pp.
Curry JU.A. and Ebert E.E., 1992, 'Annual cycle of radiation fluxes over the Arctic ocean: Sensitivity to cloud optical properties', *J. Climate*, **5**, 1267–1280.
Dahlgren L., 1974, *Solar Radiation Climate Near Sea Level in the Canadian Arctic Archipelago*, Arctic Institute of North America Devon Island Expedition 1961–62, Meddelande nr 121, Meteorologiska Institutionen, Uppsala Universitet, Uppsala, 119 pp.
Dahl-Jensen D., Mosegaard K., Gundestrup N., Clow C.D., Johnsen S.J., Hansen A.W. and Balling N., 1998, 'Past temperatures directly from the Greenland Ice Sheet', *Science*, **282**, 268–271.
Dansgaard W., Johnsen S.J., Clausen H.B., Langway C.C. Jr, 1971, 'Climatic record revealed by the Camp Century ice core', in: K.K. Turekian (Ed.), *The Late Cenozoic Glacial Ages*, Yale Univ. Press, New Haven, CN, pp. 37–56.
Defant F., 1951, 'Local winds', in: Malone T.F. (Ed.), *Compendium of Meteorology*, Amer. Met. Soc., Boston, Mass., pp. 655–672.
Delworth T.L., 1996, 'North Atlantic interannual variability in a coupled ocean-atmosphere model', *J. Climate*, **9**, 2356–2375.
Delworth T.L. and Dixon K.W., 2000, 'Implications of the recent trend in the Arctic/North Atlantic Oscillation for the North Atlantic thermohaline circulation', *J. Climate*, **13**, 3721–3727.
Denton G.H. and Karlén W., 1973, 'Holocene climatic variations – their pattern and possible causes', *Quat. Res.*, **3**, 155–205.
Deser C., 2000, 'On the teleconnectivity of the "Arctic Oscillation"', *Geophys. Res. Lett.*, **27**, 779–782.
Dethloff K., Abegg C., Rinke A., Hebestadt I. and Romanov V.F., 2001, 'Sensitivity of Arctic climate simulations to different boundary-layer parametrizations in a regional climate model', *Tellus*, **53A**, 1–26.
Dethloff K., Rinke A. and Lehmann R., 1996, 'Regional climate model of the Arctic atmosphere', *J. Geophys. Res.*, **101**, 23,401–23,422.
Dewey K.F., 1987, 'Satellite-derived maps of snow cover frequency for the Northern Hemisphere', *J. Clim. Appl. Meteorol.*, **26**, 1210–1229.
Diamond M., 1958, *Air Temperature and Precipitation on the Greenland Ice Cap*, U.S. Army Corps. Engrs., Snow, Ice, Permafrost Res. Estab., Res. Rept., 43, 9 pp.
Diamond M., 1960, 'Air temperature and precipitation on the Greenland ice sheet', *J. Glaciol.*, **3**, 558–567.

Dickinson R.E., Errico R.M., Giorgi F. and Bates G.T., 1989, 'A regional climate model for the western U.S.', *Clim. Change*, **15**, 383–422.

Dickson R.R., Osborn T.J., Hurrell J.W., Meincke J., Blindheim J., Adlandsvik B., Vinje T., Alekseev G. and Maslowski W., 2000, 'The Arctic ocean response to the North Atlantic Oscillation', *J. Climate*, **13**, 2671–2696.

Dickson R.R., Osborn T.J., Hurrell J.W., Meincke J., Blindheim J., Adlandsvik B., Vinje T., Alekseev G., Maslowski W. and Cattle H., 1997, 'The Arctic ocean response to the North Atlantic Oscillation', in: *Proceedings Conference on Polar Processes and Global Climate*, Part II of II, Rosario, Orcas Island, Washington, USA, 3–6 November 1997, pp. 46–47.

Dmitriev A.A., 1994, *Variability of Atmospheric Processes in the Arctic and Their Application in Long-term Forecasts*, Gidrometeoizdat, St. Petersburg, 207 pp. (in Russian).

Dolgin I.M., 1960, 'Arctic aero-climatological studies', *Probl. Arkt.*, **4**, 64–75 (in Russian).

Dolgin I.M., 1962, 'Some results of atmospheric investigation over the Arctic Ocean', *Probl. Arkt. i Antarkt.*, **11**, 31–36 (in Russian).

Dolgin I.M., Bryazgin N.N. and Petrov L.S., 1975, 'Snow cover in the Arctic', *Trudy AANII*, **326**, 165–170 (in Russian).

Dolgin I.M. and Gavrilova L.A. (Eds.), 1974, *Climate of the Free Atmosphere of the non-Soviet Arctic*, Gidrometeoizdat, Leningrad, 320 pp. (in Russian).

Donina S.M., 1971, 'Air temperature', in: Dolgin I.M. (Ed.), *Meteorological Conditions of the non-Soviet Arctic*, Gidrometeoizdat, Leningrad, pp. 83–104 (in Russian).

Dorn W., Dethloff K., Rinke A. and Botzet M., 2000, 'Distinct circulation states of the Arctic atmosphere induced by natural climate variability', *J. Geophys. Res.*, **105**, 29,659–29,668.

Dorsey H.G., 1945, 'Some meteorological aspects of the Greenland ice cap', *J. Meteorol.*, **2**, 135–142.

Dorsey H.H., 1949, *Meteorological Characteristics of Northern Arctic America*, Mass. Inst. of Tech., unpublished.

Drozdov O.A., Sorochan O.G., Voskresenskii A.I., Burova L.P. and Kryshko O.V., 1976, 'Characteristics of the atmospheric water budget over Arctic Ocean drainage basins', *Trudy AANII*, **327**, 15–34 (in Russian).

Duce R.P., Hoffman G.L. and Zoller W.H., 1975, 'Atmospheric trace elements at remote Northern and Southern Hemispheric sites: pollution or natural?', *Science*, **187**, 59–61.

Dudley J.F., Jr. and Davy R.D., 1989, 'Global snow depth climatology', in: *Amer. Met. Soc., Sixth Conference on Applied Clim.*, March 7–10 1989, Charleston, S. Carolina, Boston, MA, Amer. Met. Soc., pp. 145–148.

Duguay C.R., 1993, 'Modelling the radiation budget of alpine snowfields with remotely sensed data: model formulation and validation', *Ann. Glaciol.*, **17**, 288–294.

Dutton E.G., Stone R.S., Nelson D.W. and Mendonca B.G., 1991, 'Recent interannual variations in solar radiation, cloudiness, and surface temperature at the South Pole', *J. Climate*, **4**, 848–858.

Dyke A.S., England J., Reimnitz E. and Jetté H., 1997, 'Changes in driftwood delivery to the Canadian Arctic Archipelago: the hypothesis of postglacial oscillations of the Transpolar Drift', *Arctic*, **50**, 1–16.

Dyke A.S., Hooper J.E. and Savelle J.M., 1996, 'A history of sea ice in the Canadian Arctic Archipelago based on postglacial remains of the Bowhead Whale (*Balaena mysticetus*)', *Arctic*, **49**, 235–255.

Dyke A.S. and Morris T.E., 1990, 'Postglacial history of the Bowhead whale and of driftwood penetrations; implications for paleoclimate, central Canadian Arctic', *Geological Survey of Canada Paper*, **89–24**, 17 pp.

Dzerdzeevskii B.L., 1941–1945, 'Circulation of the atmosphere in the Central Basin', in: *Trudy Dreifuyushchei Stantsii "Severnyi Polyus"*, vol. 2, izd. GUSMP, 64–200 (in Russian), Transl. Univ. of Calif., Dep. of Meteorol., Scient. Rep. No. 3, 1954, variously paged.

Dzerdzeevskii B. L., 1945, *Circulation Models in the Troposphere of the Central Arctic*, Moscow–Leningrad, (in Russian), Transl. Univ. of Calif., Dep. of Meteorol., Scient. Rep. No. 3, 1954, 1–40.

Elverhøi A., Svendsen J.I., Solheim A., Andersen E.S., Milliman J.D., Mangerud J. and Hook Leb. R., 1995, 'Late Quaternary sediment yield from the high Arctic Svalbard area', *J. Geol.*, **103**, 1–17.

Evans D.J.A., 1988, *Glacial Geomorphology and Late Quaternary History of Phillips Inlet and the Wootton Peninsula, Northwest Ellesmere Island, Canada*, PhD thesis, University of Alberta.

Evans D.J.A. and England J., 1992, 'Geomorphological evidence of Holocene climatic change from northwest Ellesmere Island, Canadian High Arctic', *The Holocene*, **2**, 148–158.

Ewert A., 1997, 'Thermic continentality of the climate of the Polar regions', *Probl. Klimatol. Polar.*, **7**, 55–64 (in Polish).

Ferrel W., 1882, 'The cause of low barometer in the Polar regions in the central part of cyclones', *Prof. Papers of the Signal Service*, U.S. War Dept., **No. VIII**, 5–51.

Ferrel W., 1889, *A Popular Treatise on the Winds*, Wigley, New York, 505 pp.

Fett R.W., 1989, 'Polar low development associated with boundary layer fronts in the Greenland, Norwegian and Barents Seas', in: Twitchell P.F., Rasmussen E.A. and Davidson K.L. (Eds.), *Polar and Arctic Lows*, A. Deepak Publishing, Hampton, pp. 313–322.

Feyling-Hanssen R.W. and Olsson I., 1960, 'Five radiocarbon datings of post-glacial shorelines in central Spitsbergen', *Norsk Geografisk Tidsskrift*, **17**, 1–4.

Fischer H., Werner M., Wagenbach D., Schwager M., Thorsteinnson T., Wilhelms F., Kipfstuhl J. and Sommer S., 1998, 'Little ice age clearly recorded in northern Greenland ice cores', *Geophys. Res. Lett.*, **25**, 1749–1752.

Fisher D.A. and Koerner R.M., 1980, 'Some aspects of climatic change in the high arctic during the Holocene as deduced from ice cores', in: Mahaney W.C. (Ed.), *Quaternary Paleoclimate*, Geobooks, Norwich, pp. 249–271.

Fisher D.A. and Koerner R.M., 1983, 'Ice core study: a climatic link between the past, present and future', in: Harington C.R. (Ed.), *Climatic Change in Canada*, Syllogeus 49, National Museums of Canada, Ottawa, pp. 50–69.

Flato G.M., Boer G.J., Lee W.G., McFarlane N.A., Ramsden D., Reader M.C. and Weaver A.J., 2000, 'The Canadian Centre for Climate Modelling and Analysis global coupled model and its climate', *Clim. Dyn.*, **16**, 451–467.

Fletcher J. O., 1965, *The Heat Budget of the Arctic Basin and Its Relation to Climate, A Report Prepared for United States Air Force Project RAND*, The RAND Corporation, Santa Monica, 180 pp.

Folland C.K., 1988, 'Numerical models of the raingauge exposure problem field experiments and an improved collector design', *Q.J.R. Meteorol. Soc.*, **114**, 1485–1516.

Førland E.J., Hanssen-Bauer I. and Nordli P.Q., 1997, 'Climate statistics & longterm series of temperature and precipitation at Svalbard and Jan Mayen', *KLIMA DNMI Report*, **21/97**, Norwegian Met. Inst., 72 pp.

Francis J.A., 1994, 'Improvements to TOVS retrievals over sea ice and applications to estimating Arctic energy fluxes', *J. Geophys. Res.*, **99**(D5), 10,395–10,408.

Franklin J., 1828, *Narrative of a Second Expedition to the Polar Sea in the Years 1825, 1826 and 1827*, John Murray, London, 320 pp.

Fyfe J.C., Boer G.J. and Flato G.M., 1999, 'The Arctic and Antarctic oscillations and their projected changes under global warming', *Geophys. Res. Lett.*, **26**, 1601–1604.

Gaigerov S.S., 1962, *Problems of the Aerological Structure Circulation and Climate of the Free Atmosphere of the Central Arctic and Antarctic*, Izd. AN SSSR, Moskva, 316 pp. (in Russian).

Gaigerov S.S., 1964, *Aerology of the Polar Regions*, Gidrometeoizdat, Moskva (in Russian), Translated also by Israel Program for Scientific Translations, Jerusalem, 1967, 280 pp.

Gates W.L., 1992, 'AMIP: The Atmospheric Model Intercomparison Project, *Bull. Amer. Meteor. Soc.*, **73**, 1962–1970.

Gates W.L. and 74 coauthors, 1996, 'Climate models – Evaluation', in: Houghton J. T., Meila Filho L.G., Callander B. A., Harris N., Kattenberg A. and Maskell K. (Eds.), *Climate Change 1995: The Science of Climate Change*, Cambridge University Press, pp. 233–284.

Gates W. L. and 15 coauthors, 1999, 'An overview of the results of the Atmospheric Model Intercomparison Project (AMIP I)', *Bull. Amer. Meteor. Soc.*, **80**, 29–55.

Gavrilova M.K., 1959, 'Radiation balance of the Arctic', *Trudy GGO*, **92**, pp. (in Russian).

Gavrilova M.K., 1963, *Radiation Climate of the Arctic*, Gidrometeoizdat, Leningrad, 225 pp. (in Russian), Translated also by Israel Program for Scientific Translations, Jerusalem, 1966, 178 pp.

Georgi J., 1935, 'Die Eismittestation. Deutsche Grönland-Expedition A. Wegener 1929 und 1930–31', *Wiss. Ergebnisse*, **4**, 1, Leipzig.

Georgi J., Holzapfel R. and Kopp W., 1935, 'Meteorologie. Das Beobachtungsmaterial', *Wiss. Ergebnisse. Deutchen Grönland-Expedition Alfred Wegener 1929 und 1930/1931*, **4**, Leipzig.

Giorgi F. and Francisco R., 2000a, 'Uncertainties in regional climate change prediction: a regional analysis of ensemble simulations with the HADCM2 coupled AOGCM', *Clim. Dyn.*, **16**, 169–182.

Giorgi F. and Francisco R., 2000b, 'Ewaluating uncertainties in the prediction of regional climate change', *Geophys. Res. Lett.*, **27**, 1295–1298.

Girs A.A., 1948, 'Some aspects concerning basic forms of atmospheric circulation', *Meteorol. i Gidrol.*, **3**, 9–11 (in Russian).
Girs A.A., 1971, *Many Years Fluctuations of Atmospheric Circulation and Long-term Hydrometeorological Forecast*, Gidrometeoizdat, Leningrad, 279 pp. (in Russian).
Girs A.A., 1981, 'Some forms of atmospheric circulation and their utilisation in forecasts', *Trudy AANII*, **373**, 4–13 (in Russian).
Gledonova N.K., 1971, 'Isobaric fields and wind', in: Dolgin I.M. (Ed.) *Meteorological Conditions of the non-Soviet Arctic*, Gidrometeoizdat, Leningrad, pp. 69–83, (in Russian).
Gluza A.F. and Piasecki J., 1989, 'The role of atmospheric circulation in formation of climatic features at South Bellsund in spring-summer season of 1987', in: Repelewska-Pękalowa J. and Pękala K. (Eds.), *Wyprawy Geograficzne UMCS w Lublinie na Spitsbergen 1986–1988*, Sesja Polarna 1989, UMCS Lublin, pp. 9–28 (in Polish).
Gobeil Ch., Macdonald R.W., Smith J.N. and Beaudin L., 2001, 'Atlantic water flow pathways revealed by lead contamination in Arctic Basin sediments', *Science*, **293**, 1301–1304.
Gol'cman M.I., 1939, 'About measurements of air humidity in low negative temperatures', *Probl. Arkt.*, **1**, 39–53 (in Russian).
Gol'cman M.I., 1948, 'Problem of air humidity measurement in the Arctic', *Probl. Arkt.*, **3** (in Russian).
Gordiyenko F.G., Kotlyakov V.M., Punning Ya.-K.M. and Vaikmäe R. A., 1981, 'Study of a 200-m core from the Lomonosov Ice Plateau on Spitsbergen and the paleoclimatic implications', *Polar Geogr. and Geol.*, **5**, 242–251.
Görgen K., Bareiss J., Helbig A., Rinke A. and Dethloff K., 2001, 'An observational and modelling analysis of Laptev Sea (Arctic Ocean) ice variations during summer', *Ann. Glaciol.*, **33**, 533–538.
Gorshkov S.G. (Ed.), 1980, *Military Sea Fleet Atlas of Oceans: Northern Ice Ocean*, USSR: Ministry of Defense, 184 pp. (in Russian).
Götz F.W.P., 1931, 'Zum Strahlungsklima des Spitzbergensommers; Strahlungs- und Ozonmessungen in der Königsbucht', *Beit. Z. Geophys.*, Bd. **31**.
Groisman P.Ya. Easterling D.R., Quayle R.G., Golubev V.S. and Peck E.L., 1997, 'Adjustment methodology for the U.S. precipitation data', in: Barry R.G., Fuchs T. and Rudolf B. (Eds.), *Proceedings of the Workshop on the Implementation of the Arctic Precipitation Data Archive (APDA) at the Global Precipitation Climatology Centre (GPCC)*, Offenbach, Germany 10–12 July, 1996, WCRP-98, WMO/TD No. 804, pp. 80–83.
Groisman P.Ya., Karl T.R. and Knight R.W., 1994, 'Observed impact of snow cover on the heat balance and the rise of continental spring temperatures', *Science*, **263**, 198–200.
Grootes P.M., Stuiver M., White J.W.C., Johnsen S. and Jouzcl J., 1993, 'Comparison of oxygen isotope records from the GISP2 and GRIP Greenland ice cores', *Nature*, **366**, 552–554.
Grossvald M.G., 1973, 'History of glaciers on the archipelago in the late Pleistocene and Holocene', in: *Glaciers of Franz Josef Land*, Izd. "Nauka", Moskva, pp. 290–305 (in Russian).

Grove J.M., 1988, *The Little Ice Age*, Methuen, London, 498 pp.
Grześ M. and Sobota I., 1999, 'Winter balance of Waldemar Glacier in 1996–1998', in: *Polish Polar Studies*, XXVI Polar Symposium, Lublin, pp. 87–98.
Gupta S.K., Wilber A.C., Ritchey N.A., Whitlock C.H. and Stackhouse P.W., 1997, 'Comparison of surface radiative fluxes in the NCEP/NCAR reanalysis and the Langley 8-year SRB dataset', in: *Proc. First Int. Conf. Reanal.*, pp. 77–80.
Haefliger M., 1998, 'Radiation balance over the Greenland Ice Sheet derived by NOAA AVHRR satellite data and in situ observations', *Zürcher Geographische Schriften*, **72**, Zürcher, 92 pp.
Haggblom A., 1982, 'Driftwood in Svalbard as an indicator of sea ice conditions', *Geogr. Ann.*, **64A**, 81–94.
Hahn C.J., Warren S.G. and London J., 1995, 'The effect of moonlight on observation of cloud cover at night, and application to cloud climatology', *J. Climate*, **8**, 1429–1446.
Hanssen-Bauer I., Førland E.J. and Nordli P.Q, 1996, 'Measured and true precipitation at Svalbard', *KLIMA DNMI Report*, **31/96**, Norwegian Met. Inst., 49 pp.
Hare F.K., 1951, 'Some climatological problems of the Arctic and sub-Arctic', in: *Compedium of Meteorol.*, Amer. Met. Soc., Boston, MA, pp. 952–964.
Harris J.M. and Kahl J., 1994, 'An analysis of ten day isentropic flow patterns for Barrow, Alaska', *J. Geophys. Res.*, 25,845–25,856.
Harris R., 1987, *Satellite Remote Sensing: An Introduction*, Routledge & Kegan Paul Ltd., London and New York, 220 pp.
Harvey L.D.D., 1980, 'Solar variability as a contributary factor to Holocene climatic change', *Prog. Phys. Geogr.*, **4**, 487–530.
Heintzenberg J., 1989, 'Arctic haze: air pollution in Polar regions', *Ambio*, **18**, 51–55.
Helmholtz H. von, 1888, 'Über atmosphärische Bewegungen', *Meteor. Zeit.*, **5**, 329–340.
Hergessell H., 1906, 'Die Erforschung der freien Atmosphäre über dem Polarmeer', *Beitr. Phys. Frei. Atmos.*, **2**, 96–98.
Herman G.F., 1986, 'Atmospheric modelling and air-sea interaction', in: Untersteiner N. (Ed.), *The Geophysics of Sea Ice*, Plenum Press, New York, pp. 713–754.
Hesselberg Th. and Birkeland B. J., 1940, 'Säkuläre Schwankungen des Klimas von Norwegen, Teil I: Die Lufttemperatur', *Geophys. Publik.*, **14**, 4.
Hilmer M. and Jung T., 2000, 'Evidence for recent change in the link between the North Atlantic Oscillation and Arctic sea ice export', *Geophys. Res. Lett.*, **27**, 989–992.
Hobbs W.H., 1910, 'Characteristics of the inland ice of the Arctic regions', *Proc. Amer. Phil. Soc.*, **49**, 57–129.
Hobbs W.H., 1926, *The Glacial Anticyclones, the Poles of the Atmospheric Circulation*, Macmillan, New York, 198 pp.
Hobbs W.H., 1945, 'The Greenland glacial anticyclone', *J. Meteorol.*, **2**, 143–153.
Hobbs W.H., 1948, 'The climate of the Arctic as viewed by the explorer and the meteorologist', *Science*, **108**, 193–201.
Houghton J.T., Callander B. A. and Varney S. K. (Eds.), 1992, *Climate Change 1992: The Supplementary Report to the IPCC Scientific Assessment*, Cambridge University Press, 200 pp.

Houghton J.T, Ding Y., Griggs D.J., Noguer M., van der Linden P.J., Dai X., Maskell K. and Johnson C.A. (Eds.), 2001, *Climate Change 2001: The Scientific Basis*, Cambridge University Press, Cambridge, 881 pp.

Houghton J.T., Jenkins G.J. and Ephraums J.J. (Eds.), 1990, *Climate Change: The IPCC Scientific Assessment*, Cambridge University Press, Cambridge, 365 pp.

Houghton J.T., Meira Filho L.G., Callander B.A., Harris N., Kattenberg A. and Maskell K. (Eds.), 1996, *Climate Change 1995: The Science of Climate Change*, Cambridge University Press, Cambridge, 572 pp.

Hughes N.A., 1984, 'Global cloud climatologies: a historical review', *J. Clim. Appl. Met.*, **23**, 724–751.

Hulme M., 1992, 'A 1951–1980 global land precipitation climatology for the evaluation of general circulation models', *Clim. Dyn.*, **7**, 57–72.

Hurrell J. W., 1995, 'Decadal trends in the North Atlantic Oscillation: Regional temperatures and precipitation', *Science*, **269**, 676–679.

Hurrell J.W., 1996, 'Influence of variations in extratropical wintertime teleconnections on Northern Hemisphere temperature', *Geophys. Res. Lett.*, **23**, 665–668.

Hurrell J.W. and van Loon H., 1997, 'Decadal variations in climate associated with the North Atlantic Oscillation', *Clim. Change*, **36**, 301–326.

Huschke R.E., 1969, *Arctic Cloud Statistics from "Air Calibrated" Surface Weather Observations*, RAND Corp. Mem. RM-6173-PR, RAND, Santa Monica, CA, 79 pp.

Hyvarinen H., 1972, 'Pollen-analytical evidence for Flandrian climatic change in Svalbard', in: Vasari Y., Hyvarinen H. and Hick S. (Eds.), *Climatic Change in Arctic Areas During the Last Ten Thousand Years*, Ouluensis Acta Univ., Oulu, 225–237.

Ice Thickness Climatology 1961–1990 Normals, 1992, Environment Canada, Atmospheric Environment Service, Publ. by Minister of Supply and Services, variously paged.

Jaeger L., 1983, 'Monthly and areal patterns of mean global precipitation', in: Street-Perrot A., Beran M. and Ratcliffe R. (Eds.), *Variations in the Global Water Budget*, D. Reidel, Boston, pp. 129–140.

Jäger J. and Kellogg W.W., 1983, 'Anomalies in temperature and rainfall during warm Arctic seasons', *Clim. Change*, **5**, 39–60.

Jackson C.I., 1969, 'The summer climate of Tanquary Fiord, N.W.T.', *Arctic Meteor. Res. Group, Publ. Meteor.*, No. **95**, McGill Univ., Montreal, 65 pp.

Jaworowski Z., 1989, 'Pollution of the Norwegian Arctic: A review', *Rapportserie*, Nr. **55**, Norsk Polarinstitutt, Oslo, 93 pp.

Jaworowski Z., Hoff P., Lund W., Hagen J.O. and Segalstad T.V., 1990, *Radial Migration of Impurities in the Glacier Ice Cores*, Norwegian Polar Research Institute, Project LH-386, Final Report, 71 pp.

Jaworowski Z., Kownacka L. and Bysiek M, 1981, 'Flow of metals into the global atmosphere', *Geochim. Cosmochim. Acta*, **45**, 2185–2199.

Jaworowski Z., Segalstad T.V. and Ono N., 1992, 'Do glaciers tell a true atmospheric CO_2 history', *The Science of the Total Environment*, **114**, 227–284.

Johnsen S.J., Clausen H.B., Dansgaard W., Fuhrer K., Gundestrup N., Hammer C.U., Iversen P., Jouzel J., Stauffer B. and Steffensen J.P., 1992, 'Irregular glacial interstadials recorded in a new Greenland ice core', *Nature*, **359**, 311–313.

Johnsen S.J., Dansgaard W., Clausen H.B., Ørsted H.C. and Langway C.C., 1970, 'Climatic oscillations 1200–2000 AD', *Nature*, **227**, 482–483.

Jones P. D., 1985, 'Arctic temperatures 1851–1984', *Clim. Monit.*, **14**, 2, 43–50.

Jones P.D., 1987, 'The early twentieth century Arctic high – fact or fiction?', *Clim. Dyn.*, **1**, 63–75.

Jones P.D., 1994, 'Hemispheric surface air temperature variations: a reanalysis and an update to 1993', *J. Climate*, **7**, 1794–1802.

Jones P.D., Jonsson T. and Wheeler D., 1997, 'Extension to the North Atlantic Oscillation using early instrumental pressure observations from Gibraltar and southwest Iceland', *Int. J. Climatol.*, **17**, 1433–1450.

Jönson P. and Bärring L., 1994, 'Zonal index variations, 1899–1992: Links to air temperature in southern Scandinavia', *Geogr. Ann.*, **76A**, 207–219.

Kahl J.D., 1990, 'Characteristics of the low-level temperature inversion along the Alaskan Arctic coast', *Int. J. Climatol.*, **10**, 537–548.

Kahl J. D., Charlevoix D. J., Zaitseva N. A., Schnell R. C. and Serreze M. C., 1993a, 'Absence of evidence for greenhouse warming over the Arctic Ocean in the past 40 years', *Nature*, **361**, 335–337.

Kahl J.D., Serreze M.C. and Schell R.C., 1992a, 'Low-level tropospheric temperature inversions in the Canadian Arctic', *Atmos.-Ocean*, **30**, 511–529.

Kahl J.D., Serreze M.C., Shoitani S.M., Skony S.M. and Schnell R.C., 1992b, 'In-situ meteorological sounding archives for Arctic studies', *Bull. Am. Meteor. Soc.*, **73**, 1824–1830.

Kahl J. D., Serreze M. C., Stone R. S., Shiotani S., Kisley M. and Schell R. C., 1993b, 'Tropospheric temperature trends in the Arctic: 1958–1986', *J. Geophys. Res.*, **98**(D7), 12,825–12,838.

Kalicki T., 1985, 'The foehnic effects of the NE winds in Palffyodden region (Sörkappland)', *Prace Geogr.*, **63**, 99–105.

Kalitin N.N., 1921, 'Radiation and polarimetric observations conducted in the town of Arkhangelsk and in the White Sea in the summer of 1920', *Meteorol. Vestn*, Nos. **1–12** (in Russian).

Kalitin N.N., 1924, 'Radiation, polarimetric and cloud observations conducted in August and September 1921 by the Hydrographic Expedition of the Arctic Ocean', *Zap. po Gidrografii*, **48** (in Russian).

Kalitin N.N., 1929, 'Some data on the incoming and outgoing of radiant energy for Matochkin Shar', *Izv. GGO*, **4** (in Russian).

Kalitin N.N., 1940, 'Global radiation in the Arctic', *Probl. Arkt.*, **1**, 36–43 (in Russian).

Kalitin N.N., 1945, 'The amounts of warmth of solar radiation in the territory of the USRR', *Priroda*, **2**, 37–42 (in Russian).

Kanevskiy Z.M., 1962, 'Climatological characteristics of the Russkaya Gavan' region (Novaya Zemlya)', in: *Sb. Issledovania Lednikov i Lednikovykh Rayonov*, vyp. 2, Moskva, Izd. AN SSSR, pp. 112–143 (in Russian).

Kanevskiy Z.M. and Davidovich N.N., 1968, 'Climate', in: *Glaciation of the Novaya Zemlya*, Izd. "Nauka", Moskva, pp. 41–78 (in Russian).

Karlqvist A. and Heintzenberg J., 1992, 'Arctic pollution and the greenhouse effect', in: Griffiths F. (Ed.), *Arctic Alternatives: Civility or Militarism in the Circumpolar North*, Science for Peace / Samuel Stevens (Canadian Papers in Peace Studies 3), Toronto, pp. 156–169.

Kattenberg A. and 82 coauthors, 1996, 'Climate models – Projection of future climate', in: Houghton J. T., Meila Filho L.G., Callander B. A., Harris N., Kattenberg A. and Maskell K. (Eds.), *Climate Change 1995: The Science of Climate Change*, Cambridge University Press, pp. 285–357.
Keegan T.J., 1958, 'Arctic synoptic activity in winter', *J. Meteorol.*, **15**, 513–521.
Kelly P. M. and Jones P. D., 1981a, 'Winter temperatures in the Arctic, 1882–1981', *Clim. Monit.*, **10**, 1, 9–11.
Kelly P. M. and Jones P. D., 1981b, 'Spring temperatures in the Arctic, 1882–1981', *Clim. Monit.*, **10**, 2, 40–41.
Kelly P. M. and Jones P. D., 1981c, 'Summer temperatures in the Arctic, 1881–1981', *Clim. Monit.*, **10**, 3, 66–67.
Kelly P. M. and Jones P. D., 1981d, 'Autumn temperatures in the Arctic, 1881–1981', *Clim. Monit.*, **10**, 4, 94–95.
Kelly P. M. and Jones P. D., 1982, 'Annual temperatures in the Arctic, 1881–1981', *Clim. Monit.*, **10**, 5, 122–124.
Kelly P. M., Jones P. D., Sear C. B., Cherry B. S. G., Tavakol R. K., 1982, 'Variations in surface air temperatures: Pt. 2, Arctic regions, 1881–1980', *Mon.Wea. Rev.*, **110**, 71–83.
Kergomard C., Bonnel B. and Fouquart Y., 1993, 'Retrieval of surface radiative fluxes on the marginal zone of sea ice from operational satellite data', *Ann. Glaciol.*, **17**, 201–206.
Key J.R. and Barry R.G., 1989, 'Cloud cover analysis with Arctic AVHRR data. 1. Cloud detection', *J. Geophys. Res.*, **94**, 8521–8535.
Key J.R. and Haefliger M., 1992, 'Arctic ice surface temperature retrieval from AVHRR thermal channels', *J. Geophys. Res.*, **97**, 5885–5893.
Khalil M.A.K. and Rasmussen R.A., 1993, 'Arctic haze: Patterns and relationships to regional signatures of trace gases', *Global Biogeoch. Cycles*, **7**, 27–36.
Khrol V.P., 1976, 'Evaporation from the surface of the Arctic Ocean', *Trudy AANII*, **323**, 148–155 (in Russian).
Khrol V.P. (Ed.), 1992, *Atlas of the Energy Balance of the Northern Polar Region*, Gidrometeoizdat, St. Petersburg, 10 pp. + 72 maps (in Russian).
Knipovich I. M., 1921, 'Thermic conditions in Barents Sea at the end of May, 1921', *Byull. Rossiisk. Gidrol. Instituta*, **9**, 10–12 (in Russian).
Koch J.P. and Wegener A., 1930, 'Wissenschaftliche Ergebnisse der Danischen Expedition nach Drounting Louises Land und guer über das Inlandeis, von Nordgrönland 1912–1913', *Medd. om Grönland*, Bd. **75**.
Koerner R.M., 1977a, 'Devon Island Ice Cap: core stratigraphy and paleoclimate', *Science*, **196**, 15–18.
Koerner R.M., 1977b, 'Ice thickness measurements and their implications with respect to past and present ice volumes in the Canadian high arctic ice caps', *Canadian J. Earth Sci.*, **14**, 2697–2705.
Koerner R.M., 1979, 'Accumulation, ablation and oxygen isotope variations on the Queen Elizabeth Island ice caps, Canada', *J. Glaciol.*, **22**, 25–41.
Koerner R.M., 1992, 'Past climate changes as deduced from Canadian ice cores', in: Woo M.-K. and Gregor D.J. (Eds.), *Arctic Environment: Past, Present & Future*, Proceedings of a Symposium held at McMaster Univ., Nov. 14–15, 1991, McMaster Univ., Hamilton, Ontario, Canada, pp. 61–70.

Koerner R.M., 1999, 'Climate and the ice core record: Arctic examples', in: Lewkowicz A.G. (Ed.), *Poles Apart: A Study in Contrasts*, Proceedings of an International Symposium on Arctic and Antarctic Issues, University of Ottawa, Canada, September 25–27, 1997, pp. 71–87.

Koerner R.M., Alt B.T., Bourgeois J.C. and Fisher D.A., 1990, 'Canadian Ice Caps as sources of environmental data', in: Weller G., Wilson C.L. and Severin B.A.B. (Eds.), *International Conference on the Role of the Polar Regions in Global Change: Proceedings of a Conference Held June 11–15, 1990 at the University of Alaska*, vol. II, Fairbanks, 576–581.

Koerner R.M. and Fisher D.A., 1981, 'Studying climatic change from Canadian high Arctic ice cores', in: Harington C.R. (Ed.), *Climatic Change in Canada 2*, Syllogeus No. 33, National Museum of Natural Sciences, Ottawa, 1981, pp. 195–218.

Koerner R.M. and Fisher D.A., 1985, 'The Devon Island ice core and the glacial record', in: Andrews J.T. (Ed.), *Quaternary Environments: Eastern Canadian Arctic, Baffin Bay, and West Greenland*, Allen and Uwin, London, pp. 309–327.

Koerner R.M. and Fisher D.A., 1990, 'A record of Holocene summer climate from a Canadian high-arctic ice core', *Nature*, **343**, 630–631.

Koerner R.M. and Paterson W.S.B., 1974, 'Analysis of a core through the Meighen Ice Cap, Arctic Canada and its paleoclimatic implications', *Quat. Res.*, **4**, 253–263.

Kononova N. K., 1982, 'Natural and anthropogenic factors of climate dynamic', *Mat. Meteorol. Issled.*, **5**, 7–16 (in Russian).

Konzelmann T., 1994, 'Radiation conditions on the Greenland Ice Sheet', *Zürcher Geographische Schriften*, **56**, Zürcher, 124 pp.

Konzelmann T. and Ohmura A., 1995, 'Radiative fluxes and their impact on the energy balance of the Greenland Ice Sheet', *J. Glaciol.*, **41**, 490–502.

Kopanev I.D., 1978, *Snow Cover in the USSR*, Gidrometeoizdat, Leningrad, 182 pp. (in Russian).

Kopp W., 1939, 'Diskussion der Ergebnisse der Oststation in Scoresbysund. Deutsche Grönland-Expedition A. Wegener 1929 und 1930–31', *Wiss. Ergebnisse*, Bd. **4**, Hf. 2, Leipzig.

Kosiba A., 1960, 'Some results of glaciological investigations in SW Spitsbergen', *Zesz. Nauk. Uniw. Wrocł.*, Ser. B, Nauki Przyr., **4**, 30 pp.

Kotlyakov V.M., 1968, *Snow Cover of Earth and Glaciers*, Gidrometeoizdat, Leningrad, 479 pp. (in Russian).

Kotlyakov V.M., Kravtsova V.I. and Dreyer N.N. (Eds.), 1997, *World Atlas of Snow and Ice Resources, Palaegeography, Palaeoclimatology, Palaeoecology (Global and Planetary Change Section)*, 90, Russian Academy of Sciences, Moscow, 392 pp.

Kożuchowski K., 1993, 'Variations of hemispheric zonal index since 1899 and its relationship with air temperature', *Int. J. Climatol.*, **8**, 191–199.

Kożuchowski K. and Przybylak R., 1995, *Greenhouse Effect*, Wiedza Powszechna, Warszawa, 220 pp. (in Polish).

Krenke A.N. and Markin V.A., 1973a, 'Climate of the archipelago in accumulation season', in: *Glaciers of Franz Joseph Land*, Izd. "Nauka", Moskva, pp. 44–59 (in Russian).

Krenke A.N. and Markin V.A., 1973b, 'Climate of the archipelago in ablation season', in: *Glaciers of Franz Joseph Land*, Izd. "Nauka", Moskva, pp. 59–69 (in Russian).

Kukla G., 1979, 'Climate role of snow covers', in: Allison I. (Ed.), *Sea Level, Ice and Climate Change*, Int. Assoc. of Scient. Hydrol., Publication no. 131, pp. 79–107.

Kukla G. and Robinson D.A., 1988, 'Variability of summer cloudiness in the Arctic Basin', *Meteorol. Atmos. Phys.*, **39**, 42–50.

Kwok R., 2000, 'Recent changes in Arctic Ocean sea ice motion associated with the North Atlantic Oscillation', *Geophys. Res. Lett.*, **27**, 775–778.

Lachenbruch A.H. and Marshall B.V., 1986, 'Changing climate: geothermal evidence from permafrost in the Alaskan Arctic', *Science*, **234**, 689–696.

Lamb H.H., 1977, *Climate: Present, Past, and Future. Vol. 2. Climatic History and the Future*, Methuen, London, 835 pp.

Lamb H.H., 1984, 'Climate and history in northern Europe and elsewhere', in: Mörner N.A. and Karlén W. (Eds.), *Climatic Changes on a Yearly to Millennial Basis: Geological, Historical and Instrumental Records*, D. Reidel, Boston, pp. 225–240.

Lamb H.H. and Johnsson A.I., 1959, 'Climatic variation and observed changes in the general circulation', *Geogr. Ann.*, **41**, 94–134.

Lamb H.H. and Morth H.T, 1978, 'Arctic ice, atmospheric circulation and world climate', *Geogr. J.*, **144**, 1–22.

Lambert S.J., 1995, 'The effect of enhanced greenhouse warming on winter cyclone frequencies and strengths', *J. Climate*, **8**, 1447–1452.

Lambert S.J. and Boer G.J., 2001, 'CMIP1 evaluation and intercomparison of coupled climate models', *Clim. Dyn.*, **17**, 83–106.

Larsson P., 1963, 'The distribution of albedo over arctic surfaces', *Geogr. Rev.*, **53**, 572–579.

Larsson P. and Orvig S., 1961, *Atlas of Mean Monthly Albedo of Arctic Surfaces*, Scient. Rep. No. 2, Publ. in Meteorol., 45, McGill Univ., Montreal (pages are not numbered).

Larsson P. and Orvig S., 1962, *Albedo of Arctic Surfaces*, Scient. Rep. No. 6, Publ. in Meteorol., 54, McGill Univ., Montreal, 33 pp.

Laurent C., Le Treut H., Li Z.X., Fairhead L. and Dufresne J.L., 1998, *The Influence of Resolution in Simulating Inter-annual and Inter-decadal Variability in a Coupled Ocean-atmosphere GCM with Emphasis over the North Atlantic*, IPSL report N8.

Lean J., Beer J. and Bradley R., 1995, 'Reconstruction of solar irradiance since 1610: Implications for climate change', *Geophys. Res. Lett.*, **22**, 3195–3198.

LeDrew E.F., 1983, 'The dynamic climatology of the Beaufort to Laptev sectors of the Polar Basin for the summer of 1975 and 1976', *J. Climatol.*, **3**, 335–359.

LeDrew E.F., 1984, 'The role of local heat sources in synoptic activity within the Polar Basin', *Atmos.-Ocean*, **22**, 309–327.

LeDrew E.F., 1985,'The dynamic climatology of the Beaufort to Laptev sectors of the Polar Basin for the winter of 1975 and 1976', *J. Climatol.*, **5**, 253–272.

Legates D.R. and Willmott C.J., 1990, 'Mean seasonal and spatial variability in gauge-corrected global precipitation', *Int. J. Climatol.*, **10**, 111–127.

Lemmen D.S., 1988, *Glacial History of Marvin Peninsula, Northern Ellesmere Island, and Ward Hunt Island*, PhD thesis, University of Alberta.

Leszkiewicz J. and Pulina M., 1996, 'Analysis of winter snow cover from the point of view of snow falling phases (Hans Glacier, Hornsund region, Spitsbergen)', *Probl. Klimatol. Polar.*, **5**, Toruń, 43–65 (in Polish).

Lindner L., Marks L. and Ostaficzyk S., 1982, 'Evolution of the marginal zone and the forefield of the Torell, Nann and Tone glaciers in Spitsbergen', *Acta Geol. Polonica*, **32**, 267–278.

Liestøl O., 1969, 'Glacier surges in west Spitsbergen', *Canadian J. Earth Sci.*, **6**, 895–897.

Lo C.P., 1986, *Applied Remote Sensing*, Longman Scientific & Technical, New York, 394 pp.

Locatelli J.D., Hobbs P.V. and Werth J.A., 1982, 'Mesoscale structures of vortices in polar air streams', *Mon. Wea. Rev.*, **107**, 1417–1433.

Loewe F., 1935, 'Das Klima des Grönlandischen Inlandeises', in: Köppen W. and Geiger R. (Eds.), *Handbuch der Klimatologie, Bd. II, Teil K, Klima des Kanadischen Archipels und Grönlands*, Verlag von Gebrüder Borntraeger, Berlin, pp. K67–K101.

Loshchilov V.S., 1964, 'Snow cover on the ice of the central Arctic', *Probl. Arkt. Antarkt.*, **17**, 36–45 (in Russian).

Lubinski D.J., Forman S.L. and Miller G.H., 1999, 'Holocene glacier and climate fluctuations on Franz Josef Land, Arctic Russia, 80°N', *Quat. Sci. Rev.*, **18**, 85–108.

Lynch A.H., Chapman W.L., Walsh J.E. and Weller G., 1995, 'Development of a regional climate model of the western Arctic', *J. Climate*, **8**, 1555–1570.

Lysgaard L., 1949, 'Recent climatic fluctuations, *Folia Geographica Danica*, **V**, Kobenhavn, 1–86.

Majorowicz J.A., Šafanda J., Harris R.N. and Skinner W. R., 1999, 'Large ground surface temperature changes of the last three centuries inferred from borehole temperatures in the Southern Canadian Prairies, Saskatchewan', *Global and Planet. Change*, **20**, 227–241.

Makshtas A.P., 1984, *The Heat Budget of Arctic Ice in the Winter*, Gidrometeoizdat, Leningrad, 67 pp. (in Russian). English version published by International Glaciological Society, Cambridge, 1991, 77 pp.

Malmgren F., 1926, 'Studies of humidity and hoar-frost over the Arctic Ocean', *Geophys. Public.*, **IV**, 6, Oslo.

Manabe S., Spelman M.J. and Stouffer R.J., 1992, 'Transient responses of a coupled ocean-atmosphere model to gradual changes of atmospheric CO_2. Part II: Seasonal responses', *J. Climate*, **5**, 105–126.

Manabe S., Stouffer R.J, Spelman M.J. and Bryan K., 1991, 'Transient responses of a coupled ocean-atmosphere model to gradual changes of atmospheric CO_2. Part I: Annual mean response', *J. Climate*, **4**, 785–817.

Marciniak K. and Przybylak R., 1985, 'Atmospheric precipitation of the summer season in the Kaffiöyra region (North–West Spitsbergen)', *Pol. Polar Res.*, **6**, 543–559.

Marciniak K., Marszelewski W. and Przybylak R., 1985, 'Air temperature on the Elise and Waldemar glaciers /NW Spitsbergen/ in the summer season – comparative study', in: *XII Sympozjum Polarne, Materiały*, Szczecin, pp. 31–42 (in Polish).

Markin V.A., 1975, 'The climate of the contemporary glaciation area', in: *Glaciation of Spitsbergen (Svalbard)*, Izd. Nauka, Moskva, pp. 42–105 (in Russian).

Marks L., 1983, 'Late Holocene evolution of the Treskelen Peninsula (Hornsund, Spitsbergen)', *Acta Geol. Polonica*, **33**, 159–167.

Marshunova M.S., 1961, 'Principal characteristics of the radiation balance of the underlying surface and of the atmosphere in the Arctic', *Trudy ANII*, **226**, 109–112 (in Russian).

Marshunova M.S. and Chernigovskii N.T., 1971, *Radiation Regime of the Foreign Arctic*, Gidrometeoizdat, Leningrad, 182 pp., (in Russian), Translated also by Indian National Scientific Documentation Centre, New Delhi, 1978, 189 pp.

Marsz A., 1994, 'Precipitation in the Arctowski Station', *Probl. Klimatol. Polar.*, **4**, Gdynia, 65–76 (in Polish).

Martyn D., 1985, *Climates of the Earth*, PWN Warszawa, 667 pp. (in Polish).

Matson M., 1991, 'NOAA satellite snow cover data', *Palaeogeogr., Palaeoclim., Palaeoecol.*, **90**, 213–218.

Matveev Y.L. and Titov V.I., 1985, *Data on the Climate Structure and Variability. Global Cloudiness Field*, ASRIHMI – MDC, Obninsk, 248 pp.

Maxwell J.B., 1980, 'The climate of the Canadian Arctic islands and adjacent waters', vol. 1, *Climatological studies*, No. **30**, Environment Canada, Atmospheric Environment Service, pp. 531.

Maxwell J.B., 1982, 'The climate of the Canadian Arctic islands and adjacent waters', vol. 2, *Climatological studies*, No. **30**, Environment Canada, Atmospheric Environment Service, pp. 589.

Mayewski P. A., Meeker L.D., Whitlow S., Twickler M.S., Morrison M.C., Alley R.B., Bloomfield P., and Taylor K., 1993, 'The atmosphere during the Younger Dryas', *Science*, **261**, 195–197.

Mayewski P.A., Meeker L.D., Whitlow S., Twickler M.S., Morrison M.C., Bloomfield P., Bond G.C., Alley R.B., Gow A.J., Grootes P.M., Meese D.A., Ram M., Taylor K.C, and Wumkes W., 1994, 'Changes in atmospheric circulation and ocean ice cover over the North Atlantic during the last 41,000 years, *Science*, **263**, 1747–1751.

McGuffie K., Barry R.G., Schweiger A., Robinson D.A. and Newell J., 1988, 'Intercomparison of satellite-derived cloud analyses for the Arctic Ocean in spring and summer', *Int. J. Remote Sensing*, **9**, 447–467.

McKay D.C. and Morris R.J., 1985, 'Solar radiation data analyses for Canada 1967–1976', vol. 6: *The Yukon and Northwest Territories*, Minister of Supply and Services Canada, Ottawa, variously paged.

McKay G.A., Findlay B.F. and Thompson H.A., 1970, 'A climatic perspective of tundra areas', in: Fuller W.A. and Kevan P.G. (Eds.), *Productivity and Conservation in Northern Circumpolar Lands*, IUCN Publ., No. 16, Morges, pp. 10–23.

McLaren A.S., Serreze M.C. and Barry R.G., 1988, 'Seasonal variations of atmospheric circulation and sea ice motion in the Arctic', in: *Amer. Met. Soc., Second Conference on Polar Meteorology and Oceanography*, Boston, pp. 20–23.

Mecking L., 1928, 'The Polar Regions: A regional geography', in: *The Geography of the Polar Regions*, Amer. Geogr. Soc. Special Publ. No. 8, New York, pp. 93–281.

Meehl G.A. and Washington W.M., 1990, 'CO_2 climate sensitivity and snow-sea-ice albedo parametrization in an atmospheric GCM coupled to a mixed-layer ocean model', *Clim. Change*, **16**, 283–306.

Meese D.A., Gow A.J., Grootes P., Mayewski P.A., Ram M., Stuiver M., Taylor K.C., Waddington E.D. and Zielinski G.A., 1994, 'The accumulation record from the GISP2 core as an indicator of climate change throughout the Holocene', *Science*, **266**, 1680–1682.

Metcalfe J.R. and Goodison B.E., 1993, 'Correction of Canadian winter precipitation data', in: *Eighth Symposium on Meteorological Observations and Instrumentations...*, Jan. 17–23 1993, Anaheim, California, Amer. Met. Soc., Boston, MA, pp. 338–343.

Meteorological Office, 1962, *A Course in Elementary Meteorology*, Met. 0.707, Her Majesty's Stationary Office, London, WC1V 6HB, 189 pp.

Meteorology of the Canadian Arctic, 1944, Department of Transport, Met. Div., Canada, 85 pp.

Mills W. and Speak P., 1998, *Keyguide to Information Sources on – the Polar and Cold Regions*, Mansell, London and Washington, 330 pp.

Mitchell J.F.B., Davis R.A., Ingram W.J. and Senior C.A., 1995, 'On surface temperature, greenhouse gases and aerosols: models and observations', *J. Climate*, **10**, 2364–2386.

Mitchell J.M., 1956, 'Visual range in the polar regions with particular reference to the Alaskan Arctic', *Atmos. Terr. Phys.*, **Special Supplement**, 195–211.

Mock S.J., 1967, *Accumulation Patterns on the Greenland Ice Sheet*, U.S. Army, Corp. Engr. Cold. Regions Res. Eng. Lab., Res. Rept., 233 pp.

Mohn H., 1905, *Meteorology. The Norwegian North Polar Exped. 1893–1896*, Scient. Res., vol.VI, Christiania–London–New York–Bombay–Leipzig, 659 pp.

Mokhov I.I. and Schlesinger M.E., 1993, 'Analysis of global cloudiness. 1. Comparison of Meteor, Nimbus-7, and International Satellite Cloud Climatology Project (ISCCP) satellite data', *J. Geophys. Res.*, **98** (D7), 12,849–12,868.

Mokhov I.I. and Schlesinger M.E., 1994, 'Analysis of global cloudiness. 2. Comparison of ground-based and satellite-based cloud climatologies', *J. Geophys. Res.*, **99** (D8), 17,045–17,065.

Moodie D.W. and Catchpole A.J.W., 1975, 'Environmental data from historical documents by content analysis: freeze-up and break-up of estuaries on Hudson Bay, 1714–1871', *Manitoba Geogr. Stud.*, **5**, 119 pp.

Mosby H., 1932, 'Sunshine and radiation. The Norwegian North Polar Expedition with the "Maud" 1918–1925', *Scient. Res.*, **1a**, 7, Geofysisk Instituett, Bergen, 1–110.

Murashova A.B., 1986, 'Computation of monthly values of turbulent fluxes over the ocean', *Trudy GGO*, **504**, 80–85 (in Russian).

Müller F. and Roskin-Sharlin N., 1967, *A High Arctic Climate Study on Axel Heiberg Island*, Axel Heiberg Island Research Reports, Meteorology, No. 3, McGill Univ., Montreal, 82 pp.

Mysak L.A., 2001, 'Patterns of Arctic Circulation', *Science*, **293**, 1269–1270.

Nagurnyi A.P., Timerev A.A. and Egorova S.A., 1991, 'Space-time inversion variability in the lower Arctic troposphere', *Dokl. RAS*, **319** (in Russian).

Niedźwiedź T., 1987, 'The influence of the atmospheric circulation on the air temperature in Hornsund region (Spitsbergen)', in: Repelewska-Pękala J. , Harasimiuk M. and Pękala K. (Eds.), *Proceedings of XIV Polar Symposium: Actual Research Problems of Arctic and Antarctic*, Lublin, Poland, May 7–8, pp. 174–180 (in Polish).

Niedźwiedź T., 1993, 'Long-term variability of the atmospheric circulation over Spitsbergen and its influence on the air temperature', in: Repelewska-Pękalowa J. and Pękala K. (Eds.), *Proceedings of XX Polar Symposium*, Lublin, Poland, pp. 17–30.

Niedźwiedź T., 1997, 'The climates of the "Polar regions"', in: Yoshino M., Douguedroit A., Paszyński J. and Nkemdirim L. (Eds.), *Climates and Societes – A Climatological Perspective*, pp. 309–324.

Niewiarowski W., 1982, 'Morphology of the forefield of the Aavatsmark Glacier (Oscar II Land, NW Spitsbergen) and phases of its formation', *Acta Univ. Nicolai Copernici, Geogr.*, **16**, 15–43.

Nordenskjöld O., 1928, 'The delimitation of the Polar regions and the natural provinces of the Arctic and Antarctic', in: *The Geography of the Polar Regions*, Amer. Geophys. Soc., New York, pp. 72–90.

Nordenskjöld O. and Mecking L., 1928, *The Geography of the Polar Regions*, Amer. Geogr. Soc. Special Publ. No. 8, New York, 359 pp.

O'Brien S.R., Mayewski P.A., Meeker L.D., Meese D.A., Twickler M.S. and Whitlow S.I., 1995, 'Complexity of Holocene climate as reconstructed from a Greenland ice core', *Science*, **270**, 1962–1964.

Observations of the International Polar Expeditions 1882–1883, Fort Rae, 1886, London, Trübner & CO., 326 pp.

Ohmura A., 1981, 'Climate and energy balance of Arctic tundra', *Zürcher Geographische Schriften*, **3**, Zürcher, 448 pp.

Ohmura A., 1982, 'A historical review of studies on the energy balance of Arctic tundra', *J. Climatol.*, **2**, 185–195.

Ohmura A., 1984, 'On the cause of "Fram" type seasonal change in diurnal amplitude of air temperature in polar regions', *J. Climatol.*, **4**, 325–338.

Ohmura A., 1987, 'New temperature distribution maps for Greenland', *Zeit. für Gletscherkunde und Glazialgeologie*, **23**, 1–45.

Ohmura A., Calanca P., Wild M. and Anklin M., 1999, 'Precipitation, accumulation and mass balance of the Greenland Ice Sheet', *Zeit. für Gletscherkunde und Glazialgeologie*, **35**, 1–20.

Ohmura A. and Reeh N., 1991, 'New precipitation and accumulation maps for Greenland', *J. Glaciol.*, **37**, 140–148.

Ohmura A., Steffen K., Blatter H., Greuell W., Rotach M., Konzelmann T., Forrer J., Abe-Ouchi A., Steiger D., Stober M. and Niederbäumer G., 1992, *Energy and Mass Balance During the Melt Season at the Equilibrium Line Altitude, Paakitsoq, Greenland Ice Sheet*, ETH Greenland Expedition Progress Report No. 2, Dept. of Geogr., ETH Zürich, 94 pp.

Ohmura A., Steffen K., Blatter H., Greuell W., Rotach M., Konzelmann T., Laternser M., Abe-Ouchi A., and Steiger D., 1991, *Energy and Mass Balance During the Melt Season at the Equilibrium Line Altitude, Paakitsoq, Greenland Ice Sheet*, ETH Greenland Expedition Progress Report No. 1, Dept. of Geogr., ETH Zürich, 108 pp.

Osborn T.J., Briffa K.R., Tett S.F.B., Jones P.D. and Trigo R.M., 1999, 'Evaluation of the North Atlantic Oscillation as simulated by a coupled climate model', *Clim. Dyn.*, **15**, 685–702.

Ottar B., Gotaas Y., Hov O., Iversen T., Joranger E., Oehme M., Pacyna J., Semb A., Thomas W. and Vitols V., 1986, *Air Pollutants in the Arctic*, Norwegian Institute for Air Research, NILU OR 30/86, Lillestrőm, Norway, 81 pp.

Palutikof J.P., 1986, 'Scenario construction for regional climatic change in a warmer world', *Proceedings of a Canadian Climatic Program Workshop*, March 3–5, Geneva Park, Ontario, pp. 2–14.

Palutikof J.P., Wigley T.M.L. and Lough J.M., 1984, *Seasonal Climate Scenarios for Europe and North America in a High-CO_2, Warmer World*, U. S. Dept. of Energy, Carbon Dioxide Res. Division, Tech. Report TRO12, 70 pp.

Parlow E., 1992, 'Klimaökologie und Fernerkundung: Integration von Messergebnissen und Fernerkundungsdaten zur Erstellung klimarelevanter Flächendatensätze', *Stuttgarter Geographische Studien*, **117**, 73–87.

Parker D. E., Jones P. D., Folland C. K. and Bevan A., 1994, 'Interdecadal changes of surface temperature since the late nineteenth century', *J. Geophys. Res.*, **99** (D7), 14,373–14,399.

Parker M.N., 1989, 'Polar lows in the Beaufort Sea', in: Twitchell P.F., Rasmussen E.A. and Davidson K.L. (Eds.), *Polar and Arctic Lows*, A. Deepak Publishing, Hampton, pp. 323–330.

Parkinson C.L., Comiso J.C., Zwally H.J., Cavalieri D.J., Gloersen P. and Campbell W.J., 1987, 'Arctic sea-ice, 1973–1976: Satellite passive-microwave observations', Technical Information Branch, NASA, Washington, DC, NASA SP-489, 296 pp.

Paterson W.S.B., Koerner R.M., Fisher D.A., Johnsen S.J., Dansgaard W., Bucher P. and Oescheger H., 1977, 'An oxygen isotope climatic record from the Devon Island ice cap, Arctic Canada', *Nature*, **266**, 508–511.

Peck E.L., 1993, 'Biases in precipitation measurements: An American experience', in: *Eighth Symposium on Meteorological Observations and Instrumentation...*, Jan. 17–23 1993, Anaheim, California, Amer. Met. Soc., Boston, MA, pp. 329–334.

Pereyma J., 1983, 'Climatological problems of the Hornsund area – Spitsbergen', *Acta Univ. Wratisl.*, **714**, 134 pp.

Petrov L.S., 1971, 'The Arctic boundary and principles of its determination', in: Govorukha L.S. and Kruchinin Yu.A. (Eds.), Problems of Physiographic Zoning of Polar Lands, *Trudy AANII*, **304**, 18–35 (in Russian). Translated and published also by Amerind Publishing Co., Pot. Ltd, New Delhi, 1981, 15–34.

Petterssen S., 1949, 'Changes in the general circulation associated with the recent climatic variations', *Geogr. Ann.*, **31**, 212–221.

Petterssen S., Jacobs W.C. and Hayness B.C., 1956, *Meteorology of the Arctic*, Washington, D.C., 207 pp.

Pękala K. and Repelewska-Pękalowa J., 1990, 'Relief and stratigraphy of quaternary deposits in the region of Recherche Fiord and southern Bellsund (Western Spitsbergen)', in: Repelewska-Pękalowa J. and Pękala K. (Eds.), *Wyprawy Geograficzne na Spitsbergen*, UMCS Lublin, pp. 9–20.

Pickard G.L. and Emery W.J., 1982, *Descriptive Physical Oceanography: An Introduction*, Pergamon Press, Oxford–New York–Sydney–Paris–Frankfurt, 249 pp.

Pietroń Z., 1987, 'Frequency and conditions of fog occurrence in Hornsund, Spitsbergen', *Pol. Polar Res.*, **8**, 277–291.
Pollack H.N. and Chapman D.S., 1993, 'Underground records of changing climate', *Scient. Amer.*, **268**, 44–50.
Prik Z.M., 1959, 'Mean position of surface pressure and temperature distribution in the Arctic', *Trudy ANII*, **217**, 5–34 (in Russian).
Prik Z.M., 1960, 'Basic results of the meteorological observations in the Arctic', *Probl. Arkt. Antarkt.*, **4**, 76–90 (in Russian).
Prik Z.M., 1965, 'Precipitation in the Arctic', *Trudy AANII*, **273**, 5–25 (in Russian).
Prik Z.M., 1969, 'To the problem of relative humidity in the Arctic in winter', *Trudy AANII*, **287**, Gidrometeoizdat, Leningrad, 98–109 (in Russian).
Prik Z.M., 1971, 'Climatic zoning of the Arctic', in: Govorukha L.S. and Kruchinin Yu.A. (Eds.), Problems of Physiographic Zoning of Polar Lands, *Trudy AANII*, **304**, 72–84 (in Russian). Translated and published also by Amerind Publishing Co., Pot. Ltd, New Delhi, 1981, 76–88.
Przybylak R., 1992a, 'Thermal-humidity relations against the background of the atmospheric circulation in Hornsund (Spitsbergen) over the period 1978–1983', *Dokumentacja Geogr.*, **2**, 105 pp. (in Polish).
Przybylak R., 1992b, 'Spatial differentiation of air temperature and relative humidity on western coast of Spitsbergen in 1979–1983', *Pol. Polar Res.*, **13**, 113–130.
Przybylak R., 1993, 'Climatic models and their utilising in forecast of climate change', *Przegl. Geogr.*, **1–2**, 163–176 (in Polish).
Przybylak R., 1995, 'Scenarios of Arctic air temperature and precipitation in a warmer world based on instrumental data', in: Heikinheimo P. (Ed.), *International Conference on Past, Present and Future Climate, Proceedings of the SILMU conference held in Helsinki, Finland, 22–25 August 1995*, pp. 298–301.
Przybylak R., 1996a, *Variability of Air Temperature and Precipitation over the Period of Instrumental Observations in the Arctic*, Uniwersytet Mikołaja Kopernika, Rozprawy, 280 pp. (in Polish).
Przybylak R., 1996b, 'Thermic and precipitation relations in the Arctic over the period 1961–1990', *Probl. Klimatol. Polar.*, **5**, 89–131 (in Polish).
Przybylak R., 1997a, 'Spatial and temporal changes in extreme air temperatures in the Arctic over the period 1951–1990', *Int. J. Climatol.*, **17**, 615–634.
Przybylak R., 1997b, 'Spatial variations of air temperature in the Arctic in 1951–1990', *Pol. Polar Res.*, **18**, 41–63.
Przybylak R., 1997c, 'Spatial relations of atmospheric precipitation changes in the Arctic in 1951–1990', *Probl. Klimatol. Polar.*, **7**, 41–54 (in Polish).
Przybylak R., 1999, 'Influence of cloudiness on extreme air temperatures and diurnal temperature range in the Arctic in 1951–1990', *Pol. Polar Res.*, **20**, 149–173.
Przybylak R., 2000a, 'Temporal and spatial variation of air temperature over the period of instrumental observations in the Arctic', *Int. J. Climatol.*, **20**, 587–614.
Przybylak R., 2000b, 'Diurnal temperature range in the Arctic and its relation to hemispheric and Arctic circulation patterns', *Int. J. Climatol.*, **20**, 231–253.
Przybylak R., 2002a, *Variability of Air Temperature and Atmospheric Precipitation During a Period of Instrumental Observation in the Arctic*, Kluwer Academic Publishers, Boston–Dordrecht–London, 330 pp.

Przybylak R., 2002b, 'Variability of total and solid precipitation in the Canadian Arctic from 1950 to 1995', *Int. J. Climatol.* **22**, 395–420.
Punning Ya. -M.K., Troitskii L.S., 1977, 'Glacier advances on Svalbard in the Holocene', *Mat. glyatsiol. issled.*, **29**, 211–216 (in Russian).
Punning Ya.-M.K., Troitskii L. S. and Rajamae R., 1976, 'The genesis and age of the Quaternary deposits in the eastern part of Van Mijenfjorden, West Spitsbergen', *Geologisk Föreningens Stockholm Forhandlingar*, **98**, 343–347.
Putnins P., 1970, 'The climate of Greenland', in: Orvig S. (Ed.), *Climates of the Polar Regions*, World Survey of Climatology, vol. 14, Elsevier Publ. Comp., Amsterdam–Londyn–New York, pp. 3–128.
Raatz W.E., 1981, 'Trends in cloudiness in the Arctic since 1920', *Atmos. Environ.*, **15**, 1503–1506.
Raatz W.E., 1991, 'The climatology and meteorology of Arctic air pollution', in: Sturges W.T. (Ed.), *Pollution of the Arctic Atmosphere*, Elsevier Science Publ., London and New York, pp. 13–42.
Radionov V.F., Bryazgin N.N. and Alexandrov E.I., 1997, *The Snow Cover of the Arctic Basin*, University of Washington, Technical Report APL-UW TR 9701, variously paged.
Rae R.W., 1951, *Climate of the Canadian Arctic Archipelago*, Department of Transport, Met. Div., Toronto, 90 pp.
Ragozin A.I. and Chukanin K.I., 1959, 'Mean trajectories and velocities of pressure systems in the European Arctic and in the Subarctic', in: Sbornik statei po meteorologii, *Trudy ANII*, **217**, 36–64 (in Russian).
Rahn K.A., Borys R.D. and Shaw G.E., 1977, 'The Asian source of arctic haze bands', *Nature*, **268**, 713–715.
Rahn K.A. and McCaffrey R.J., 1980, 'On the origin and transport of the winter arctic aerosol', *Ann. New York Acad. Sci.*, **338**, 486–503.
Rahn K.A. and Shaw G.E., 1982, 'Sources and transport of arctic pollution aerosols: a chronicle of six years of ONR research', *Naval Research Rev.*, **March**, S2-26.
Ramsden D. and Fleming G., 1995, 'Use of a coupled ice-ocean model to investigate the sensitivity of the Arctic ice cover to doubling atmospheric CO_2', *J. Geophys. Res.*, **100**, 6817–6828.
Randall D., Curry J., Battisti D., Flato G., Grumbine R., Hakkinen S., Martinson D., Preller R., Walsh J. and Weatherly J., 1998, 'Status of and outlook for large-scale modeling of atmosphere-ice-ocean interactions in the Arctic', *Bull. Amer. Met. Soc.*, **79**, 197–219.
Rasmussen E., 1981, 'An investigation of polar low with a spiral cloud structure', *J. Atmos. Sci.*, **38**, 1785–1792.
Rasmussen E., 1983, 'A review of mesoscale disturbances in cold air masses', in: Lilly D.K. and Gal-chen T. (Eds.), *Mesoscale, Meteorology-theories, Observations and Models*, Reidel, Boston, pp. 247–283.
Rasmussen E., 1985a, 'A case study of a polar low development over the Barents Sea', *Tellus*, **37A**, 407–418.
Rasmussen E., 1985b, 'Paskestormen et baroklink polart lavtryk', *Vejret*, **4–7 Argang**, 3–17.

Rasmusson E.M., 1977, *Hydrological Application of Atmospheric Vapour-flux Analyses*, Operational Hydrology Report 11, WMO, Geneva, 50 pp.
Ratzki E., 1962, 'Contribution to the climatology of Greenland', *Exped. Polaires Franc.*, Publ. No. **212**, Paris.
Reed R.J., 1979, 'Cyclogenesis in polar air streams', *Mon. Wea. Rev.*, **107**, 38–52.
Reed R.J. and Kunkel B.A., 1960, 'The Arctic circulation in summer', *J. Meteorol.*, **17**, 489–506.
Rempp G. and Wagner A., 1917, 'Die Hydrodynamik des Föhns und die "lokalen" Winde in Spitzbergen, Deutsches Observatorium, Ebeltofthafen-Spitzbergen', *Veröffentlichungen*, **7**, 12 pp.
Report of the International Polar Expedition to Point Barrow, Alaska, 1885, Meteorology, Washington, pp. 203–260.
Rigor I.G., Colony R.L. and Martin S., 2000, 'Variations in surface air temperature observations in the Arctic, 1979–1997', *J. Climate*, **13**, 896–914.
Rigor I.G. and Heiberg A., 1997, *International Arctic Buoy Program Data Report: 1 January 1996 – 31 December 1996*, Advance Copy Technical Memorandum APL-UW TM5-97, Seattle, 163 pp.
Rinke A. and Dethloff K., 2000, 'On the sensitivity of a regional Arctic climate model to initial and boundary conditions', *Clim. Res.*, **14**, 101–113.
Rinke A., Dethloff K., Christensen J.H., 1999a, 'Arctic winter climate and its interannual variations simulated by a regional climate model', *J. Geophys. Res.*, **104**, 19,027–19,038.
Rinke A., Dethloff K., Christensen J.H., Botzet M. and Machenhauer B., 1997, 'Simulation and validation of Arctic radiation and clouds in a regional climate model', *J. Geophys. Res.*, **102** (D25), 29,833–29,847.
Rinke A., Dethloff K., Spekat A., Enke W. and Christensen J.H., 1999b, 'High resolution climate simulations over the Arctic', *Polar Research*, **18**, 143–150.
Rinke A., Lynch A.H. and Dethloff K., 2000, 'Intercomparison of Arctic regional climate simulations: Case studies of January and June 1990', *J. Geophys. Res.*, **15**, 29,669–29,683.
Robinson D.A., 1991, 'Merging operational satellite and historical station snow cover data to monitor climate change', *Palaeogeogr., Palaeoclim., Palaeoecol. (Global and Planetary Change Section)*, **90**, 235–240.
Robinson D.A., Dewey K.F. and Heim R.R., Jr., 1993, 'Global snow cover monitoring: An update', *Bull. Amer. Met. Soc.*, **74**, 1689–1696.
Robinson D.A., Serreze M.C., Barry R.G., Scharfen G. and Kukla G., 1992, 'Large-scale patterns and variability of snow melt and parameterized surface albedo in the Arctic Basin', *J. Climate*, **5**, 1109–1119.
Rodewald M., 1950, 'Zur Frage der Luftdruckverhältnisse in der Arktis', *Ann. Meteorol.*, **3**, 284–290.
Romanov I.P., 1991, *Ice Cover of the Arctic Basin*, Arkt. i Antarkt. Nauchno-Issled. Inst., Leningrad, 212 pp. (in Russian).
Ropelewski Ch. F., 1991, 'Real-time monitoring of global snow cover', *Palaeogeogr., Palaeoclim., Palaeoecol. (Global and Planetary Change Section)*, **90**, 225–229.
Rossow W.B., 1992, 'Polar cloudiness: Some results from ISCCP and other cloud climatologies', in: *Amer. Met. Soc., Third Conference on Polar Meteorology and Oceanography*, 29 Sept. – 2 Oct. 1992, Portland, Oregon, Boston, pp.1–3.

Rossow W.B., 1995, 'Another look at the seasonal variation of polar cloudiness with satellite and surface observations', in: *Amer. Met. Soc., Fourth Conference on Polar Meteorology and Oceanography*, Jan. 15–20 1995, Dallas, Texas, Boston, (J10)1 – (J10)4.

Rossow W.B. and Garder L.C., 1992, 'Cloud detection using satellite measurements of infrared and visible radiances for ISCCP', *J. Climate*, **6**, 2341–2369.

Rossow W.B. and Schiffer R.A., 1991, 'ISCCP Cloud Data Products', *Bull. Amer. Meteorol. Soc.*, **72**, 2–20.

Salinger M.J. and Pittock A.B., 1991, 'Climate scenarios for 2010 and 2050 AD Australia and New Zealand', *Clim. Change*, **18**, 259–269.

Santer B. D., Taylor K. E., Wigley T. M. L., Penner J. E., Jones P. D. and Cubash U., 1995, 'Towards the detection and attribution of an anthropogenic effect on climate', *Clim. Dyn.*, **12**, 77–100.

Saravanan R., 1998, 'Atmospheric low-frequency variability and its relationship to midlatitude SST variability: Studies using the NCAR climate system model', *J. Climate*, **11**, 1386–1404.

Sater J.E. (Ed.), 1969, *The Arctic Basin*, The Arctic Inst. of North America, Washington, 337 pp.

Sater J.E., Ronhovde A.G. and Van Allen L.C. (Eds.), 1971, *Arctic Environment and Resources*, The Arctic Inst. of North America, Washington, 310 pp.

Schatz H., 1951, 'Ein Föhnsturm in Nordostgrönland', *Polarforschung*, **3**, 13–14.

Scherer D., 1992, 'Klimaökologie und Fernerkundung: Erste Ergebnisse der Messkampagne 1990/1991', *Stuttgarter Geographische Studien*, **117**, 89–104.

Scherhag R., 1931, 'Eine bemerkungswerte Klimaänderung Über Nord-Europa', *Ann. Hydr. Mar. Met.*, 57–67.

Scherhag R., 1937, 'Die Erwärmung der Arktis', *ICES Journal*.

Scherhag R., 1939, 'Die Erwärmung der Arktis', *Ann. Hydr. Mar. Met.*

Scherr P.E., Glasser A.M., Barnes J.C. and Willard J.M., 1968, *World-wide Cloud Distribution for Use in Computer Simulations*, Final Report Contract NAS-8-21040, Allied Research Associetes, Inc., Baltimore, Maryland, 272 pp.

Schlesinger M.E. and Mitchell J.F.B., 1987, 'Climate model simulations of the equilibrium climatic response to increased carbon dioxide', *Rev. Geophys.*, **25**, 760–798.

Schneider G., Paluzzi P. and Oliver J.P., 1989, 'Systematic error in synoptic sky cover record of the South Pole', *J. Climate*, **2**, 295–302.

Schweiger A.J. and Key J.R., 1992, 'Arctic cloudiness: Comparison of ISCCP-C2 and *Nimbus-7* satellite-derived cloud products with a surface-based cloud climatology', *J. Climate*, **5**, 1514–1527.

Schweiger A.J., Serreze M.C. and Key J.R., 1993, 'Arctic sea ice albedo: A comparison of two satellite-derived data sets', *Geophys. Res. Lett.*, **20**, 41–44.

Schwerdtfeger W. 1931, 'Zur Theorie polarer Temperatur- und Luftdruckwellen', *Veroff. des Geophysikalischen Instituts der Universitat Leipzig*, II, Serie, Bd. **IV**, H. 5.

Seaman N.L., Otten H. and Anthers R.A., 1981, 'A rapidly developing polar low in the North Sea of January 2, 1979', in: *First International Conference on Meteorology and Air/Sea Interaction of Coastal Zone*, May 10–14, 1981, The Hague, Netherlands.

Serreze M.C. and Barry R.G., 1988, 'Synoptic activity in the Arctic Basin in summer, 1979–1985', in: *Amer. Met. Soc., Second Conference on Polar Meteorology and Oceanography*, March 29–31, 1988, Madison, Wisc., Boston, pp. 52–55.

Serreze M.C., Barry R.G., Rehder M.C. and Walsh J.E., 1995a, 'Variability in atmospheric circulation and moisture over the Arctic', *Phil. R. Soc. Lond.*, 215–225.

Serreze M.C., Barry R.G. and Walsh J.E., 1994a, 'Atmospheric water vapor characteristics at 70°N', *J. Climate*, **8**, 719–731.

Serreze M.C., Box J.E., Barry R.G. and Walsh J.E., 1993, 'Characteristics of Arctic synoptic activity, 1952–1989', *Meteorol. Atmos. Phys.*, **51**, 147–164.

Serreze M.C., Carse F. and Barry R.G., 1997, 'Icelandic low cyclone activity: Climatological features, linkages with the NAO, and relationships with recent changes in the Northern Hemisphere circulation', *J. Climate*, **10**, 453–464.

Serreze M.C., Kahl J.D. and Schnell R.C., 1992, 'Low-level temperature inversions of the Eurasian Arctic and comparisons with Soviet drifting stations', *J. Climate*, **5**, 615–630.

Serreze M.C. and Rehder M.C., 1990, 'June cloud cover over the Arctic Ocean', *Geophys. Res. Lett.*, **17**, 2397–2400.

Serreze M.C. Rehder M.C., Barry R.G. and Kahl J.D., 1994b, 'A climatological database of Arctic water vapor characteristics', *Polar Geogr. and Geol.*, **18**, 63–75.

Serreze M.C., Rehder M.C., Barry R.G., Kahl J.D. and Zaitseva N.A., 1995b, 'The distribution and transport of atmospheric water vapour over the Arctic Basin', *Int. J. Climatol.*, **15**, 709–727.

Sevruk B., 1982, *Methods of Correction for Systematic Error in Point Precipitation Measurement for Operational Use*, Operational Hydrology Report No. 21, Publ. 589, WMO, Geneva, 91 pp.

Sevruk B., 1986, 'Correction of precipitation measurements: Swiss experience', in: Sevruk B. (Ed.), Correction of Precipitation Measurements, *Zürscher Geographische Schriften*, **23**, 187–196.

Shapaev W.M., 1959, 'Basic data about local disturbances of wind and about representatives of the meteorological stations in the Soviet Arctic', *Trudy AANI*, **217**, 87–98 (in Russian).

Shapiro M.A., Fedor L.S. and Hampel T., 1987, 'Research aircraft measurements within a polar low over the Norwegian Sea', *Tellus*, **37**, 272–306.

Shaw G.E., 1995, 'The Arctic haze phenomenon', *Bull. Amer. Met. Soc.*, **76**, 2403–2413.

Shaw N., 1927, 'The influence of the north polar region upon the meteorology of the northern hemisphere', in: Breitfuss L. (Ed.), *Internat. Studiengesellschaft zur Erforschung der Arktis mit dem Luftschiff (Aeroarctic)*, Gotha: Justus Perthes, 25–30.

Shaw N., 1928, *Manual of Meteorology*, vol. II, Cambridge.

Shindell D.T., Miller R.L., Schmidt G. and Pandolfo L., 1999, 'Simulation of recent northern winter climate trends by greenhouse-gas forcing', *Nature*, **399**, 452–455.

Shuleykin V.V., 1935, 'Elements of the thermal regime of the Kara Sea', *Trudy Taymyrskoy Gidrograficheskoy Ekspeditsii*, **2**, Izd. Gidrogr. Otdela UMS, Leningrad, 7–48 (in Russian).

Slonosky V.C. and Yiou P., 2001, 'The North Atlantic Oscillation and its relationship with near surface temperature', *Geophys. Res. Lett.*, **28**, 807–810.

Solar Radiation at Polaris Bay. Terrestrial Radiation, 1876, Scientific Results of the U.S. Arctic Expedition. Steamer "Polaris", C.F. Hall Commanding, vol. I, Physical observations, Washington.

Spinnangr G., 1968, 'Global radiation and duration of sunshine in northern Norway and Spitsbergen', *Meteorol. Ann.*, **5**, Oslo, 137 pp.

Stanhill G., 1995, 'Solar irradiance, air pollution and temperature changes in the Arctic', *Phil. Trans. R. Soc. Lond. A*, **352**, 247–258.

Starkel L., 1984, 'The reflection of abrupt climatic changes in the relief and sequence of continental deposits', in: Mörner N.A. and Karlén W. (Eds.), *Climatic Changes on a Yearly to Millennial Basis; Geological, Historical and Instrumental Records*, D. Reidel, Boston, 135–146.

Steffensen E., 1969, 'The climate and its recent variations at the Norwegian arctic stations', *Meteorol. Ann.*, **5**, 349 pp.

Steffensen E., 1982, 'The climate at Norwegian Arctic stations', *KLIMA DNMI Report*, **5**, Norwegian Met. Inst., 44 pp.

Stepanova N.A., 1965, *Some Aspects of Meteorological Conditions in the Central Arctic: A review of U.S.S.R. Investigations*, U.S. Department of Commerce, Weather Bureau, Washington, 136 pp.

Stewart T.G. and England J., 1983, 'Holocene sea ice variations and paleoenvironmental change, northernmost Ellesmere Island, NWT, Canada', *Arctic and Alpine Res.*, **15**, 1–17.

Stonehouse B. (Ed.), 1986, *Arctic Air Pollution*, Cambridge University Press, Cambridge, 328 pp.

Stonehouse B., 1990, *North Pole South Pole. A Guide to the Ecology and Resources of the Arctic and Antarctic*, PRION, London, 216 pp.

Stowe L.L., Wellemeyer C.G. Eck T.F., Yeh H.Y.M. and the NIMBUS-7 Cloud Data Processing Team, 1988, 'NIMBUS-7 global cloud climatology. Part I: Algorithms and validation', *J. Climate*, **1**, 445–470.

Stowe L.L., Yeh H.Y.M., Eck T.F., Wellemeyer C.G., Kyle H.L., and the NIMBUS-7 Cloud Data Processing Team, 1989, 'NIMBUS-7 global cloud climatology. Part II: First year results', *J. Climate*, **2**, 671–709.

Stuiver M. and Braziunas T.F, 1989, 'Atmospheric ^{14}C and century-scale solar oscillations', *Nature*, **338**, 405–408.

Sturges W.T. (Ed.), 1991, *Pollution of the Arctic Atmosphere*, Elsevier Science Publ., London and New York, 334 pp.

Sugden D., 1982, *Arctic and Antarctic. A Modern Geographical Synthesis*, Basil Blackwell, Oxford, 472 pp.

Supan A., 1879, 'Die Temperaturzonen der Erde', *Petermanns Mitt.*, H. **25**, 349–358.

Supan A., 1884, *Grundzüge der Physischen Erdkunde*, Leipzig.

Surova T.G., Troitskii L.S. and Punning Ya.-M.K., 1982, 'The history of glaciation in Svalbard during the Holocene on the basis of palynological investigations', *Mat. glyatsiol. issled.*, **42**, 100–106 (in Russian).

Süring R., 1895, 'Temperatur und Feuchtigkeitsbeobachtungen über und auf der Schneedecke des Brockengipfels', *Met. Zeitschr.*, Bd. **12**.

Svendsen J.I. and Mangerud J., 1997, 'Holocene glacial and climatic variations on Spitsbergen, Svalbard', *The Holocene*, **7**, 45–57.

Sverdrup H.U., 1933, *Meteorology. The Norwegian North Polar Expedition with the "Maud" 1918–1925*, Scient. Res., vol. II, part 1, Discussion, Bergen, 331 pp.
Sverdrup H.U., 1935, 'Übersicht über das Klima des Polarmeeres und des Kanadischen Archipels', in: Köppen W. and Geiger R. (Eds.), *Handbuch der Klimatologie, Bd. II, Teil K, Klima des Kanadischen Archipels und Grönlands*, Verlag von Gebrüder Borntraeger, Berlin, pp. K3-K30.
Sychev K.A. (Ed.), 1959, *Observations of Research Drifting Stations "North Pole 4", "North Pole 5" and "North Pole 6", 1956/57, vol. 2, Meteorology, Actinometry*, Izd. "Morskoy Transport", Leningrad, 648 pp. (in Russian).
Szupryczyński J., 1968, 'Some problems of the Quaternary on Spitsbergen', *Prace Geogr.*, **71**, 2–128 (in Polish).
Tao X., Walsh J.E. and Chapman W.L., 1996, 'An assessment of global climate model simulations of Arctic air temperature', *J. Climate*, **9**, 1060–1076.
Tarussov A., 1988, 'Accumulation changes on Arctic glacier during 1656–1980 A.D.', *Mat. glyatsiol. issled.*, **14**, 85–89 (in Russian).
Tarussov A., 1992, 'The Arctic from Svalbard to Severnaya Zemlya: climatic reconstructions from ice cores', in: Bradley R.S. and Jones P.D. (Eds.), *Climate Since A.D. 1500*, Routledge, London, pp. 505–516.
Taylor K.C., Mayewski P.A., Alley R.B., Brook E.J., Gow A.J., Grootes P.M., Meese D.A., Saltzman E.S., Severinghaus J.P., Twickler M.S., White J.W.C., Whitlow S. and Zielinski G.A., 1997, 'The Holocene-Younger Dryas transition recorded at Summit, Greenland', *Nature*, **278**, 825–827.
The Polar Group, 1980, 'Polar-atmosphere-ice-ocean processes: A review of polar problems in climate research', *Rev. Geophys. Space Phys.*, **18**, 525–543.
The Times Atlas of the World, 1992, Times Books, London, 222 pp.
Thompson D.W. and Wallace J.M., 1998, 'The Arctic Oscillation signature in the wintertime geopotential height and temperature fields', *Geophys. Res. Lett.*, **25**, 1297–1300.
Thompson D.W. and Wallace J.M., 2000, 'Annular modes in the extratropical circulation. Part I: Month-to-month variability', *J. Climate*, **13**, 1000–1016.
Thompson D.W., Wallace J.M. and Hegerl G.C., 2000, 'Annular modes in the extratropical circulation. Part II: Trends', *J. Climate*, **13**, 1018–1036.
Thorndike A.S. and Colony R., 1980, *Arctic Ocean Buoy Program Data Report: 1 January 1979 – 31 December 1979*, Polar Science Center, University of Washington, Seattle, 131 pp.
Timerev A.A. and Egorova S.A., 1991, 'Spatial-temporal variability of surface inversions in the Arctic', *Soviet Meteorology and Hydrology*, **7**, 39–44.
Trenberth K.E. and Hurrell J.W., 1994, 'Decadal atmosphere-ocean variations in the Pacific', *Clim. Dyn.*, **9**, 303–319.
Turner J., Lachlan-Cope T. and Rasmussen E.A., 1991, 'Polar lows', *Weather*, **46** (4), 107–114.
Twitchell P.F., Rasmussen E.A. and Davidson K.L. (Eds.), 1989, *Polar and Arctic Lows*, A. Deepak Publishing, Hampton, 421 pp.
Ukhanova E.V., 1971, 'Fogs and visibility', in: Dolgin I.M. (Ed.), *Meteorological Conditions of the non-Soviet Arctic*, Gidrometeoizdat, Leningrad, pp. 142–151 (in Russian).

Untersteiner N., 1964, 'Calculations of temperature regime and heat budget of sea ice in the central Arctic', *J. Geophys. Res.*, **69**, 4755–4766.
Vahl M., 1911, 'Zones et biochores géographiques', in: *Oversigt over der Kgl. Danske Vidensk. Selskabs. Forhandl.*, Copenhagen, pp. 269–317.
Vaikmäe R. A., 1990, 'Paleonvironmental data from less-investigated polar regions', in: Weller G., Wilson C.L. and Severin B.A.B. (Eds.), *International Conference on the Role of the Polar Regions in Global Change*, vol. II, Proceedings of a Conference held June 11–15, 1990 at the University of Alaska, Fairbanks, pp. 611–616.
Vaikmäe R. A. and Punning Ya.-M.K., 1982, 'Isotope and geochemical studies of the Vavilov Ice Dome, Severnaya Zemlya', *Mat. glyatsiol. issled.*, **44**, 145–149 (in Russian).
Valero F.P.J., Ackerman T.P., Gore W.J.Y. and Weil M.L., 1988, 'Radiation studies in the Arctic', in: Hobbs P.V. and McCormick M.P. (Eds.), *Aerosols and Climate*, A. Deepak Publishing, Hampton, Virginia, pp. 271–275.
Vangengeim G. Ya., 1937, 'Meteorological conditions of the region of Franz Joseph Land in the warm season of the year (April–August)', *Trudy Arkt. Inst.*, **103**, 3–64 (in Russian).
Vangengeim G. Ya., 1952, 'Bases of the macrocirculation method for long-term weather forecasts for the Arctic', *Trudy ANII*, **34**, 314 pp. (in Russian).
Vangengeim G. Ya., 1961, 'Degree of the atmospheric circulation homogeneity in different parts of Northern Hemisphere under the existence of main macrocirculation types W, E and C', *Trudy AANII*, **240**, 4–23 (in Russian).
Vinje T., 2001, 'Anomalies and trends of sea-ice extent and atmospheric circulation in the Nordic seas during the period 1864–1998', *J. Climate*, **14**, 255–267.
Vize V. Yu., 1925, 'Novaya Zemlya's bora', *Izv. Central'nogo Gidrometeorologicheskogo biuro*, vyp. **5** (in Russian).
Vize V. Yu., 1940, *Sea Climate in Russian Arctic*, Izd-vo Glavsevmorputi, Leningrad–Moskva, 124 pp. (in Russian).
Voieikov A.I., 1889, 'Snow cover: its influence on soil, climate and weather, and methods of its investigation', *Zap. Russk. Geogr. Ob.-va po Obshchey Geogr.*, **18** (in Russian). Also published in Izbr. Soch., 2, Izd. Akad. Nauk SSSR, Moskva, 1948 (in Russian).
Volkov N.A. (Ed.), 1958, *Results of Scientific Research Work of the Drifting Stations "North Pole 4" and "North Pole 5", 1955–1956*, vol. 3, Meteorology, Actinometry, Izd. "Morskoy Transport", Leningrad (in Russian).
Voskresenskiy A. I., Baranov G. I., Dolgin M. I., Nagurnyi A.P., Aleksandrov E. I. Bryazgin N. I., Dementev A. A., Marshunova M. S., Burova L. P. and Kotova N. M., 1991, 'Estimation of possible changes of atmospheric climate in the Arctic up to 2005 including anthropogenic factors', in: Krutskich B.A. (Ed.), *Klimaticheskii Rezhim Arktiki na Rubezhe XX i XXI vv.*, Gidrometeoizdat, St.-Petersburg, pp. 30–61 (in Russian).
Vowinckel E., 1962, 'Cloud amount and type over the Arctic', *Scient. Rep.*, No. **4**, Publ. in Meteorol., 51, McGill Univ., Montreal, 27 pp.
Vowinckel E., 1964a, 'Heat flux through the polar ocean ice', *Scient. Rep.*, No. **12**, Publ. in Meteorol., 70, McGill Univ., Montreal, 15 pp.

Vowinckel E., 1964b, 'Atmospheric energy advection in the Arctic', *Scient. Rep.*, No. **13**, Publ. in Meteorol., 71, McGill Univ. Montreal, 14 pp.

Vowinckel E., 1965, 'The inversion over the Polar Ocean', *Scient. Rep.*, No. **14**, Publ. in Meteorol., 72, McGill Univ., Montreal, 30 pp.

Vowinckel E. and Orvig S., 1962, 'Insolation and absorbed solar radiation at the ground in the Arctic', *Scient. Rep.*, No. **5**, Publ. in Meteorol., 53, McGill Univ., Montreal, 32 pp.

Vowinckel E. and Orvig S., 1963, 'Long wave radiation and total radiation balance at the surface in the Arctic', *Scient. Rep.*, No. **8**, Publ. in Meteorol., 62, McGill Univ., Montreal, 33 pp.

Vowinckel E. and Orvig S., 1964a, 'Radiation balance of the troposphere and of the Earth-Atmosphere system in the Arctic', *Scient. Rep.*, No. **9**, Publ. in Meteorol., 63, McGill Univ., Montreal, 23 pp.

Vowinckel E. and Orvig S., 1964b, 'Incoming and absorbed solar radiation at the ground in the Arctic', *Arch. Meteorol. Geophys. Bioklim.*, **13**, 352–377.

Vowinckel E. and Orvig S., 1964c, 'Long wave radiation and total radiation balance at the surface in the Arctic', *Arch. Meteorol. Geophys. Bioklim.*, **13**, 451–479.

Vowinckel E. and Orvig S., 1964d, 'Radiation balance of the troposphere and of the earth-atmosphere system in the Arctic', *Arch. Meteorol. Geophys. Bioklim.*, **13**, 480–502.

Vowinckel E. and Orvig S., 1965, 'The heat budget over the Arctic Ocean', Final Rep., *Publ. in Meteorol.*, **74**, McGill Univ., Montreal, variously paged.

Vowinckel E. and Orvig S., 1966, 'The heat budget over the Arctic Ocean', *Arch. Meteorol. Geophys. Bioklim.*, **14**, 303–325.

Vowinckel E. and Orvig S., 1967, 'The inversion over the Polar Ocean', in: Orvig S. (Ed.), *W.M.O. – S.C.A.R. – J.C.P.M. Symp. Polar Meteorol.*, Proc. – W.M.O. Tech. Note, 87, pp. 39–59.

Vowinckel E. and Orvig S., 1970, 'The climate of the North Polar Basin', in: Orvig S. (Ed.), *Climates of the Polar Regions*, World Survey of Climatology, 14, Elsevier Publ. Comp., Amsterdam–London–New York, pp. 129–252.

Vowinckel E. and Taylor B., 1964, 'Evaporation and sensible heat flux over the Arctic Ocean', *Scient. Rep.*, No. **10**, Publ. in Meteorol., 66, 30 pp.

Vowinckel E. and Taylor B., 1965, 'Evaporation and sensible heat flux over the Arctic Ocean', *Arch. Meteorol. Geophys. Bioklim.*, **14**, 36–52.

Wadhams P., 1981, 'Sea ice topography of the Arctic Ocean in the region 70°W to 25°E', *Phil. Trans. R. Soc. Lond. A*, **302**, 45–85.

Wadhams P., 1995, 'Arctic sea ice extent and thickness', *Phil. Trans. R. Soc. Lond. A*, **352**, 301–319.

Wagner G., 1965, *Klimatologische Beobachtungen in Südostspitzbergen 1960*, Franz Steiner Verlag, Wiesbaden, 69 pp.

Walsh J.E., 1995, 'Recent variations of Arctic climate: The observations evidence', in: *Amer. Met. Soc., Fourth Conference on Polar Meteorology and Oceanography*, Jan. 15–20 1995, Dallas, Texas, Boston, pp. (J9)20 – (J9)25.

Walsh J.E. and Crane R.G., 1992, 'A comparison of GCM simulations of Arctic climate', *Geophys. Res. Lett.*, **19**, 29–32.

Walsh J.E., Kattsov V., Portis D. and Meleshko V., 1998, 'Arctic precipitation and evaporation: Model results and observational estimates', *J. Climate*, **11**, 72–87.

Walsh J.E., Lynch A., Chapman W. and Musgrave D., 1993, 'A regional model for studies of atmosphere-ice-ocean interaction in the western Arctic', *Meteorol. Atmos. Phys.*, **51**, 179–194.

Wang J. and Ikeda M., 2000, 'Arctic Oscillation and Arctic Sea-Ice Oscillation', *Geophys. Res. Lett.*, **27**, 1287–1290.

Warren S.G., Hahn C.J., London J., Chervin R.M. and Jenne R.L., 1986, *Global Distribution of Total Cloud Cover and Cloud Type Amounts over Land*, NCAR/TN-273 + STR, National Center for Atmospheric Research, Boulder, CO, 29 pp. + 199 maps.

Warren S.G., Hahn C.J., London J., Chervin R.M. and Jenne R.L., 1988, *Global Distribution of Total Cloud Cover and Cloud Type Amounts over the Oceans*, NCAR/TN-317 + STR, National Center for Atmospheric Research, Boulder, CO, 42 pp. + 170 maps.

Warren S.G., Rigor I.G., Untersteiner N., Radionov V.F., Bryazgin N.N., Aleksandrov Y.I. and Colony R., 1999, 'Snow Depth on Arctic sea ice', *J. Climate*, **12**, 1814–1829.

Washington W.M. and Meehl G.A., 1984, 'Seasonal cycle experiment on the climate sensitivity due to a doubling of CO_2 with an Atmospheric General Circulation Model coupled to a Simple Mixed Layer Ocean Model', *J. Geophys. Res.*, **89**, 9475–9503.

Weatherly J.W., Briegleb B.P., Large W.G. and Maslanik J.A., 1998, 'Sea ice and polar climate in the NCAR CSM', *J. Climate*, **11**, 1472–1486.

Wegener K., 1939, 'Ergänzungen für Eismitte. Deutsche Grönland-Expedition A. Wegener 1929 und 1930–31', *Wiss. Ergebnisse*, Bd. **4**, Hf. 2, Leipzig.

Weickmann L., 1942, *Zur Diskussion der Arktis zugeführten Wärmemenge. Die Erwärmung der Arktis*, Veröff. Deutschen Wiss. Inst. Kopenhagen.

Weiss W., 1975, *Arktis*, Verlag Anton Schroll & Co., Wien und München, 188 pp.

Werner A., 1990, 'Lichen growth rates for the northwest coast of Spitsbergen, Svalbard', *Arctic and Alpine Res.*, **22**, 129–140.

Werner A., 1993, 'Holocene moraine chronology, Spitsbergen, Svalbard: lichenometric evidence for multiple Neoglacial advances in the Arctic', *The Holocene*, **3**, 128–137.

Westman J., 1903, *Measures de l'intensité de la radiation solaire faites en 1899 et en 1900 à la station d'hivernage suédoise à la baie de Treurenberg, Spitzberg Missions Scientifiques pour la Mesure d'un Arc de Méridien au Spitzberg Entreprises en 1899–1902*, 2, sec. 8, B. Radiation Solaire, Stockholm, 59 pp.

Whitlock C.H., Charlock T.P., Staylor W.F., Pinker R.T., Laszlo I., DiPasquale R.C., and Ritchey N.A., 1993, *WCRP Surface Radiation Budget Shortwave Data Product Description* – Version 1.1. Technical report, NASA Technical Memorandum 107747.

Whittaker L.M. and Horn L.H., 1984, 'Northern Hemisphere extratropical cyclone activity', *J. Climatol.*, **4**, 297–310.

Wigley T.M.L., Jones P.D. and Kelly P.M., 1986, 'Empirical climate studies. Warm world scenarios and the detection of climate change induced by radiatively active gases', in: Bolin B., Döös Bo.R., Jäger J. and Warrick R.A. (Eds.), *The Greenhouse Effect, Climate Change and Ecosystems*, SCOPE 29, John Wiley & Sons, Chichester, pp. 271–322.

Wild M., Ohmura A. and Cubasch U., 1997, 'GCM-simulated surface energy fluxes in climate change experiments', *J. Climate*, **10**, 3093–3110.
Wilson C. 1988, 'The summer season along the east coast of Hudson Bay during the nineteenth century. Part III. Summer thermal and wetness indices. B. The indices, 1800 to 1900', *Canadian Climate Centre Report*, **88-3**, 1–42.
Wilson C., 1992, 'Climate in Canada, 1809–1820: three approaches to the Hudson's Bay Company Archives as an historical database', in: Harington C.R. (Ed.), *The Year Without a Summer? World Climate in 1816*, Canadian Museum of Nature, Ottawa, pp. 162–184.
Woo M-K. and Young K.L., 1996, 'Summer solar radiation in the Canadian High Arctic', *Arctic*, **49**, 170–180.
Wójcik G., 1976, *Problems of Climatology and Glaciology in Iceland*, Uniwersytet Mikołaja Kopernika, Rozprawy, 226 pp. (in Polish).
Wójcik G. and Kejna M., 1991, 'Annual distribution of wind direction frequencies and annual and daily course of wind velocity in Hornsund (SW Spitsbergen)', *Acta Univ. Wratisl.*, **1213**, Prace Inst. Geogr., Ser. A, t. V, 351–363 (in Polish).
Wójcik G., Marciniak K. and Przybylak R., 1983, 'Air humidity during the summer on the coastal Kaffiöyra Plain and Waldemar Glacier (NW Spitsbergen)', in: Olszewski A. and Wójcik G. (Eds.), *Polskie Badania Polarne 1970–1982*, Rozprawy UMK, X Sympozjum Polarne, Toruń, pp. 187–199 (in Polish).
Wójcik G. and Przybylak R., 1991, 'Meteorological conditions on the Kaffiöyra Plain (NW Spitsbergen) in the period 14th July – 9th September 1982', *Acta Univ. Nicolai Copernici, Geografia*, **22**, Toruń, 109–124 (in Polish).
Yi D., Mysak L.A. and Venegas S.A., 1999, 'Singular Value Decomposition of Arctic sea ice cover and overlying atmospheric circulation fluctuations', *Atmos.-Ocean*, **37**, 389–415.
Young S.B., 1989, *To the Arctic: An Introduction to the Far Northern World*, John Wiley & Sons, Inc., New York, 354 pp.
Zaitseva N.A., Skony S.M. and Kahl J.D., 1996, 'Temperature inversions over the Western Arctic from radiosonde data', *Russian Meteorol. and Hydrol.*, **6**, 6–17.
Zavyalova I.N., 1971, 'Air humidity', in: Dolgin I.M. (Ed.), *Meteorological Conditions of the non-Soviet Arctic*, Gidrometeoizdat, Leningrad, pp. 104–113 (in Russian).
Zhang J., Rothrock D.A. and Steele M., 1998, 'Warming of the Arctic Ocean by a strengthened Atlantic inflow: Model results', *Geophys. Res. Lett.*, **25**, 1745–1748.
Zhang Y. and Hunke E.C., 2001, 'Recent Arctic change simulated with a coupled ice-ocean model', *J. Geophys. Res.*, **106**, 4369–4390.

COPYRIGHT ACKNOWLEDGEMENTS

I gratefully acknowledge the following copyright holders, who have kindly provided permission to reproduce the figures and tables indicated. Sources of all figures and tables are given in each figure and table caption and the full publication details are found in the *References* section (pp. 225–257):

Figs. 1.2, 2.2 (January and July), 2.5, 3.1, 6.4–6.5, 7.11: from *Atlas Arktiki* 1985, by permission of the Arctic and Antarctic Research Institute; Fig. 1.3: from the Central Intelligence Agency, 1978, *Polar Regions Atlas*, National Foreign Assessment Center, C.I.A. Washington, DC, 66 pp.; Figs. 1.4–1.5 and Table 1.1: from *Descriptive Physical Oceanography: An Introduction* by G.L. Pickard and W.J. Emery, reprinted by permission of Elsevier Science; Fig. 2.1: from H.J. Crutcher H.J. and J.M. Meserve, 1970, *Selected Level Height, Temperatures, and Dew Points for the Northern Hemisphere*, reprinted by permission of NAVAIR, Washington, D.C.; Fig. 2.2 (April and October): from S.G. Gorshkov (Ed.), 1980, *Military Sea Fleet Atlas of Oceans: Northern Ice Ocean*, USSR: Ministry of Defense, 184 pp.; Figs. 2.3 and 2.4: from M.C. Serreze, J.E. Box, R.G. Barry and J.E. Walsh, Characteristics of Arctic synoptic activity, 1952–1989, *Meteorol. Atmos. Phys.*, 51, 147–164, © 1993, with kind permission of Springer–Verlag KG and the authors; Fig. 2.6: from S.G. Gorshkov (Ed.), 1980, *Military Sea Fleet Atlas of Oceans: Northern Ice Ocean*, USSR: Ministry of Defense, 184 pp.; Figs. 3.2–3.3, 3.8–3.14: from V.P. Khrol (Ed.), *Atlas of the Energy Balance of the Northern Polar Region*, © 1992, with kind permission of the Arctic and Antarctic Research Institute; Figs. 3.4–3.6: from the publication: M.S. Marshunova and N.T. Chernigovskii, 1971, *Radiation Regime of the Foreign Arctic*, Gidrometeoizdat, Leningrad, 182 pp.; Fig. 3.7 and Table 3.1: from E. Vowinckel and S. Orvig, 1970, The climate of the North Polar Basin, in: S. Orvig (Ed.), *Climates of the Polar Regions*, World Survey of Climatology, 14, Elsevier Publ. Comp., Amsterdam–London–New York, pp. 129–252.; Fig. 4.2: with the kind permission of the publisher and the editor; Fig. 4.6a–d: reprinted with the kind permission of P. Calanca; Fig. 4.6e: from A. Ohmura, New temperature distribution maps for Greenland, *Zeit. für Gletscherkunde und Glazialgeologie*, 23, 1–45, © 1987, with the kind permission of the publisher and the author; Fig. 4.14: from the publication N.A. Zaitseva, S.M. Skony and J.D. Kahl, 1996, Temperature inversions over the Western Arctic from radiosonde data, *Russian Meteorol. and Hydrol.*, 6, 6–17, reprinted with the permission of the publisher and the authors; Figs. 5.1 and 5.2: from R.E. Huschke, 1969, *Arctic Cloud Statistics from "Air Calibrated" Surface Weather Observations*, RAND Corp. Mem. RM-6173-PR, RAND, Santa Monica, CA, 79 pp., reprinted with permission; Fig. 5.4: from E. Vowinckel, 1962, Cloud amount and type over the Arctic, *Scient. Rep.*, No. 4, Publ. in Meteorol., 51, McGill Univ., Montreal, 27 pp.; Fig. 5.5: from the publication S.G. Gorshkov (Ed.), 1980, *Military Sea Fleet Atlas of Oceans: Northern Ice Ocean*, USSR: Ministry of Defense, 184 pp.; Figs. 5.7 and 6.3: from I.M. Dolgin (Ed.), *Meteorological Conditions of the non-Soviet Arctic*, with kind permission of the Arctic and Antarctic Research Institute; Fig. 6.2: from M.C. Serreze,

R.G. Barry, M.C. Rehder and J.E. Walsh, Variability in atmospheric circulation and moisture flux over the Arctic, *Phil.Trans. R. Soc. Lond. A.*, 215–225, © 1995, with the kind permission of The Royal Society and the authors; Fig. 7.1: reprinted with permission from *Polar Geography and* Geology, Vol. 18, No. 1, pp. 63–75. © V.H. Winston & Son, Inc., 360 South Ocean Boulevard, Palm Beach, FL 33480. Al rights reserved; Fig. 7.4: from A. Ohmura, P. Calanca, M. Wild and M. Anklin, Precipitation, accumulation and mass balance of the Greenland Ice Sheet, *Zeit. für Gletscherkunde und Glazialgeologie*, 35, 1–20, © 1999, with the kind permission of the publisher and the authors; Fig. 7.5: reprinted from the Journal of Glaciology with permission of the International Glaciological Society and A. Ohmura; Figs. 7.12–7.14 and Table 4.1: from V.F. Radionov, N.N. Bryazgin and E.I. Alexandrov, *The Snow Cover of the Arctic Basin*, University of Washington, Technical Report APL-UW TR 9701. © 1997, with the kind permission of the University of Washington; Figs. 7.15–7.16: from S.G. Warren, I.G. Rigor, N. Untersteiner, V.F. Radionov, N.N. Bryazgin, Y.I. Aleksandrov and R. Colony, Snow Depth on Arctic sea ice, *J. Climate*, 12, 1814–1829. © 1999, with the kind permission of the American Meteorological Society; Figs. 8.1 and 8.2: from J. Heintzenberg, Arctic haze: air pollution in Polar regions, *Ambio*, 18, 51–55. © 1989, with the kind permission of the Royal Swedish Academy of Science; Fig. 8.3: from L.A. Barrie, Arctic air chemistry: an overview, in: Stonehouse B. (Ed.), *Arctic Air Pollution*, Cambridge University Press, pp. 5–23. © 1986, with the kind permission of Cambridge University Press; Fig. 8.4: reprinted with the kind permission of K.A. Rahn; Fig. 8.5: from WMO Bulletin 41 (2), by permission of the publisher; Fig. 10.1: from S.J. Johnsen, H.B. Clausen, W. Dansgaard, K. Fuhrer, N. Gundestrup, C.U. Hammer, P. Iversen, J. Jouzel, B. Stauffer and J.P. Steffensen, 1992, Irregular glacial interstadials recorded in a new Greenland ice core, *Nature*, 359, 311–313, with the kind permission of the publisher and the authors; Figs. 10.2, 10.8 and 10.10: reprinted with the permission of D.A. Meese, A.J. Gow, P. Grootes, P.A. Mayewski, M. Ram, M. Stuiver, K.C. Taylor, E.D. Waddington and G.A. Zielinski, The accumulation record from the GISP2 core as an indicator of climate change throughout the Holocene, *Science*, 266, 1680–1682. © 1994 American Association for the Advancement of Science; Figs. 10.3 and 10.5: reprinted with the permission of S.R. O'Brien, P.A. Mayewski, L.D. Meeker, D.A. Meese, M.S. Twickler and S.I. Whitlow, Complexity of Holocene climate as reconstructed from a Greenland ice core, *Science*, 270, 1962–1964. © 1995 American Association for the Advancement of Science; Fig. 10.4: reprinted with the permission of D. Dahl-Jensen, K. Mosegaard, N. Gundestrup, C.D. Clow, S.J. Johnsen, A.W. Hansen. and N. Balling, Past temperatures directly from the Greenland Ice Sheet, *Science*, 282, 268–271. © 1998 American Association for the Advancement of Science; Fig. 10.6: from D.J.A. Evans and J. England, 1992, Geomorphological evidence of Holocene climatic change from northwest Ellesmere Island, Canadian High Arctic, *The Holocene*, 2, 148–158, with the kind permission of the publisher and the authors; Figs. 10.7 and 10.13 from R.A. Vaikmäe, Paleoenvironmental data from less-investigated polar regions, in: G. Weller, C.L. Wilson and B.A.B. Severin (Eds.), *International Conference on the Role of the Polar Regions in Global Change*, vol. II, Proceedings of a Conference held June 11–15, pp. 611–616. © 1990, with the kind permission of the

University of Alaska.; Fig. 10.9: from H. Fischer, M. Werner, D. Wagenbach, M. Schwager, T. Thorsteinnson, F. Wilhelms, J. Kipfstuhl and S. Sommer, Little ice age clearly recorded in northern Greenland ice cores, *Geophys. Res. Lett.*, 25, 1749–1752, © 1998 American Geophysical Union, reproduced by permission of American Geophysical Union; Fig. 10.11: from B.T. Alt, D.A. Fisher and R.M. Koerner, Climatic conditions for the period surrounding the Tambora signal in ice core from the Canadian High Arctic Islands, in: C.R. Harington (Ed.), *The Year Without a Summer? World Climate in 1816*, Ottawa, pp. 309–325. © Canadian Museum of Nature 1992, with the kind permission of the publisher and the authors; Fig. 10.12: from R.S. Bradley and P.D. Jones, 1993, 'Little Ice Age' summer temperature variations: their nature and relevance to recent global warming trends, *The Holocene*, 3, 367–376, with the kind permission of the publisher and the authors; Fig. 11.1: J.E. Walsh and R.G. Crane, A comparison of GCM simulations of Arctic climate, *Geophys. Res. Lett.*, 19, 29–32, © 1992 American Geophysical Union, reproduced by permission of American Geophysical Union; Figs. 11.2 and 11.4: from X. Tao, J.E. Walsh and W.L. Chapman, An assessment of global climate model simulations of Arctic air temperature, *J. Climate*, 9, 1060–1076. © 1996, with the permission of the American Meteorological Society; Fig. 11.3: from J.E. Walsh, V. Kattsov, D. Portis and V. Meleshko, Arctic precipitation and evaporation: Model results and observational estimates, *J. Climate*, 11, 72–87. © 1998, with the permission of the American Meteorological Society; Figs. 11.5 and 11.9: from S. Manabe, R.J. Stouffer, M.J. Spelman and K. Bryan, Transient responses of a coupled ocean-atmosphere model to gradual changes of atmospheric CO_2. Part I: Annual mean response, *J. Climate*, 4, 785–817. © 1991, with the permission of the American Meteorological Society; Figs. 11.6, 11.8 and 11.10 from H. Cattle and J. Crossley, Modelling Arctic climate change, *Phil. Trans. R. Soc. Lond. A*, 352, 201–213. © British Crown Copyright 1995, with the kind permission of the Royal Society and the authors; Fig. 11.7: from A. Kattenberg and 82 co-authors, Climate models – Projection of future climate, in: J.T. Houghton, L.G. Meila Filho, B.A. Callander, N. Harris, A. Kattenberg and K. Maskell (Eds.), *Climate Change 1995: The Science of Climate Change*, Cambridge University Press, pp. 285–357, reprinted with the kind permission of the Intergovernmental Panel on Climate Change.

INDEX

A

accumulation
  annual, 119
  rate, 138, 165–168, 173
actinometric stations, 34, 39, 45, 48
Advanced Very High Resolution Radiometer (AVHRR), 34
aerosol
  forcing, 216, 223
  optical depth, 148
air
  circulation, 14
  humidity, 47, 54, 109, 111–112, 114–115
  masses, 25, 29–31, 71, 76–78, 91, 101, 111, 125, 138, 151, 197
  pollution, 105, 141–145, 147
  pressure, 16–18, 20–21, 95, 152–155
  stream, 25–26, 28, 138, 157
  temperature, 8, 16, 47, 54, 63–78, 81–92, 104–105, 109, 124, 127, 129, 150–156, 181–203, 206–207, 210–216, 221–223
Alaska, 20, 30, 38, 46, 76, 85, 88, 106, 130, 138, 142, 187, 190, 193–194, 202, 214, 221–222
albedo, 5, 7, 34, 43–45, 186
Aleutian low, 154, 204
analogue
  method, 211, 221, 223
  scenarios, 221–223
annual cycle of air temperature
  coastal, 66–67
  continental, 66–68
  maritime, 65–66
Antarctic, 1, 5
anticyclonic activity, 20, 25, 28, 36, 122, 124, 150, 157
anti-greenhouse effect, 186
Arctic
  anticyclone, 13, 16, 20
  boundary, 1–5
  Canadian, 7, 16–17, 20, 22–23, 25, 28, 30, 35–36, 38–40, 47, 51, 70, 72, 76, 78, 83–88, 99, 101, 104, 109, 111–113, 118–119, 124, 127–129, 131, 133–134, 136, 138, 150, 154–156, 163, 169–170, 172, 177–179, 181, 187, 190, 193–195, 200, 202, 221–222
  central, 7, 13–14, 16–17, 20–23, 25, 34, 45, 54–55, 59, 63–64, 72, 77, 84–86, 88, 99, 101, 136, 190, 214
  Circle, 1–2, 5, 38, 45, 49, 148
  climate system, 8, 12, 98, 196, 217, 222
  front, 20
  Gas and Aerosol Sampling Program (AGASP), 147
  haze, 141–144, 147–148, 203, 222
  lows, 29, 31
  Norwegian, 82–83, 85–88, 187, 190, 195
  Ocean, 5, 7–9, 11–13, 16–17, 20, 22–23, 25, 28, 38–40, 43–44, 46, 51, 55, 59, 63–64, 76–77, 81, 92, 101, 103–106, 113–114, 120, 122, 124, 127, 129, 134, 138, 149, 151, 201, 206, 208–210, 216–217
  Ocean Buoy Program, 17, 21, 63
  Oscillation (AO), 196–199
  regions, 1, 3–4, 29, 78, 144
  Russian, 20, 25, 28, 33, 38, 46, 72, 77, 84–88, 110, 134, 136, 138, 187, 190, 194–195, 216, 221–222
synoptic activity, 17
temperature inversions, 91–92
  advection, 91
  base, 92–93
  frequency, 91–95
  intensity, 94
  subsidence, 91
  surface/lower tropospheric, 7, 91–95, 145, 147, 157
  thickness, 92–94

263

upper tropospheric, 92, 94
Western, 22, 23, 28, 92, 210
artificial contamination, 165
Atlantic
　Ocean, 85, 197, 212–213
　region, 20, 25, 28, 36, 40, 51, 63, 65, 71–72, 76, 78–82, 94, 99, 101, 103–104, 106, 111, 113, 120, 124, 128–131, 134, 136, 149–154, 158, 182, 184, 187, 189, 192, 197, 200–201, 222
　sector, 14, 20–21, 197, 214
　Water, 9–10, 12, 201
atmospheric
　circulation, 12–14, 25, 29, 40, 65, 70, 72, 78, 81, 86, 88, 111, 122, 128, 149, 154, 157, 169, 184, 186, 190, 195–197, 200, 202
　moisture balance, 118
　precipitation, 117, 120, 122–124, 127, 198, 211
　refraction, 5
　sublimation, 105
Atmospheric Model Intercomparison Project (AMIP), 206–209

B

Baffin Bay, 8, 12, 20–21, 23, 25, 28, 36, 45–47, 51, 54–55, 65, 71–72, 76, 78–81, 89, 99, 101, 103–104, 106, 110, 113–114, 120, 122, 124, 128–129, 134, 136, 149, 153, 156–157, 187, 192, 196–197, 200, 202
Baffin Island, 5, 12, 138, 156–157
Barents Sea, 20, 22–23, 25, 28, 31, 38–40, 46–48, 51, 54–55, 59, 101, 104, 106, 134, 151, 201, 208, 216, 221–222
Beaufort Sea, 10, 20, 23, 40, 54–55, 92–94
Beringia, 7
Bering Strait, 5, 12, 20, 23, 28, 36, 55, 153–154
biosphere, 98
black element carbon (soot), 148

Bölling/Alleröd, 164–165
Boothia Peninsula, 84
bora, 29
borehole temperature, 165–167, 169, 172–174, 176
Bottom Water, 9–10
British Isles, 30

C

calm, 29
Canadian
　Arctic Archipelago, 5, 12, 29, 43, 46, 55, 64, 70, 72, 76, 84, 103, 106, 138, 154
　Basin, 10, 23
　high, 16, 20
chemical
　composition, 143, 165
　flux, 163, 166–167
　models, 203
　transport model, 146
Chukchi
　Peninsula, 28, 194
　Sea, 23, 28–31, 36, 40, 55, 59, 104, 106, 154
Chukotka region, 38
clear days, 29, 81–84, 89, 91, 103, 111, 114, 159
Climate and Cryosphere (CLIC), 161
climate
　global, 98, 161, 203
　present–day, 203, 209, 211–212, 222
　scenarios, 203, 211
　sensitivity, 203–204
climatic
　barrier, 151,
　changes, 161, 169, 171–174, 203
　classification, 154–155
　deterioration, 172, 179
　events, 165
　models, 128, 142, 161, 190, 195
　optimum, 165, 167, 172
　regionalisation, 3, 149–150
　regions, 3, 68, 78–81, 122, 131–132, 149–150, 182, 187, 192, 196

sub-regions, 149–150
variability, 169
cloudiness, 35–36, 38–40, 47, 49, 69, 81–84, 86, 88–89, 91, 93, 97–101, 103–104, 110–111, 114, 116, 128, 150–159, 208–209, 212, 217
cloudy days, 82–84, 89, 91, 103–104, 111, 114, 152, 155–156, 159
cluster analyses, 147
coagulation, 143
condensation, 54, 112, 118
Contemporary Global Warming (CGW), 173, 182, 185
continental climate, 7, 122, 130, 134, 170
convergence of sea ice, 8
correlation
 analysis, 152
 coefficients, 78–79, 185, 197–200, 202
CO_2 doubling, 142, 203, 211–215, 217–218, 220, 223
cryosphere, 98, 161
Current
 Baffin Island, 12
 East Greenland, 11, 55
 Irminger, 12
 Labrador, 12
 Murmansk, 55
 Norwegian, 55
 West Greenland, 12, 55, 156
 West Spitsbergen, 12, 49, 55
cyclonic
 activity, 20–21, 25, 28, 36, 38, 40, 69–72, 78–79, 86, 88–89, 91–92, 95, 99, 101, 103, 111, 122, 124–125, 133–134, 138, 149–150, 152–153, 156
 vortex, 14

D

daily course of temperature
 normal, 68
 reverse, 68–69
data
 glaciological, 126
 meteorological, 63, 126

Denmark Strait, 51, 54
Devon
 Island, 38, 169
 Ice Cap, 169, 177–178
diurnal temperature range (DTR), 85–90
divergence of sea ice, 8
driftwood, 169–170, 177
drizzle, 117
dust content, 165

E

Earth's rotation, 14
earth–atmosphere system, 5, 34–35
Earth Radiation Budget Experiment, 34
East Siberian Sea, 21–23, 28–30, 54, 59, 104, 106, 153
electricity conductivity, 165
elemental carbon, 143
Ellesmere Island, 5, 84, 103–104, 113, 177–179
El Niño–Southern Oscillation (ENSO), 202
emissivity of snow, 7
empirical orthogonal function, 197
Europe, 146, 148, 166, 168, 174
European Centre for Medium–Range Weather Forecasts (ECMWF), 204, 207, 210–211
evaporation, 54, 59, 117–118, 122, 206, 208, 213

F

Fast Ice, 8
feedback mechanism, 186
foehn, 29
fog
 advection, 104
 ice, 104–105
 radiation, 104, 107
 "steam", 104–105
"Fram"
 drift, 10, 13, 63
 type, 88
Fram Strait, 12, 34
frost, 118

G

general circulation model (GCM), 203–204, 209–212, 214, 217, 222–223
geomagnetic activity, 69
glacial anticyclone theory, 13–14
glacier
   advance, 168, 171–172, 179–181,
   retreat, 171
   tidewater, 9
   wind, 30
global
   cooling, 186
   solar radiation, 35, 39–40, 42–43, 45
   warming, 4, 92, 117, 186, 222
grid points/boxes, 14, 24, 182, 185
greenhouse
   effect, 148, 186
   gases, 92, 142, 165, 203, 215, 218, 220
      carbon dioxide, 142, 165
      freons, 142
      methane, 142, 165
   warming, 186
Greenland
   ice core, 164, 168
   Ice-core Project (GRIP), 163
   ice sheet/ice cap, 7, 34–36, 51, 54, 64, 68, 70, 72, 84–85
   Ice Sheet Project (GISP2), 162
   Sea, 12, 28, 31, 45–46, 51, 55, 59, 128, 172, 222
Grönfjord–Fridtjov ice divide, 179–180
Gulf Stream, 104, 150

H

heat
   balance, 35, 54, 88, 217, 219
   geothermal, 165
   latent, 8, 54, 58–59, 211, 217
   sensible, 7, 54–59, 211, 217
heavy metals, 147
Hopen, 38, 83
human activity, 117, 141, 161
hummocks, 8

I

ice core
   analyses, 161, 165, 170–171, 173, 178
   data, 162, 164, 168, 170–171, 173, 177–180
ice crystals, 105, 118, 141, 148
"ice crystals haze", 141
ice wedge polygons, 179
Iceland, 1, 12, 20, 72, 78, 134, 151, 172, 199, 216,
Icelandic low, 20, 71, 79, 125, 138
Iceland–Kara Sea trough, 20, 78
industrial age, 142
interdiurnal temperature changes, 157
interglacial, 221
International
   Polar Year, 33, 63
   Satellite Cloud Climatology Project, 34
isobaric systems, 13
isotopic data/record, 181

J

Jan Mayen
   Island, 20, 38, 104, 150, 172
   station, 26, 38, 65–66, 68, 82–83, 89, 111, 122–123, 131–133, 150

K

Kara Sea, 20, 22–23, 28–30, 38, 51, 54, 59, 72, 78, 104, 134, 151, 201–202
Kola Peninsula, 148

L

Labrador
   Peninsula, 5, 104
   Sea, 12, 25
lake
   sediment, 172, 178
   sedimentation, 179
Lake Baykal, 1

Laptev Sea, 23, 28–30, 38, 46, 51, 54–55, 104, 106, 136, 153
leads, 8, 55, 105
least geometrical distance, 78
limited–area models/ Regional Climate Models, 204, 210–211, 222–223
Little Ice Age (LIA), 172–181
litosphere, 98
logbook, 141
Lomonosov
   ice cap, 178–179
   plateau, 180–181

M

Malmgren formula, 113
mass balance, 117
"Maud" expedition, 13, 63
Medieval Warm Period, 166, 173–174, 177, 179
melting, 8–9, 72, 77, 81, 92–94, 136, 138–139, 156, 158, 178, 180–181
meltwater discharge, 179
model biases, 206, 209–211, 223
moisture
   balance, 118
   content, 120, 124
   fluxes, 8
molluscs, 172
moraines, 172, 181

N

National Centers for Environmental Prediction (NCEP) reanalyses, 211
National Center for Atmospheric Research (NCAR), 204–207, 210
natural factors, 186
Neoglacial advances, 172
New Siberian Sea, 136
nitrate, 143
Nordaustlandet Island, 180–181
Nordenskjöld line, 2–3
Nordic seas, 201
normal distribution, 133
North America, 5, 43, 146, 148, 168

North Atlantic Oscillation (NAO) index, 184, 190, 196–202, 204
Northern Hemisphere, 1, 31, 79, 84, 120, 144, 153, 182, 185, 187, 191–192, 195–197, 199, 221
North Pacific (NP) index , 201–202
North Pole/Pole, 5, 7–8, 11, 13, 16–17, 20–21, 23, 25, 28, 36, 39–40, 46–49, 51, 55, 59, 64, 68, 70–72, 84, 88, 92–94, 101, 134, 136, 138, 158
North Water polynya, 8, 54, 156
Norway, 5, 101, 141
Norwegian Sea, 20, 22, 28, 31, 38–40, 45–51, 54–55, 59, 99, 101, 106, 114, 172, 208, 216
Novaya Zemlya, 20, 29–30, 38, 76, 101, 151, 171, 180, 216
Novosibirskiye Ostrova, 136

O

oceanic circulation, 12, 149, 151, 156, 196
orographic barrier, 25, 153
Ostrov Vrangelya, 29, 85
oxygen isotope, 162–163, 167–169, 175, 177
ozone, 69

P

Pacific Ocean, 212
Pack Ice, 8
paleoclimatological information/reconstruction, 161, 221
passive microwave, 119
peat studies, 177
physical parameterisation, 210
polar
   anticyclone, 13–14
   cap ice, 8
   cell, 14
   day, 5, 29, 72, 77–78, 152, 155, 158
   easterlies, 25
   front, 145
   lows, 25, 29–31

night, 5, 12, 29, 36, 38–40, 47, 49, 68–69, 106, 111, 117, 119, 149, 196
regions, 1–2, 91, 97–98, 161, 186, 212, 222
vortex, 197
POLES (Polar Exchange at the Sea Surface), 211
pollen, 168, 172
polynyas, 8, 51, 55, 105
power plants, 148
precipitable water 120–121,
precipitation
   measurable, 134
   number of days, 134
   regime, 134
"proxy data", 161, 165–166, 168–170, 172–174, 177–181
   botanical, 180
   geomorphological, 180
   glaciological, 161
psychrometer, 109
pyranometer, 33

R

radiation
   balance, 5, 7, 34–35, 39, 43, 49–51, 54, 69, 91, 142, 148
   counter, 47
   cooling, 12, 29
   effective, 47–49, 51
   forcing, 186
   incoming, 39–40, 45, 50, 78, 88, 148, 158
   infrared, 47
   inversion, 94
   net, 2–3, 14, 34, 39, 43, 47–49, 51–54, 91, 150, 210–211, 217
   outgoing, 34, 91, 148
   regime, 33–34, 39–40, 49
   solar, 5, 33–35, 39, 43, 45, 47, 51, 78, 94, 111, 149, 155–156, 196
   terrestrial, 34, 47
   thermal, 148
radiocarbon age, 168, 170

radiosonde
   data, 92
   stations, 118
rainfall, 117
reflectivity, 7, 45
relative humidity, 109–110, 112–116
remote sensing, 34, 119
runoff, 118, 148
Russian/Soviet drifting stations, 7, 16–17, 34, 63, 92

S

sandar, 179
satellite
   images/charts, 119
   data, 34, 119
   era, 33, 120
   remote sensing, 34, 119
saturation deficit, 109
Scandinavia, 172
scenarios of the Arctic climate
   coupled model, 203
   equilibrium model, 211–213, 217, 223
   transient model, 211, 213, 217, 223
sea
   breeze, 30–31
   level, 70, 117
sea ice
   albedo, 7
   albedo–temperature feedback, 186
   circulation, 10–11
   concentration, 105
   cover, 43, 105, 149
   edge, 45, 51, 55, 59, 127
   extent, 7, 212
   model, 206
   perennial, 55
   thickness, 210, 217
sea level pressure (SLP), 14, 16–17, 158, 197, 199, 201–202, 204–205
Second International Polar Year, 33, 63
sedimentation, 143, 179
"Sedov" drift, 10
semi-permanent

Arctic high, 14, 21
inversion, 91
Severnaya Zemlya, 20, 77, 129, 136, 170–171
Siberian high, 16, 20, 25, 152–154
snow
   accumulation, 118, 122, 138
   cover, 7, 43, 45–46, 119–120, 136, 138, 149–150
      decay, 7, 136
      density, 119, 138–140
      formation, 136–137
      depth/thickness, 7, 119, 138–139
   crystals, 118
snowfall, 117, 119
snowflakes, 117
snowstorm, 117
solar
   activity, 164, 186
   irradiance, 5, 175, 186, 196
soot, 148
Southern
   Hemisphere, 168, 212
   Oscillation (SO) index, 202
South Pole, 147
Special Sensor Microwave/Imager (SSM/I), 34
Spitsbergen, 5, 12, 25, 30, 33, 36, 38, 40, 49, 51, 55, 59, 63, 68–69, 76–77, 85, 101, 104, 106, 109–111, 116, 122, 134, 138, 143, 150–152, 172, 178, 180–181, 201, 216
standard deviation, 76–77, 129, 139, 167, 175
stratification of the atmosphere, 47
stratiform clouds, 222
stratosphere
   lower, 210, 214
   middle, 214
Subarctic, 84, 182, 187, 190
sulphate
   aerosols, 186, 214
   concentration, 143
   forcing, 216, 223
summer melt/melting, 93, 178, 180–181
sunshine duration, 35–40

supersaturated with respect to ice, 112
Svalbard, 29, 77–78, 136, 151, 170–172, 178–181, 184
synoptic scale, 21, 25, 31, 91, 119–120, 122

T

Taymyr Peninsula, 46, 136, 151
teleconnection, 129
thermal inertia, 212
thermohaline circulation, 213
thermophilus marine, 172
topography, 25, 120, 150, 156
trace gases, 147, 195
Transpolar Drift Stream, 11
tree
   line, 2–4
   ring
      records, 178
      width, 168
troposphere, 7, 14, 91, 109, 120, 210, 214
tundra, 3, 45, 122, 138, 149, 210

U

U.S. Historical Weather Map, 14

V

Vahl's method, 2
variability coefficient, 129–131
vertical gradient, 122
Victoria Island, 5
Vikings, 173

W

water
   balance/budget, 118–119, 138
   vapour pressure, 109–111, 114
westerlies, 13, 126
whalebone studies, 177
wind
   local, 25, 30

bora, 29
downslope, 29
foehn, 29
katabatic, 25, 29, 157–158
mountain, 30
severe, 152–153, 156
storm, 25, 151, 154, 158
streamlines, 126
stress, 8
World War II, 63–64
Wrocław dendrite method, 78, 128

Y

Younger Dryas, 162, 165, 167

Z

Zemlya Frantsa Josifa, 29–30, 76, 129, 136, 151, 170–171, 180, 216
zonal
circulation, 186, 196
index, 196–197, 199–200

ATMOSPHERIC AND OCEANOGRAPHIC SCIENCES LIBRARY

1. F.T.M. Nieuwstadt and H. van Dop (eds.): *Atmospheric Turbulence and Air Pollution Modelling.* 1982; rev. ed. 1984
ISBN 90-277-1365-6; Pb (1984) 90-277-1807-5
2. L.T. Matveev: *Cloud Dynamics.* Translated from Russian. 1984
ISBN 90-277-1737-0
3. H. Flohn and R. Fantechi (eds.): *The Climate of Europe: Past, Present and Future.* Natural and Man-Induced Climate Changes: A European Perspective. 1984
ISBN 90-277-1745-1
4. V.E. Zuev, A.A. Zemlyanov, Yu.D. Kopytin, and A.V. Kuzikovskii: *High-Power Laser Radiation in Atmospheric Aerosols.* Nonlinear Optics of Aerodispersed Media. Translated from Russian. 1985 ISBN 90-277-1736-2
5. G. Brasseur and S. Solomon: *Aeronomy of the Middle Atmosphere.* Chemistry and Physics of the Stratosphere and Mesosphere. 1984; rev. ed. 1986
ISBN (1986) 90-277-2343-5; Pb 90-277-2344-3
6. E.M. Feigelson (ed.): *Radiation in a Cloudy Atmosphere.* Translated from Russian. 1984 ISBN 90-277-1803-2
7. A.S. Monin: *An Introduction to the Theory of Climate.* Translated from Russian. 1986
ISBN 90-277-1935-7
8. S. Hastenrath: *Climate Dynamics of the Tropics*, Updated Edition from *Climate and Circulation of the Tropics.* 1985; rev. ed. 1991
ISBN 0-7923-1213-9; Pb 0-7923-1346-1
9. M.I. Budyko: *The Evolution of the Biosphere.* Translated from Russian. 1986 ISBN 90-277-2140-8
10. R.S. Bortkovskii: *Air-Sea Exchange of Heat and Moisture During Storms.* Translated from Russian, rev. ed. 1987 ISBN 90-277-2346-X
11. V.E. Zuev and V.S. Komarov: *Statistical Models of the Temperature and Gaseous Components of the Atmosphere.* Translated from Russian. 1987
ISBN 90-277-2466-0
12. H. Volland: *Atmospheric Tidal and Planetary Waves.* 1988 ISBN 90-277-2630-2
13. R.B. Stull: *An Introduction to Boundary Layer Meteorology.* 1988
ISBN 90-277-2768-6; Pb 90-277-2769-4
14. M.E. Berlyand: *Prediction and Regulation of Air Pollution.* Translated from Russian, rev. ed. 1991 ISBN 0-7923-1000-4
15. F. Baer, N.L. Canfield and J.M. Mitchell (eds.): *Climate in Human Perspective.* A tribute to Helmut E. Landsberg (1906-1985). 1991 ISBN 0-7923-1072-1
16. Ding Yihui: *Monsoons over China.* 1994 ISBN 0-7923-1757-2
17. A. Henderson-Sellers and A.-M. Hansen: *Climate Change Atlas.* Greenhouse Simulations from the Model Evaluation Consortium for Climate Assessment. 1995
ISBN 0-7923-3465-5
18. H.R. Pruppacher and J.D. Klett: *Microphysics of Clouds and Precipitation*, 2nd rev. ed. 1997 ISBN 0-7923-4211-9; Pb 0-7923-4409-X
19. R.L. Kagan: *Averaging of Meteorological Fields.* 1997 ISBN 0-7923-4801-X
20. G.L. Geernaert (ed.): *Air-Sea Exchange: Physics, Chemistry and Dynamics.* 1999
ISBN 0-7923-5937-2
21. G.L. Hammer, N. Nicholls and C. Mitchell (eds.): *Applications of Seasonal Climate Forecasting in Agricultural and Natural Ecosystems.* 2000 ISBN 0-7923-6270-5

ATMOSPHERIC AND OCEANOGRAPHIC SCIENCES LIBRARY

22. H.A. Dijkstra: *Nonlinear Physical Oceanography.* A Dynamical Systems Approach to the Large Scale Ocean Circulation and El Niño. 2000   ISBN 0-7923-6522-4
23. Y. Shao: *Physics and Modelling of Wind Erosion.* 2000   ISBN 0-7923-6657-3
24. Yu.Z. Miropol'sky: *Dynamics of Internal Gravity Waves in the Ocean.* Edited by O.D. Shishkina. 2001   ISBN 0-7923-6935-1
25. R. Przybylak: *Variability of Air Temperature and Atmospheric Precipitation during a Period of Instrumental Observations in the Arctic.* 2002   ISBN 1-4020-0952-6
26. R. Przybylak: *The Climate of the Arctic.* 2003   ISBN 1-4020-1134-2

KLUWER ACADEMIC PUBLISHERS – DORDRECHT / BOSTON / LONDON